# OPTICAL COMMUNICATIONS SYSTEMS

# OPTICAL COMMUNICATIONS SYSTEMS

*Editor*

**Brian Ismay**

scitus
academics

**Optical Communications Systems**
Edited by **Brian Ismay**

ISBN: 978-1-68117-596-6
Library of Congress Control Number: 2016946451


616, Corporate Way, Suite 2, 4766,
Valley Cottage, NY 10989

www.scitusacademics.com



**Notice**



Printed in United States of America on Acid Free Paper

# PREFACE

Optical communications, sometimes referred to as optical telecommunications, is distant communication with light for information. It can be performed visually or electronically. For all types of networks and telecommunications, optical communication systems are really important. They consist of a transmitter encoding a message into an optical signal, a channel carrying the signal to its destination and a receiver reproducing the message from the optical signal received. Of course, the worldwide telephone network, the biggest communication system in the world, is based on fiber optics. In addition, optics is gradually finding more communications applications every day, never to mention the many sensing applications currently being implemented. In some areas optics are implemented a little slower than originally anticipated. This is largely due to the cost. In fact, hardware costs in the telephone network were no problem compared with many other costs, such as way and cable installation, and the original implementations were carried out quickly rather than cost-effectively. As a result, the telecommunications solutions, which were the first to be implemented, could never be transferred directly to various other applications. Today, however, the telecommunications market is competitive, as are the market for data communications and others, and new cost-effective solutions are emerging. There are many applications in which optics compete.

This book covers basic concepts of optical and wireless optical communications systems, optical networks and amplifiers, optical multiplexing and demultiplexing for optical communications systems and network traffic technology. This book will serve as a valuable tool for students, practitioners as well as researchers.

# TABLE OF CONTENTS

7

# A MAGNETICALLY LEVITATED PRECISE POINTING MECHANISM FOR APPLICATION TO OPTICAL COMMUNICATION TERMINALS

**Thomas Edward Donaldson Frame, Alexandre Pechev**

Surrey Space Centre, University of Surrey, Guildford, UK

## ABSTRACT

Increasing data bandwidth requirements from spacecraft systems is beginning to pressure existing microwave communications systems. Free-Space optical communications allows for larger bandwidths for lower relative power consumption, smaller size and weight when compared to the microwave equivalent. However optical communication does have a formidable challenge that needs to be overcome before the advantages of the technology can be fully utilized. In order for the communication to be successful the transmitter and receiver terminals need to be pointed with a high accuracy (generally in the order of ≤10 μradians) for the duration of communication. In this paper we present a new concept for the precise pointing of optical communications terminals (termed the Precise Pointing Mechanism). In this new concept we combine the separate pointing mechanisms of a conventional optical terminal into a single mechanism, reducing the complexity and cost of the optical bench. This is achieved by electromagnetically actuating the whole telescope assembly in 6 degrees-of-freedom with an angular resolution of less than ±3 μradians within a 10 (Az. El.) field of view and linear resolution of ±2 μm. This paper presents the new pointing mechanism and discusses the modelling, simulation and experimental work undertaken using the bespoke engineering model developed.

## KEYWORDS

Optical Communications; Magnetic Levitation; Control; Pointing; Active Anti-Vibration

## 1. INTRODUCTION

Communication is an ever growing and advancing field that is driven by many factors. As space borne instrumentation and data gathering/processing systems advance, an inherent increase in the data rate and bandwidth of the communications system follows. Optical terminals can provide advantages such as larger data bandwidths, lower relative power consumption, and smaller size/weight over microwave systems due to the nature of the medium used [1]. Also as the RF communication frequency spectrum becomes more crowded an optical link can offer an attractive alternative without the need for introduction of complex channel filters and algorithms [2] along with a greater immunity and security. However due to the narrow divergence of the medium used in optical communications, a very accurate and responsive pointing system is required in order to maintain the link between the communication terminals. A pointing error greater than a few micro radians can dramatically reduce the power seen on the receiver. Therefore a stable and precise pointing mechanism is required for the duration of communications.

Generally optical terminals consist of a fixed Telescope acting as the antenna, and a series of mirror/optics assemblies (the Coarse Pointing Assembly (CPA), Fine Pointing Assembly (FPA) and the Point Ahead Assembly (PAA)).

The CPA traditionally consists of a gimbal mechanism using stepper motors and gear boxes to achieve movement in azimuth and elevation, such as in the ESA SOUT terminal [3]. An alternative is to use a form of hybrid stepper motor as in the ISLFE optical terminals [3]. The FPA consists of a pair of mirrors that work together to focus the beam onto the receiver unit. In the SOUT terminal the FPA uses a single mirror that is manipulated using a permanently excited DC motor [3]. In the ISLFE terminal however the FPA consists of a single mirror that is connected to Lorentz force actuators that manipulate it using capacitive sensors to control the angular position [4].

Another major requirement of the pointing system is to overcome spacecraft disturbances and maintain the optical link for the duration of communication. Magnetic levitation technology provides a solution to overcoming disturbances, and offers infinite resolution that is only constrained by the sensor resolution and noise. In this research the functionality of the separate pointing mechanisms are combined into a single unit that can reject disturbances shown in [5], [6] and fulfils the requirements of the separate mechanisms [7,8]. This single unit will magnetically levitate and actuate the entire telescope assembly to reduce the number of individual optical elements and reduce complexity of the optical bench.

Magnetic levitation has been considered and implemented in many areas and differing fields, and achieved high precision positioning, as demonstrated by [9] and [10]. However this technology has not been applied to a space-borne magnetically levitated optical telescope. The system developed by [9] levitates a 6-DOF triangular platform that demonstrates a 5 nm position resolution for 6 axes but has a very small travel range in the order of micro-meters. Used effectively this technology can provide the accuracy requirements in order to maintain an effective optical communications link [9,11,12]. Applications of this technology are terrestrial inter building links, space-borne deep space communications, or a GEO-LEO relay satellite in the micro-sat range. The technology presented in this paper could also be adapted to form part of the pointing mechanism of an imaging system where a form of active anti-vibration would be advantageous and improve images [13]. Another application due to recent ESA activity is a direct Earth to Mars link that would significantly improve communications [14].

The purpose of this paper is to introduce the Precise Pointing Mechanism developed here at the Surrey Space Centre. An engineering model that can completely magnetically levitate and actuate a 60 mm telescope antenna in all 6 degrees of freedom has been developed and used to demonstrate the use of magnetic levitation to provide two pointing resolutions (±10 mRadians and ±3 μRadians) for application to an optical communications scenario.

This mechanism is developed to be an enabling technology for use in terrestrial and free-space optical communications, or applications where active anti-vibration and precise pointing would be advantageous. The design is scalable dependent upon the gain requirements of the telescope.

This paper discusses an example application of the engineering model developed, gives an overview of the control strategy used and finally presents some results to demonstrate the resolution of the terminal.

# 2. OPTICAL COMMUNICATIONS USING THE PPM

This section presents some applications that can be considered for the engineering model PPM presented in this paper. Using the PPM as the actuation mechanism for optical terminals, it is envisaged that the field of view of the PPM would require that the telescope assembly element be mounted on each spacecraft and Pre-set (As shown in **Figure 1**) to point in the direction of the next spacecraft to allow communications with the target satellite.

As can be seen from **Figure 1** the OCT is not aligned to the spacecraft axis it is aligned to point at the target spacecraft. This mounting should be configured so that it is aligned to compensate for the separation of the spacecraft at a pre-defined altitude. Assuming an orbit with an eccentricity of 0.001 (near circular) an assumption for height variations between the two spacecraft is made at ±10 km. If the link distance is in the region of 1000 km then 8% of the terminals azimuth and elevation range is used to compensate for altitude errors. Assuming that the typical attitude error is 1°, the ±5° azimuth and elevation range of the terminal of the engineering model in this research will be able to adequately compensate for altitude and attitude errors of the two spacecraft. If the link distance reduces to 200 km then the pointing requirement for the terminal changes to 77% of the ±5° limit of the engineering model.

Considering a GEO satellite to LEO satellite link the usefulness of the PPM can be demonstrated. The GEO satellite has a single PPM terminal mounted on a slip ring on the satellite that allows communication with the LEO satellite. A LEO satellite that has the same orbit properties as the UK-DMC also has a single PPM mounted on a slip ring on it to facilitate communication with the GEO satellite.

The PPM on the GEO satellite is mounted on the Earth facing side of the spacecraft and has a Field of View (FOV) of ±5°. It has an increased azimuth range of 280° due to mounting on a slip ring. By modelling this scenario in AGI's Satellite Tool Kit (STK) the access between the two satellites can be determined.

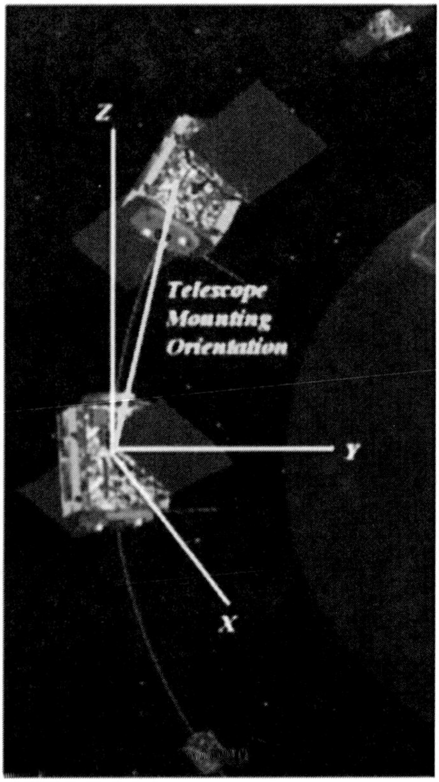


**Figure 1**. Pre-Set mounting configuration for current T1004- PPM applied to an OCT. Elevation is controlled by the PPM and the Azimuth range is controlled by the slip ring on which the PPM is mounted.

The actuation that is required from the terminal is shown in **Figure 2**. This shows the range through which the PPM terminal has to actuate in order to achieve the access between the GEO and LEO satellites.

The total access for the two week period is 80.82 minutes, at 1 Gbit/s this converts to a total of 4849 Gbit of data transferred. This limited Azimuth and Elevation range is a limitation of the engineering model and not the technology. Future work is considering increasing this range to ±10°.

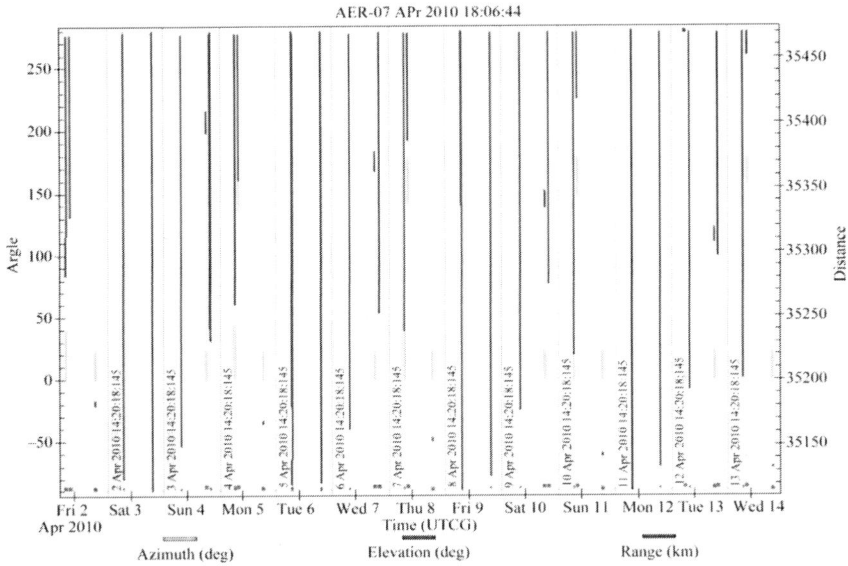


**Figure 2**. Azimuth and Elevation range for access between UK-DMC and a GEO relay sat. The azimuth range (green) is actuated by the slip ring on which the PPM is mounted. The Elevation (red) is actuated by the PPM and the change in link distance is shown in blue.

# 3. THE PRECISE POINTING MECHANISM (PPM)

This research differs from existing Optical Communications Terminals (OCT) by magnetically levitating the entire telescope assembly of the OCT using a combination of reluctance force electromagnets and permanent magnets. The telescope element is then actuated in 6 degrees of freedom using the electromagnets to compensate for optical pointing errors, thus maximizing received power on an Avalanche Photo Diode (APD). For this evaluation model the pointing controller can compensate for pointing errors within a 10° FOV. By actuating the telescope of the OCT the complexity and number of optical components required in the optical bench is reduced. This has the effect of reducing the cost and complexity of the optical bench. A trade off is an increase in power consumption, but this is offset by a saving in mass.

The antenna of this optical communications system consists of a telescope providing a gain to overcome transmission losses and maximize the amount of optical power received from the transmitted source, and thus maximizing the signal to noise ratio and reducing the probability of a bit error rate. OCT applications use reflecting telescopes that utilise the property of paraboloidal mirrors to concentrate and magnify the optical signal. In traditional systems the telescope remains fixed while mirror assemblies are manipulated in order to provide the pointing. This research differs by manipulating the entire telescope assembly in order to fulfill the requirements of the separate mirror assemblies. The operation of the optical system of the PPM is shown diagrammatically in**Figure 3**.

The type of telescope used in the PPM engineering model will be a Cassegrain configuration. The main driving force for the selection of this type of telescope is the gain and speed achieved from a very small compact terminal (when compared to other types such as Newtonian). A typical Cassegrain configuration consists of two mirrors (a primary parabolic, and a secondary mirror), where the primary has a central aperture drilled that allows the focused light to be reflected through it from the secondary mirror. The focal length of this telescope assembly has been designed to give an F-number of 5 which combines speed and brightness of image.

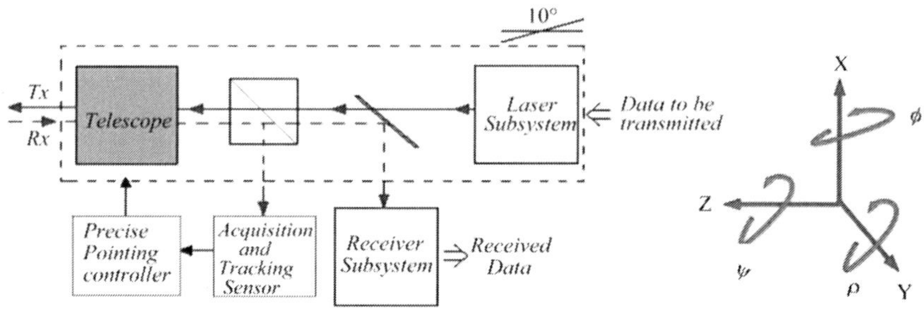


**Figure 3**. Configuration of OCT using the PPM as the actuation method for the telescope assembly. The optical axis of the telescope assembly is aligned with the Z axis shown.

## 3.1. System Configuration

The PPM developed in this research consists of eight reluctance force electromagnets and four pairs of permanent magnets that are used to levitate and actuate the telescope element of the OCT that consists of a catadioptric Schmidt Cassegrain spider configuration. A computer aided design (CAD) image of the evaluation model is shown in **Figure 4**. The model shown in **Figure 4** is only a receiver version of the PPM OCT, but we also have a version that includes a laser transmitter (with some additional optics and different sensor mounting).

The system uses two nested control loops and feedback networks to provide coarse pointing and fine tracking. Coarse pointing is achieved through the use of three low cost Eddy Current Probes (ECP) mounted underneath the telescope assembly (PRS-04 Fabricated by Sensonics). These provide positional data with accuracy around 2 μm which is not adequate to sustain an optical link (as will be discussed later). Fine tracking occurs once an optical signal is present on entrance of the telescope and thus focused onto the Position Sensing Device (PSD), the PDM- 10 produced by On-Trak Photonics. This is a dual axis PSD and tracks the laser beacon to an accuracy of 0.1 μm (±6 μm seen on the active area of the PSD) which can sustain the proposed optical link without excessive degradation of the communication channel. These measured errors in pointing are then used by the controllers to actuate the telescope to reduce the overall pointing error.

The engineering model was fabricated by the workshops of the University of Surrey with some parts. Such as the laminated electromagnet cores being sub contracted. **Figure 5** shows the completed PPM setup in the SSC laboratories, mounted on a passive anti-vibration table. This is to reduce the influence of stray vibrations on the PPM during experimentation.

The engineering model was designed to demonstrate the levitation and actuation of the telescope of an OCT using low cost COTS, and to demonstrate the use of layered control loops combining lower resolution sensors and low cost higher resolution sensors, resulting in a range of pointing accuracies required to maintain an optical link.

Although the data rate of the OCT scenario is not evaluated in experimentation the divergence of the link scenario is facilitated using a small lens mounted on the aperture of the laser. This diverges the laser at an angle of approximately 30°. The entrance pupil to the telescope assembly is then 150 mm away from the laser source which allows the optics (which were designed for a 3000 km optical link) to focus the laser source back to a fine spot on the PSD. It was found during experimentation that the quality of the low cost laser introduced a small positional error caused by the elliptical beam shape at the exit of the laser.

A system diagram that describes the configuration of the PPM experiment is shown in **Figure 6**.

The PPM OCT is divided into a number of different elements that have been developed to achieve a pointing resolution to sustain a free-space optical communications link.

## 3.2. System Model

In order to simulate the PPM system a model was developed in Matlab that represents the dynamics of the ter minal and the hardware from which the engineering model is constructed. **Figure 6**shows the system diagram that contains models of the Telescope Antenna dynamics, Sensor Amplifiers, Actuator Coils and Current Amplifier models in order to generate an accurate system model for development of the control system. This model was implemented in the Simulink environment of Matlab and used to simulate performance of the PPM before implementation on the bespoke hardware. The model initially included linearised models of the electromagnetic cores in order to allow the development of the control system. Once this was completed and demonstrated in simulation the model for the electromagnetic actuators was replaced with its non-linear version to model the real life behavior of the actuators within the system. Experimentation was also undertaken to ascertain the resolution of the ECP's and fine tracking sensor so that they could be modeled as accurately as possible. Finally noise from the bespoke precise current amplifiers that drive the current in each actuator was found experimentally for each actuator channel and then modeled for completeness.

In hardware all but the "Computer" elements (shown in the green box) are replaced with their hardware equivalents. National Instruments 6023E Analogue to Digital Converter (ADC) card and 6703 Digital to Analogue Converter (DAC) card are used as the interface components between the PPM hardware, and the host computer running XPC real time environment on which the control system is implemented.

**Telescope Assembly Dynamics**

The configuration of the Telescope Antenna of the PPM contains both translation and rotation components.

The behaviour of the PPM in translation along the X, Y, or Z axis is governed by Newton's Second Law as in Equation (1),

$$F_X = M_{PPM}\alpha_X \tag{1}$$

where $F_X$ is the force acting along the X axis, $M_{PPM}$ is the mass of the telescope assembly and $\alpha_X$ is the acceleration along the X axis. The accelerations ($\alpha_X$, $\alpha_Y$ and $\alpha_Z$) are found in the PPM by taking the second derivative of the terminal displacement along each respective axis, which is measured through the use of the coarse sensors (ECP's).

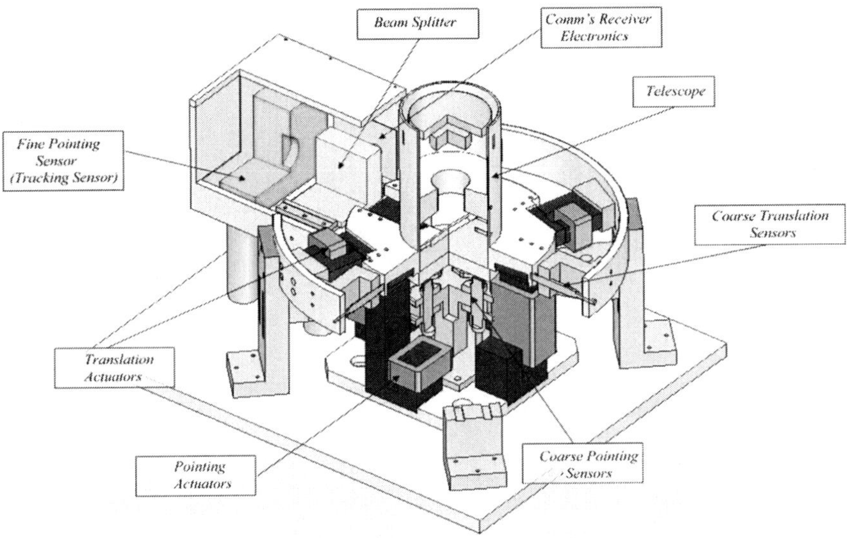


**Figure 4**. CAD image showing layout and elements of the PPM.

The Telescope Assembly (shown in **Figure 7**) is designed so that Centre of Mass (CM) is also the centre through which the Telescope Assembly rotates. This location is shown in **Figure 7**, and occurs in the centre of the 45° mirror that is used to divert the focused laser spot onto the fine tracking sensor. This reduces the complexity in the control and reduces the amount of power needed to rotate the telescope. If the CM of the Telescope Assembly was located at the entrance pupil more power would be needed in the actuators to maintain stability of the system.

The Inertia of the telescope assembly shown in **Figure 7** can be described as Equation (2).

$$I_{PPM} = \begin{bmatrix} 264 & -0.005 & -0.008 \\ -0.008 & 263 & 0.008 \\ 0.008 & -0.005 & 283 \end{bmatrix} \times 10^{-6}\ \text{kgm}^2 \tag{2}$$

**Figure 5**. Implemented PPM mounted on anti vibration table.

These align with the X, Y and Z axis of the telescope assembly. Using Equation (2) in Euler's equation of motion to describe the rotational motion of the telescope assembly, we yield Equation (3).

$$T_X = I_{XX}\dot{\omega}_X + \omega_X \times I_{XX}\omega_X \tag{3}$$

where the torque about the X axis is given by $T_X$, (and similarly for the Y and Z axis, $T_Y$ and $T_Z$respectively), $\omega_X$ is the angular acceleration about the X axis.

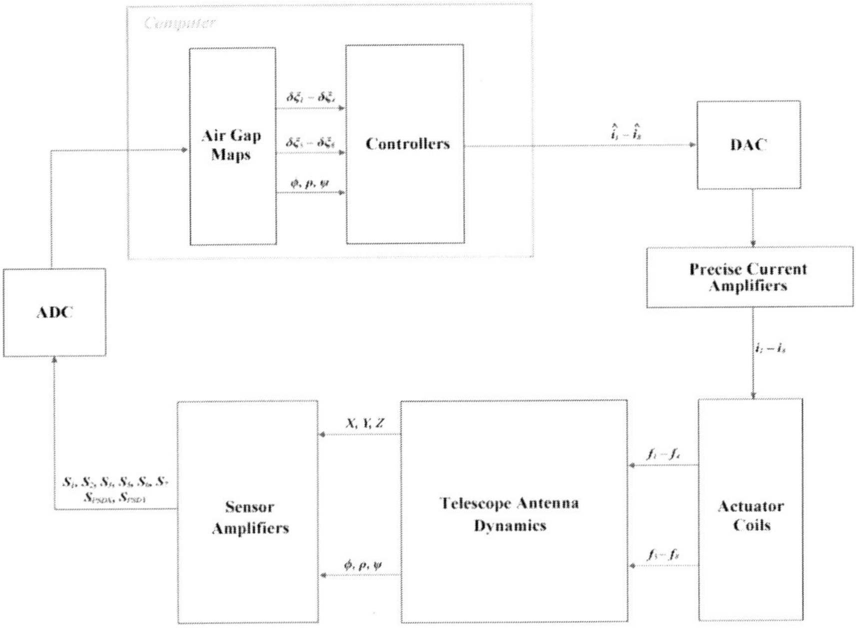


**Figure 6**. Overview of the setup of the PPM experiment that is shown in the photo of **Figure 5**.

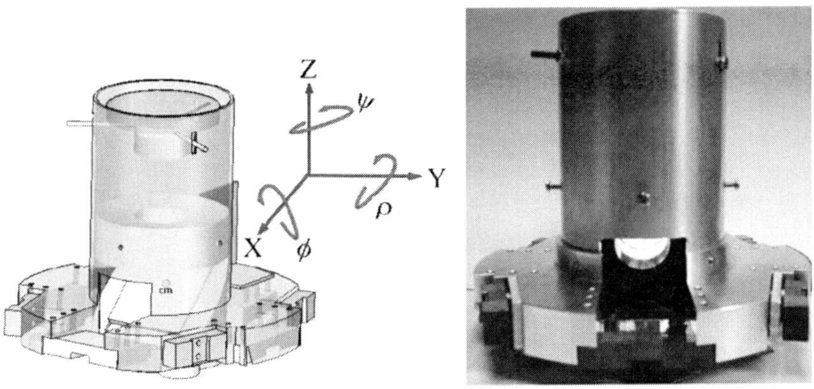


**Figure 7**. Illustration showing the location of the Centre of Mass of the Telescope Assembly (left), and a photo of the fabricated Telescope Assembly for use in the engineering model (right).

## 3.3. Optical Bench

This part considers the simulation of the antenna element of the OCT and considers some sources of error that will affect the performance of it, which will impact the fine pointing resolution of the PPM. An F5 catadioptric Cassegrain telescope has been developed as this offers a good compromise between image intensity at the focal plane and speed. A note to consider is that this Telescope Assembly would need to be used at least 15° off the sun-line in order to prevent damage to the receiver and optics. This requirement can be modified by developing baffles. The 60 mm diameter parabolic mirror and telescope aperture gives a gain of approximately 106 db generated by the telescope to offset some of the losses associated with free space optical communications.

The telescope assembly is tilted about φ and ρ (X and Y shown in **Figure 7** to correct for pointing errors, while the remaining four degrees of freedom remain fixed (X, Y, Z and rotation about Z, ψ).

**Figure 8** shows the ray trace for the engineering model with a zero degree pointing error in the incoming rays. The laser source is modelled as a Laser Diode placed 150 mm away from the entrance to the telescope with a 30° divergence angle. This is representative of the engineering model configuration and the divergence angle of the laser is achieved through the use of a diffusing lens mounted at the exit of the laser source.

In **Figure 9** the incoming rays from the laser source are focused onto the centre of the Tracking Sensor and APD (via the beam splitter hard mounted onto the stator element of the terminal). The radius of the focused laser signal spot is approximately 2 microns RMS seen on both the centre of the Tracking Sensor and the APD. An image of the spot using relative intensity values can be seen in **Figure 9**. The relative illumination on the fine tracking sensor and the spot position can be seen in Figures 10 and 11 respectively.

## 3.4. Magnetic Actuators

The PPM consists of two electromagnetic systems that provide levitation and control over the pointing and translation of the telescope antenna in 6 Degrees of Freedom (X, Y, Z, φ, ρ and ψ). Two sets of electromagnetic cores were designed and manufactured by Photofabrication.

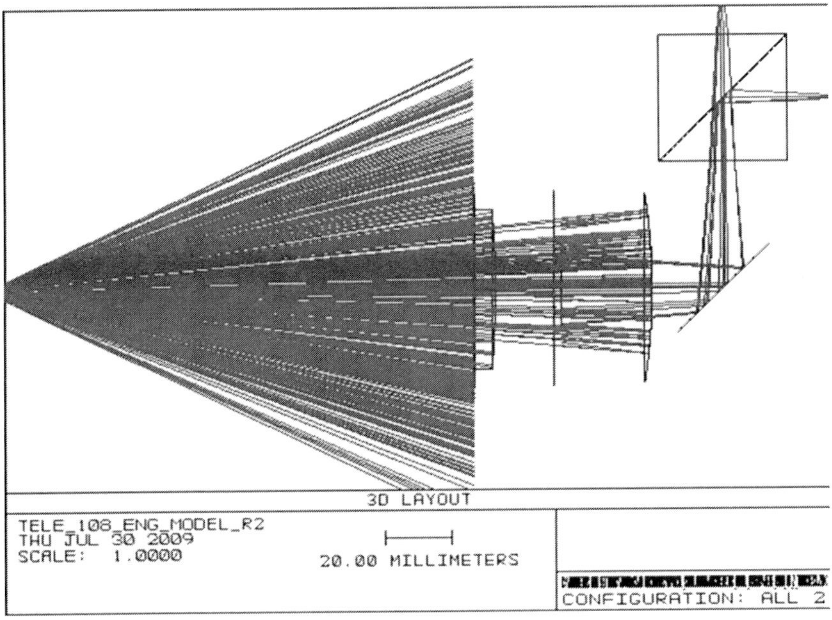


**Figure 8**. Ray trace image of terminal with 0º pointing error.

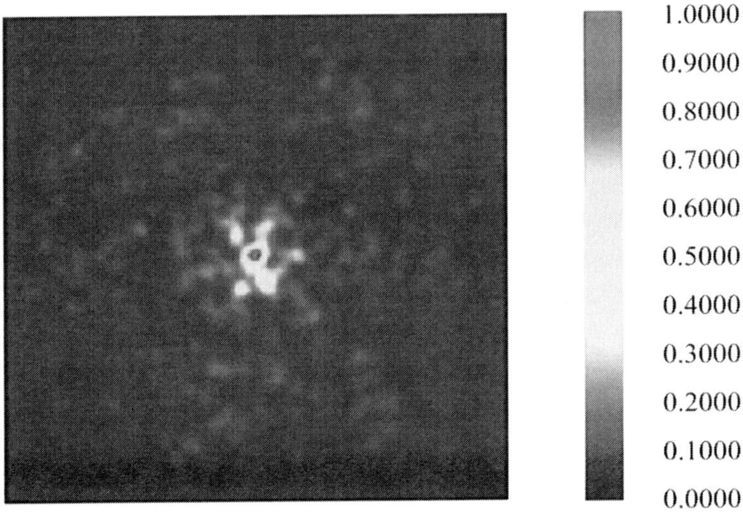


**Figure 9**. Relative intensity spot image on the tracking sensor with a 0º pointing error. This slide is 10 μm square at a resolution of 128 by 128 pixels.

The magnetic systems are illustrated graphically in **Figure 12**.

The actuators highlighted in blue control translation along Z, rotation about X (φ) and rotation about Y (ρ). The actuators highlighted in green control translations along X and Y and rotation about Z (ψ).

Pointing in this context refers to φ, ρ and ψ, and translation refers to linear motion along axis X, Y and Z.

Permanent magnets (PM's) are used in a repulsive configuration that have been sized and selected in order to provide a levitation force along the Z axis (**Figure 14**) are shown in **Figure 13**. The repulsive force generated by these permanent magnet pairs serves two purposes. For Earth testing of the PPM this force offsets the effects of gravity on the mass of the telescope antenna. It also provides a force that opposes the reluctance force generated by the four electromagnets, thus allowing control over the Z displacement. The forces produced by the pointing actuators for this engineering model total a net force of 12.04 N.

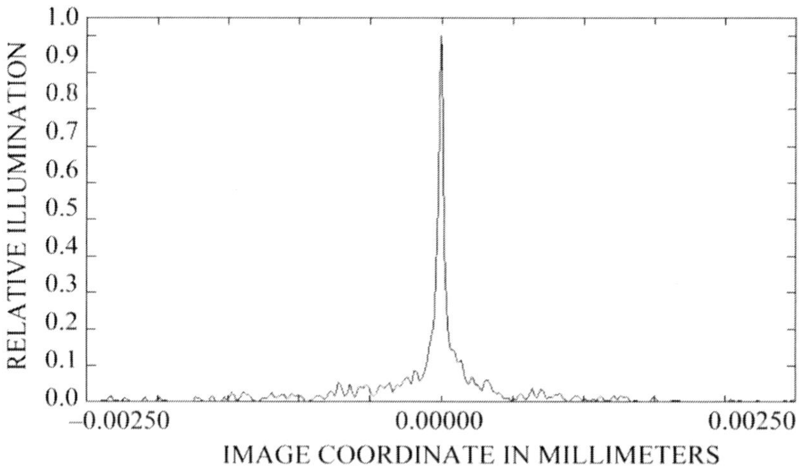


**Figure 10**. 2D relative intensity on the X axis of the tracking sensor with 0º pointing error.

The force allows the effects of gravity on the telescope assembly to be offset, as well as providing a net force of approximately 0.04 N (in reference position) for the pointing actuators ($f_1$, $f_2$, $f_3$ and $f_4$) to overcome, allowing levitation to be achieved.

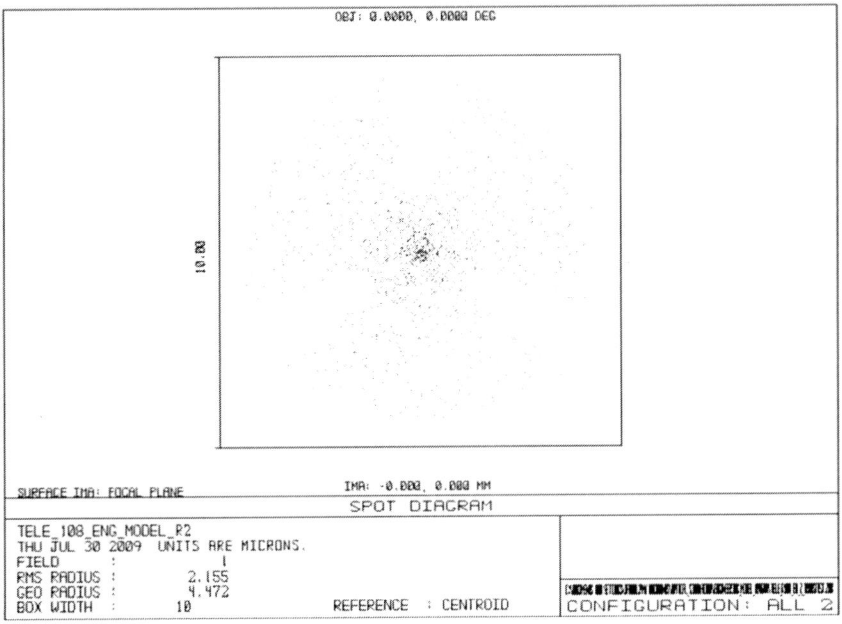


**Figure 11**. Spot size and position on the active area of the tracking sensor with 0º pointing error.

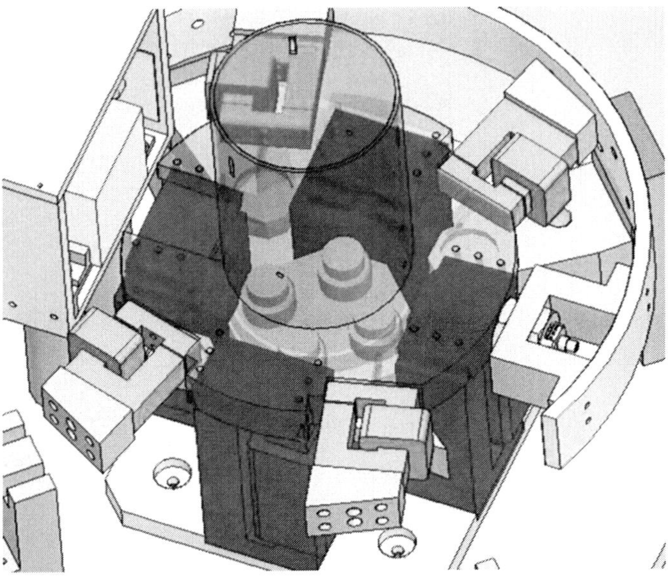

**Figure 12**. Illustration showing the magnetic systems that levitate and actuate the PPM.

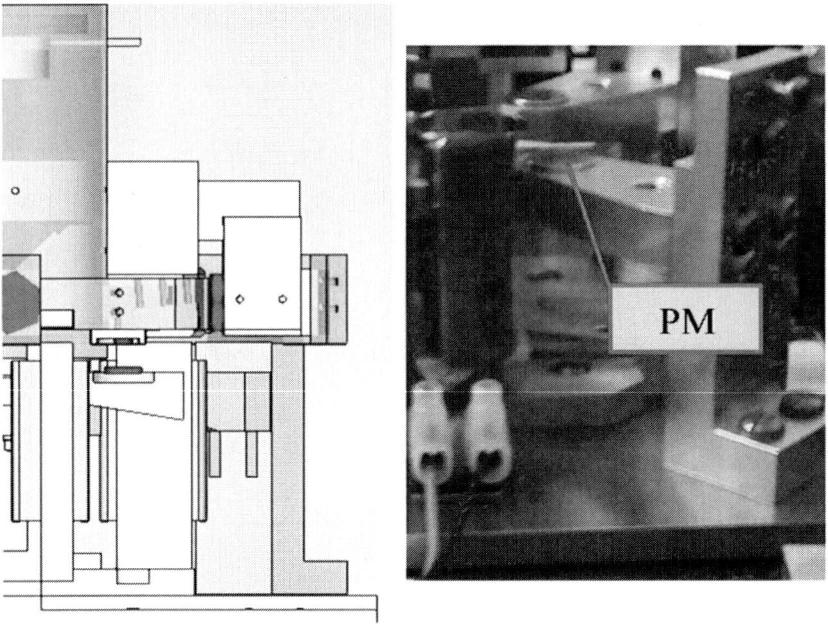


**Figure 13**. CAD image showing position of the permanent magnets highlighted in red (left) with a photo showing the position of the lower permanent magnet within the PPM. The top PM is embedded in the telescope assembly (right).

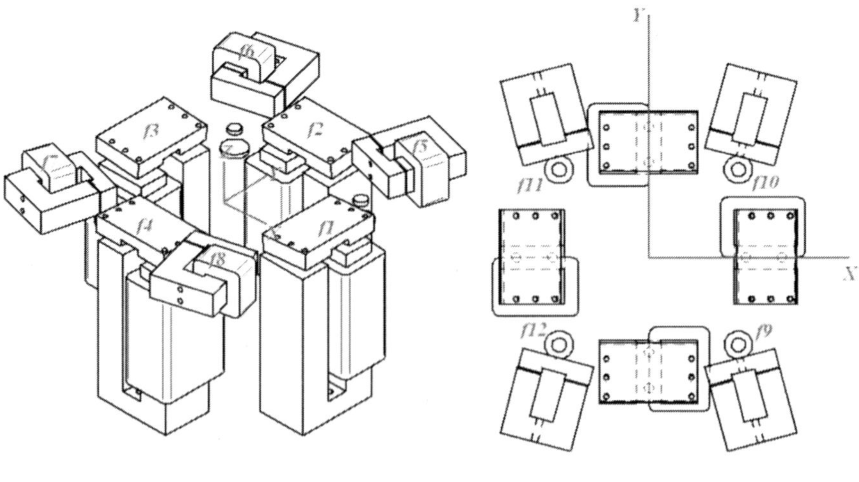


**Figure 14**. Layout showing placement of electromagnets and forces.

The repulsive forces produced by the permanent magnets is a function of the air gap between the pairs, therefore this is adjustable in order to allow these forces to be tuned on the engineering model.

The permanent magnets also produce undesired forces that induce a torque ($T_Z$) about the Z axis, and unwanted translation along X and Y.

The effects of these are reduced as much as possible through offsetting the positions of the permanent magnets. Through magnetic modelling using the Ansoft Maxwell 3D package the unwanted forces generated were found to be less than 10% of the force produced by the electromagnets when a full 5° (scenario in which worse case unwanted disturbances are experienced), rotation about $\varphi$ and $\rho$ was commanded, so there is adequate margin in the electromagnets to overcome these disturbances.

## 3.5. Control System

The control architecture of the PPM consists of six Single Input Single Output (SISO) decoupled controllers that each controls a single degree of freedom utilising a coarse sensor network that comprises Sensonics PRS04 and PRS02 Eddy Current Probes.

A further two controllers are then implemented using a fine tracking sensor that consists of the On-Trak PSM-10 position sensing device (PSD). This control strategy is summarised in **Figure 15**.

It is this combination that allows for the two resolutions achievable using the PPM (Coarse and Fine). The telescope assembly is actuated by manipulating the magnetic field in order to change the displacement of the telescope assembly.

In essence it obeys the relationship of Equation (4), where $F_X$ is the force along the X axis, $f_{bias}$ is the bias force applied to that actuator and $\delta\hat{f}$ is the demanded force from the control system to correct for disturbances on the telescope assembly.

$$F_X = f_{bias} + \delta\hat{f} \tag{4}$$

Using the relationship between the force produced by the actuators and the current required to generate that force (Equation (5)) together with the inverse of the terminal dynamics the change in each individual actuator $f_n$,

$$f_n = -\left(\frac{\mu_0 N^2 A_n}{4}\right)\left(\frac{i_n^2}{\xi_n^2}\right) \tag{5}$$

where $i_n$ is the current in the $n^{th}$ coil and $\xi_n$ is the air gap between the rotor and core parts (the rotor being attached to the telescope ring), $\mu_0$ is the permeability of free-space, N is the number of turns of conductor on the electromagnets core and $A_n$ is the area of a single actuator pole face. The force generated by the electromagnets is a reluctance force which is always attractive therefore the minus sign in Equation (5) represents this attraction.

The current demands from each controller are then mapped onto the required electromagnetic cores to generate these desired force changes to correct for disturbances and pointing errors influencing the telescope assembly. The controllers are implemented in the form of Lead-Lag compensator's in the form of Equation (6).

$$K = K_X \times \left[\frac{\sqrt{\beta_X}}{\beta_X} \times \frac{s + \sqrt{\beta_X}\,\omega_{cX}}{s + \sqrt{\beta_X}\,\frac{\omega_{cX}}{\beta_X}}\right] \times \left[\frac{s + \omega_{iX}}{s}\right] \tag{6}$$

where $K_X$ is a gain term for the axis in question, $\beta_X$ is the damping coefficient and $\omega_{cX}$ and $\omega_{iX}$ are the corner frequencies of the lead and lag terms respectively. The gain term $K_X$ was derived from the dynamic model for the telescope assembly when the gain of the controllers causes the magnitude to be 1 at the cross-over frequency. The gains are then tuned once when implemented in simulation or on the hardware. When implementing the control strategy on the simulation model or hardware the controllers did not initially include an integration term (the lag term in the controller), which was added once the system was stabilised. This made it easier to achieve levitation and actuation, although as expected the steadystate error was quite high for each axis.

It was at this point that the fine controllers were introduced to the system to provide an extra level of feedback from the fine tracking sensor that manipulates $\varphi$ and $\rho$ to compensate for pointing errors using the position error of the laser spot on the PSD.

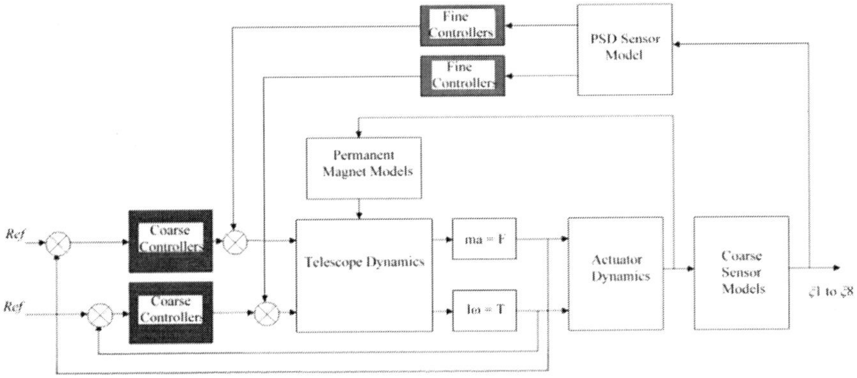


**Figure 15**. Overview of control strategy to levitate and track an optical signal. In red are the coarse controllers that levitate and actuate the telescope assembly using the eddy current probes, and in blue are the fine controllers that tilt the telescope about X and Y (φ and ρ).

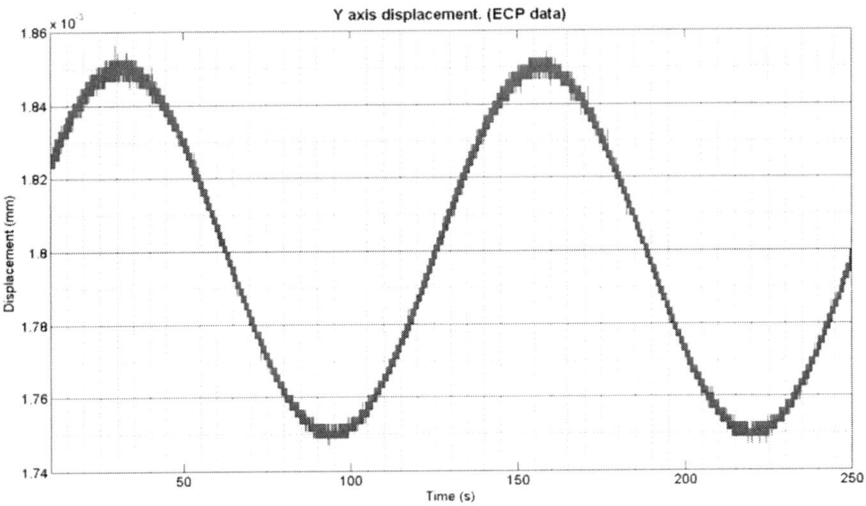


**Figure 16**. Response of the Y axis ($S_6$ ECP data) to a sine wave using the coarse control loops and sensor network.

They take the same form as shown in Equation (6), and did not include the lag term initially. A commanded sine wave was used to demonstrate the response of the terminal along the Y axis. The response of the telescope assembly, along with the quantisation noise can be seen in

Figures 16 and 17. This also shows the step sizes achievable along this axis.

A commanded rotation for φ was driven with a similar sine wave (±8 milli-radians @ 0.05 rad/s). Rotation φ is driven by a commanded torque $T_X$ about the X axis. Figure 18 shows the rotation of the telescope assembly in response to the commanded position changes.

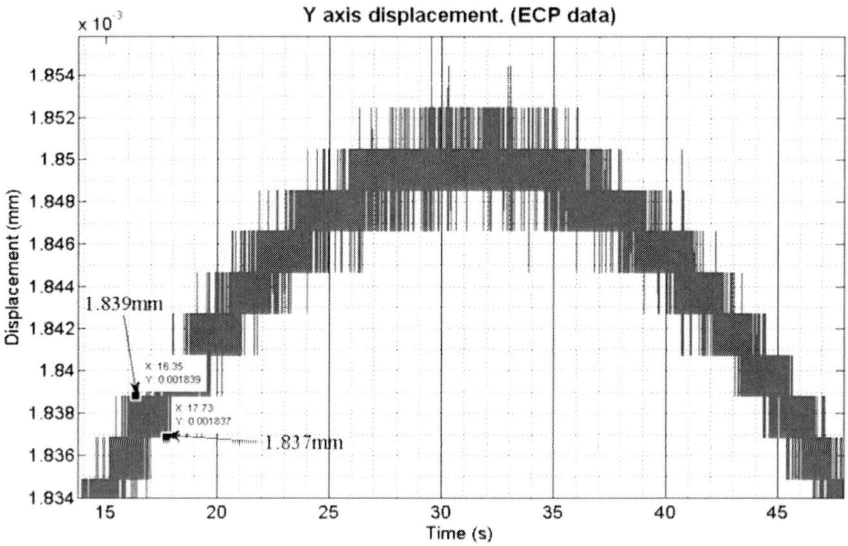


**Figure 17**. Zoomed in top section of Y axis response. The quantisation noise can clearly be seen and the positional resolution of the axis is ±2 μm.

**Figure 19** shows the position of the terminal once the Laser is passed through the telescope optics and incident on the active area of the PSD.

Due to the mechanical fixing of the PSD a positive change in angle φ corresponds to a negative peak on the active area of the fine tracking sensor (the PSD). This is illustrated in the time differences in positive peaks in Figures 18 and 20.

The positional resolution of the telescope assembly while under coarse control is within ±5 μradians for both φ and ρ, which is short of the resolution required to maintain an optical communications link.

However during simulation of the engineering model the resolution of the eddy current probes was expected to be lower than the 2 μm (expected to be around ±4 μm) that could be expected under typical operating conditions. This is shown in the histogram of **Figure 21**.

During implementation the performance of the ECP sensors was worse than modelled in simulation (In the order of ±6 μm. This was due to the metal used in fabrication of the telescope assembly and the small size of sensing area available on the telescope assembly. The performance was dramatically improved however with the addition of very thin steel sheets that were glued to the sensing surfaces of the telescope assembly.

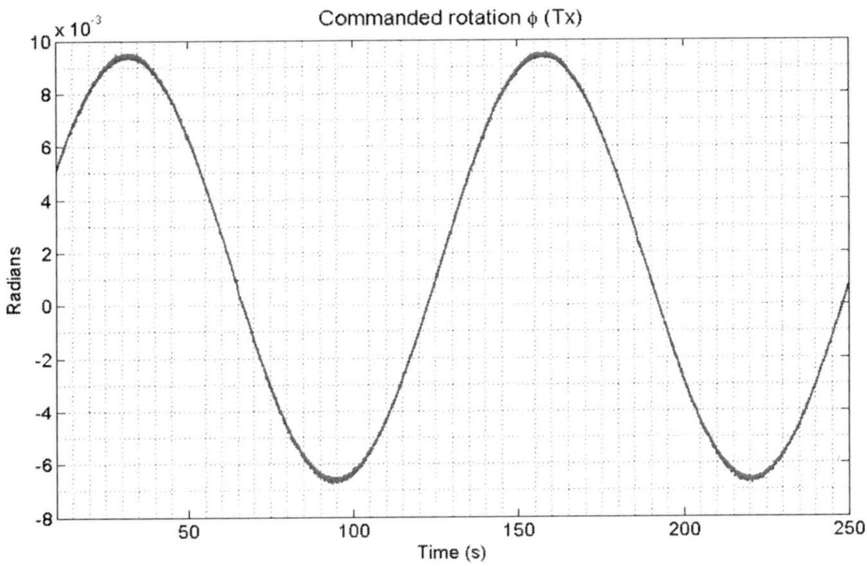


**Figure 18**. Response of the telescope assembly to a commanded rotation of φ.

This has the effect of dramatically increasing the conduction of the eddy currents which in turn improves the resolution and performance of the ECP's.

At this point if an optical signal is incident on the PSD tracking sensor then the fine control loops are activated and the terminal actuated to track the signal.

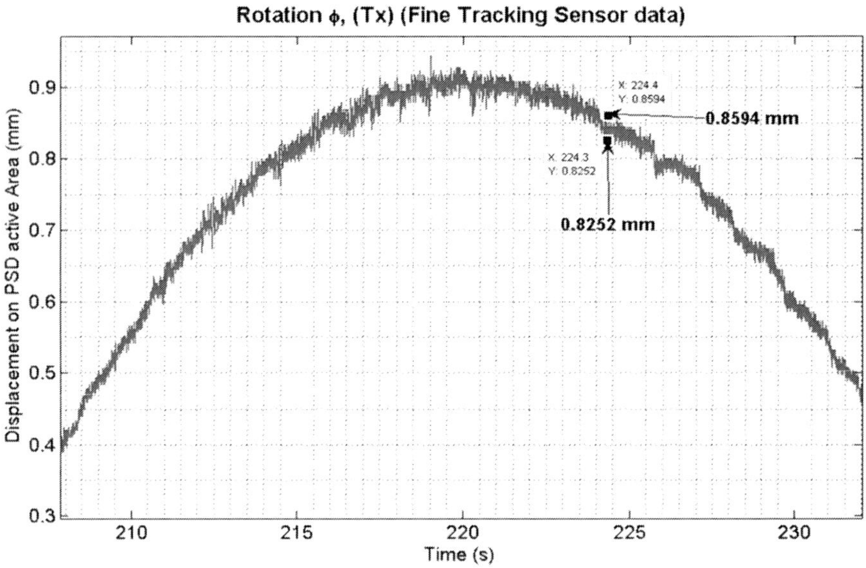


**Figure 19**. Rotation φ as seen on the respective axis of the fine tracking sensor (PSD). This corresponds to a resolution of approximately 0.18 mRadians under coarse control.

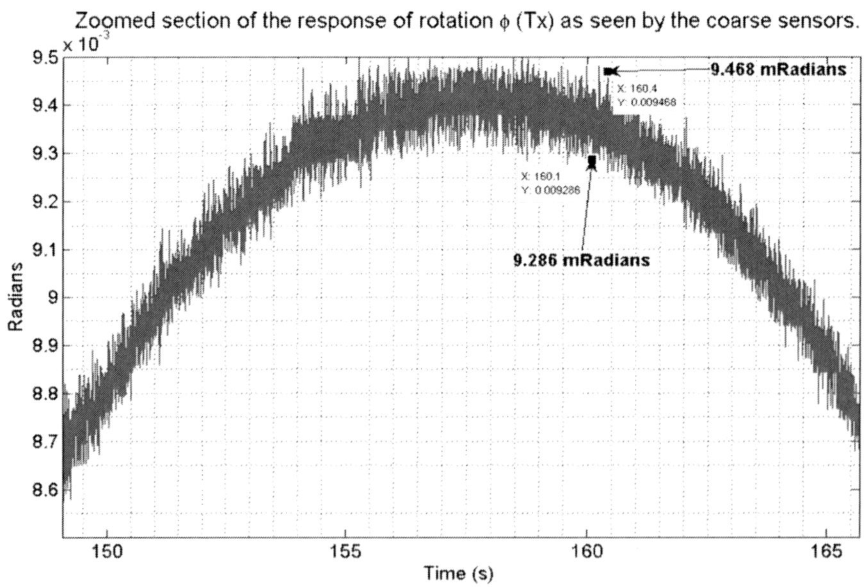


**Figure 20**. Zoomed view of the response of φ shown in **Figure 18**. This shows the positional resolution to be approximately 0.18 m Radians.

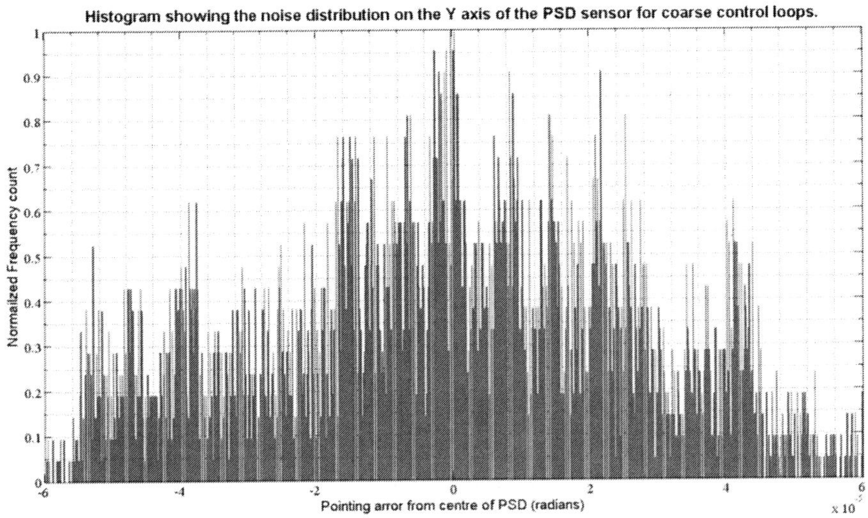


**Figure 21**. Pointing resolution of the simulated coarse control loops using the ECP sensors.

The resolution of the fine loop is plotted over the top of the coarse control loops to demonstrate the two resolutions achievable with the PPM of this research (Figures 22 and 23).

The histogram is taken over 400 seconds of steady state data. The experiment was repeated a number of times over a large range of pointing positions. The curve shown in black is the modelled pointing resolution requirements for a 3000 km optical link [15].

# 4. CONCLUSIONS

This research has seen the investigation, implementation and successful testing of a new concept for the precise pointing of, in this instance an optical antenna for communications. The realisation of the PPM allows for sub micro-meter, and sub micro-radian resolution in all six DOF over a larger displacement range than existing systems. This represents a significant step forward when compared with other mechanisms that utilise magnetic levitation, which, while achieving resolutions comparable to the PPM generally lack the displacement range.

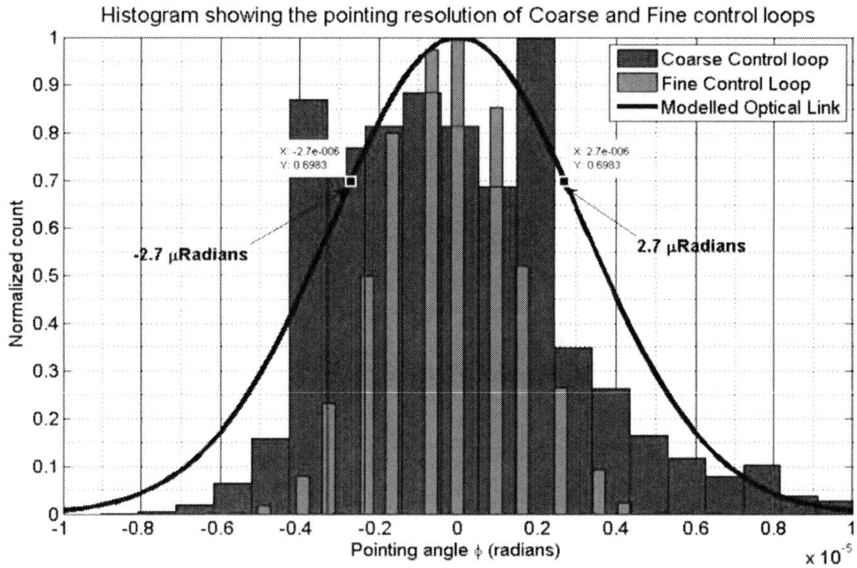


**Figure 22**. Coarse and fine pointing control loops plotted against the received optical power as a function of pointing angle error for φ.

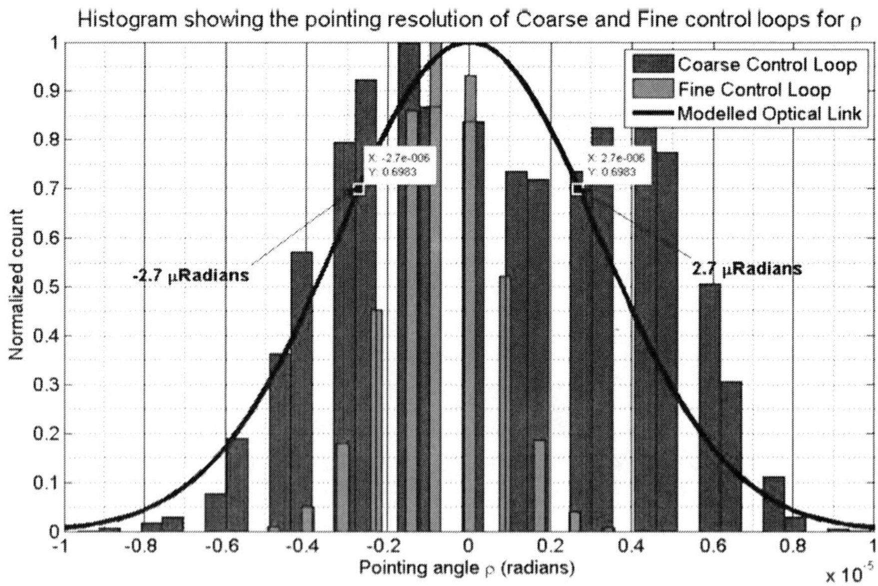


**Figure 23**. Coarse and fine pointing control loops plotted against the received optical power as a function of pointing angle error for ρ.

The control strategy presented allows the combination of low cost lower resolution sensors for measuring the position within 6 DOF, with a higher cost greater resolution sensor to achieve an overall increase in accuracy and resolution of the system. This strategy can be utilised to reduce the cost of implementation while retaining the resolution requirements within a system.

The frictionless environment, and contact less nature of the PPM means that the positional accuracy will not degrade as a function of mechanical wear. This also means that no lubrication is required, which can be a significant issue in space craft mechanism. Although the lifetime and consistency of operation cannot be evaluated in the lifetime of this research, no degradation or inconsistency in functionality could be observed during the eight months of experimentation.

## ACKNOWLEDGEMENTS

The authors would like to thank The Nuffield Foundation for their support of this project through grant NAL/32791, and ESA for their contribution under a pump priming fund (FRSF09/16).

## REFERENCES

1. S. Arnon, S. Rotman and N. S. Kopeika, "Optimum Transmitter Optics Aperture for Satellite Optical Communication," IEEE Transactions on Aerospace and Electronic Systems, Vol. 34, No. 2, 1998, pp. 590-596.

2. H. Wang and Q. Chu, "An Inline Coaxial Quasi-Elliptic Filter with Controllable Mixed Electric and Magnetic Coupling," IEEE Transactions on Microwave Theory and Techniques, Vol. 57, No. 3, 2009, pp. 667-673.

3. G. Baister and P. Gatenby, "The SOUT Optical Intersatellite Communication Terminal," IEEE Proceedings Optoelectronics, Vol. 141, No. 6, 1994, pp. 345-355.

4. G. Baister and C. Haupt, "The ISLFE Terminal Development Project Results from the Engineering Breadboard Phase", 20th AIAA

International Communication Satellite Systems Conference, 12-15 May 2002.

5. M. Toyoshima and K. Araki, "In-Orbit Measurements of Short Term Attitude and Vibrational Environment on the Engineering Test Satellite VI Using Laser Communication Equipment," Optical Engineering, Vol. 40, No. 5, 2001, pp. 827-832.

6. D. Bamber and P. Palmer, "Attitude Determination through Image Registration Model and Test-Case for Novel Attitude System in Low Earth Orbit," AIAA/AAS Astrodynamics Specialist Conference and Exhibit, Keystone, 21-14 August 2006.

7. S. Arnon and N. S. Kopeika, "Laser Satellite Communication Network-Vibration Effect and Possible Solutions," Proceedings of the IEEE, Vol. 85, No. 10, 1997, pp. 1646-1661.

8. R. Seiler and C. Allegranza, "Mechanism Noise Signatures: Identification and Modelling," 13th European Space Mechanisms and Tribology Symposium, Vienna, 23-25 September 2009.

9. S. Verma and W. J. Kim, "Six-Axis Nanopositioning Device with Precision Magnetic Levitation Technology," IEEE/ASME Transactions on Mechatronics, Vol. 9, No. 2, 2004, pp. 384-391.

10. J. Seddon and A. Pechev, "3Dwheel: 3-Axis Low Noise, High-Bandwidth Attitude Actuation from a Single Momentum Wheel Using Magnetic Bearings," 23rd Annual AIAA/USU Conference on Small Satellites, Utah, 10-13 August 2009.

11. Z. Ren and L. Stephens, "Laser Pointing and Tracking Using a Completely Electromagnetically Suspended Precision Actuator," American Institute of Aeronautics and Astronautics, Journal of Guidance, Control and Dynamics, Vol. 29, No. 5, 2006, pp. 1235-1238.

12. Y. Chen and M. Wang, "Modeling and Controller Design of a Maglev Guiding System for Application in Precision Positioning," IEEE Transactions on Industrial Electronics, Vol. 50, No. 3, 2003, pp. 493-506.

13. F. Ayoub, S. Leprince, R. Binet, K. Lewis and O. Aharonson, "Influence of Camera Distortions on Satellite Image Registration and Change Detection Applications", Geoscience and Remote Sensing Symposium, Vol. 2, No. 1, 2008, pp. II-1072-II-1075.

14. R. McKay and M. Macdonald, "Non-Keplerian Orbits Using Low Thrust, High ISP Propulsion Systems," 60th International Astronautical Congress, Daejeon, 12-16 October 2009.

15. T. Frame and A.Pechev, "Optical Communications Terminals with Precise Pointing," Proceedings of the 13th European Space Mechanisms and Tribology Symposium, Vienna, 23-25 September 2009.

2

# EFFECT OF POLARIZATION CONTROLLERS ROTATION ANGLES ON THE PERFORMANCE OF OPTICALLY GENERATED UWB-BASED COMMUNICATION SYSTEMS


**J. E. Kadum [1], R. S. Fyath [2]**

[1]Department of Electronic and Communications Engineering, Al-Nahrain University, Baghdad, Iraq

[2]Department of Computer Engineering, Al-Nahrain University, Baghdad, Iraq

Correspondence to: R. S. Fyath , Department of Computer Engineering, Al-Nahrain University, Baghdad, Iraq.



## ABSTRACT

One of the interesting methods reported in the literature for photonic generation of impulse radio-ultrawideband (IR-UWB) signals is based on a phase modulator (PM) and an asymmetric Mach–Zehnder interferometer (AMZI). The AMZI is electrically reconfigurable by employing a polarization modulator (PolM). UWB signals with different modulation formats are realized by adjusting the polarization controllers (PC) in the AMZI and the amplitude of the electrical drive signal to the PolM. This paper addresses the effect of PCs rotation angles on the performance of on-off keying (OOK) and biphase modulation (BPM) systems based on Gaussian monocycle and doublet UWB pulses. The simulation results reveal that there are optimum values of these rotation angles that yield minimum bit-error-rate (BER) of the received

UWB signals at the end of the transmission fiber link. Simulation results are presented using Optisystem Version 13.0.

## KEYWORDS

Ultrawidwband-over-fibre, Photonic generation, Photonic microwave delay-line filter

## 1. INTRODUCTION

Ultrawideband (UWB) signals are applicable for short rang wireless communication systems, such as indoor wireless communication networks, intra-vehicle networks and wireless sensor network [1-4]. One of the versions of UWB-based communication is impulse radio (IR) which attracts increasing interest due to its intrinsic advantages, such as immunity to multipath fading, being carrier free, broad bandwidth, and low-power spectral density (PSD) [5-9]. The Federal Communication Commission (FCC) has allocated 7.5 GHz of spectrum for unlicensed use of UWB devices with 3.1-10.6 GHz band, with PSD less than -41.3 dBm/MHz for indoor wireless communication [10-12]. The UWB signals are characterized by a short propagation range which is generally limited, by the extremely low PSD, to a few meters to tens of meters [13]. To overcome the short propagation, UWB-over-fiber (UWBoF) systems have been proposed to increase the transmission range [14-16]. Thus, generation and distribution of UWB signals in the optical domain have attracted considerable interest. Several methods have been proposed in the literature for photonic generation of UWB signals. Some of these methods are based on nonlinear optical loop mirror [17, 18], Mach-Zehnder modulators [19, 20], polarization modulators [21-23], nonlinear fiber optics [24-26], optical filters [27, 28], phase modulation-to-intensity modulation conversion [29], or cascaded refractive semiconductor optical amplifiers [30]. One of the interesting methods for photonic generation of IR-UWB signals has been presented by Pan and Yao [31]. They have performed a comprehensive investigation technique to implement on-off keying (OOK), bi-phase modulation (BPM), pulse position modulation, pulse amplitude modulation, and pulse shape modulation. The method is based on phase modulator (PM) and an asymmetric Mach-Zehnder interferometer (AMZI) as shown in Fig. 1.

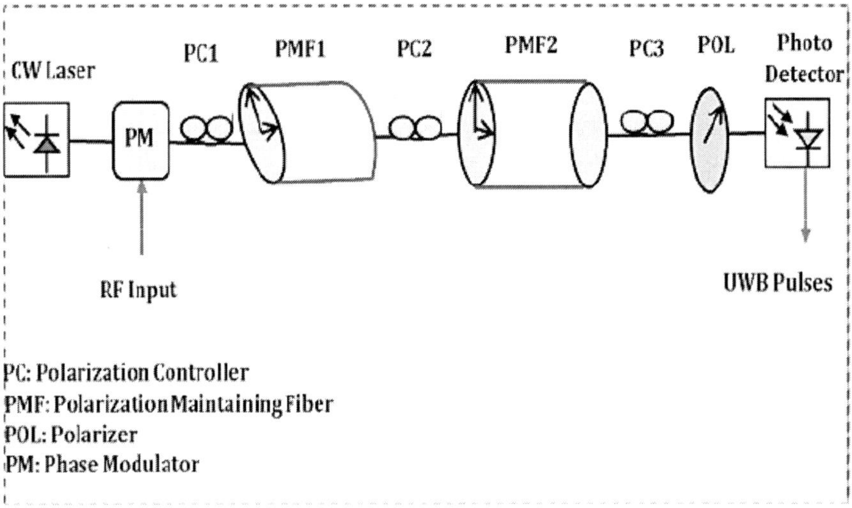


**Figure 1.** Schematic diagrams of the photonic microwave bandpass filter based on a phase modulator [31]

The AMZI consists of two sections of polarization maintaining fiber (PMF) followed by a polarizer. Three polarization controllers (PCs) are inserted to adjust the polarization states of lightwaves before PMF1, PMF2, and the polarizer. The photodiode (PD) is connected at the output of the polarizer to perform optical to electrical conversion. The AMZI is electrically reconfigurable by employing a polarization modulator (PoLM). UWB signal with different modulation formats can be realized by adjusting these polarization controllers in the AMZI and the amplitude of the electric drive signal to the PolM. The study in [31] assumes fixed values for the rotation angles of the PCs ($\alpha_1 = 0, 90°$ or $45°$, $\alpha_2 = \alpha_3 = 45°$) in order to generate the various UWB modulation formats. To the authors' knowledge, the effect of deviations from these angle values is not addressed in the literature and this issue is investigated in this paper. The analysis presented in [31] is extended to take into account the effect of PC rotation angles. The simulation results reveal that there are optimum values of these angles that yield minimum bit-error-rate (BER) for the received UWB signals at the end of the transmission fiber. It is worth to mention here that this study can be considered as a first step toward analyzing optically generated UWB signals in the presence of components impairments and addressing its effect on the performance of UWBoF systems. This topic attracts little attention in the literature and the analysis and the results reported in this paper can be used as a guideline to investigate other optically generated UWB systems.

# 2. ANALYSIS FRAMEWORK

## 2.1. System Description

Figure 2 shows a simplified block diagram of the reconfigurable UWB system under investigation. The block diagram reflects the main components of the system in Fig. 1. To simplify the analysis, each PMF is modeled here by using three subcomponents. First, a polarization beam splitter (PBS) is used to resolve the incident field into two orthogonal polarization components. Then a time delay τ is introduced to one of the polarization components before combining them using a polarization beam combiner (PBC). The system of Fig. 2 consists of a laser diode (LD) operating in continuous wave (CW) mode, a phase modulator, an asymmetric Mach-Zehnder interferometer, and a photodiode.

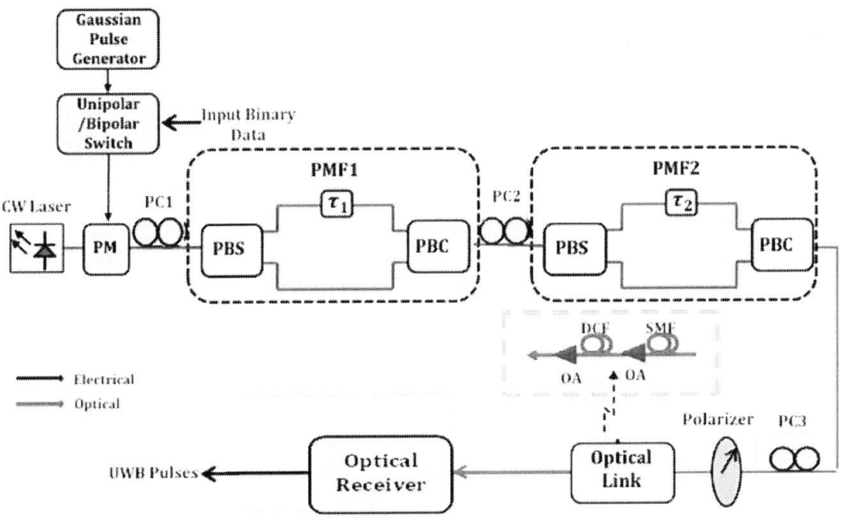


**Figure 2.** Configuration of the optical generation system. The insert shows the details of the transmission link

The AMZI consists of two sections of PMFs followed by a polarizer. Three polarization controllers PC1, PC2 and PC3 are inserted to adjust the polarization of the optical wave before PMF1, PMF2 and the polarizer. Practically, the PD is the front-side of the optical receiver which receives the optical signal after transmission through an optical link. The effect of this link will be not taken into account in the analysis and its effect will be addressed through simulation. The input binary data d(t) is used to control a unipolar or bipolar switch driven by a series of a periodic short-duration Gaussian pulses to obtain OOK or BPM signaling, respectively. The gated Gaussian pulses are used to modulate the phase of CW laser through a PM which is modeled here using Mach-Zehnder modulator. The repetition rate of the Gaussian pulses is set equal to the bit rate of the binary data. Thus each of the data bit is spread by a single short Gaussian pulse. It is worth to mention here that this generating scheme is different from that adopted in [31] where the Gaussian pulses and the binary data are applied to the PM and PolM, respectively. Note further that the system in Fig. 2 does not use a PolM and hence can be used to generate OOK and BPM signaling only.

## 2.2. System Analysis

The normalized optical field at the output of the PM along the two polarization directions can be expressed as [31]

$$e_{PM}(t) = \begin{cases} \exp[j\omega_c t + jm\, d(t) g(t)] \\ 0 \end{cases} \tag{1}$$

where $\omega_c$ is the angular frequency of the optical carrier, m is the phase modulation index, $g(t)$ is a normalized Gaussian pulse, and d(t) is the input binary data. The binary data d(t) is given by

$$d(t) = \begin{cases} 1 \\ 0 \end{cases} \text{OOK} \qquad d(t) = \begin{cases} 1 \\ -1 \end{cases} \text{BPM} \tag{2}$$

The complex transfer functions of PC and PMF can be expressed as

$$T_{PC}(\omega) = \begin{bmatrix} \cos\alpha \, e^{j\emptyset} & -\sin\alpha \\ \sin\alpha & \cos\alpha \end{bmatrix} \tag{3a}$$

$$T_{PMF}(\omega) = \begin{bmatrix} e^{j\omega\tau} & 0 \\ 0 & 1 \end{bmatrix} \tag{3b}$$

where $\alpha$ and $\emptyset$ are, respectively, the rotation angle and phase shift introduced by the PC while $\tau$ represents the time delay between the two polarization components introduced by the PMF. The complex transfer function of the cascaded optical components (three PCs and two PMFs) can be computed using the following relation

$$T(\omega) = T_{PC3}(\omega).T_{PMF2}(\omega).T_{PC2}(\omega).T_{PMF1}(\omega).T_{PC1}(\omega) \tag{4}$$

If the polarizer is adjusted to select the optical signal in the direction [1 0], the optical field at the output of the polarizer can be computed from

$$E_{POL}(\omega) = [1 \quad 0].T(\omega).E_{PM}(\omega) \tag{5}$$

where $E_{PM}(\omega)$ is the Fourier transform of the phase modulated signal and given by

$$E_{PM}(\omega) \equiv \mathcal{F}\{e_{PM}(t)\} = \begin{bmatrix} \mathcal{F}\{exp\,[j\omega_c t + jm_{eff}\,g(t)]\} \\ 0 \end{bmatrix} \tag{6}$$

where $\mathcal{F}\{.\}$ denotes the Fourier transform of the argument and $m_{eff}$ represents the effective phase modulation index; $m_{eff} \equiv md(t)$ with d(t) takes the values 0, 1, or -1 according to the logic states and signaling format. Following the analysis reported in [31], the optical field at the polarizer output can be expressed in frequency and time domains as follows

$$E_{POL}(\omega) = \{\cos\alpha_1 \cos\alpha_2 \cos\alpha_3 \exp(j[\omega(\tau_1+\tau_2)+\emptyset_1+\emptyset_2+\emptyset_3]) - \cos\alpha_1 \sin\alpha_2 \sin\alpha_3 \exp(j[\omega\tau_1+\emptyset_1]) - \cos\alpha_3 \sin\alpha_2 \sin\alpha_1 \exp(j[\omega\tau_2+\emptyset_3]) - \cos\alpha_2 \sin\alpha_1 \sin\alpha_3\} \times \mathcal{F}\{[\exp(j\omega_c t + jm_{eff}\,g(t)]$$

$$e_{POL}(t) = \mathcal{F}^{-1}\{E_{POL}(\omega)\} = -\exp(j\omega_c t)\Big\{\cos\alpha_2 \sin\alpha_1 \sin\alpha_3 \exp\big(jm_{eff}\,g(t)\big) + \cos\alpha_1 \sin\alpha_2 \sin\alpha_3 \exp\big(j[\theta_1 + m_{eff}\,g(t+\tau_1)]\big) + \cos\alpha_3 \sin\alpha_2 \sin\alpha_1 \exp\big(j[\theta_3 + m_{eff}\,g(t+\tau_2)]\big) - \cos\alpha_2 \cos\alpha_1 \cos\alpha_3 \exp\big(j[\theta_1+\theta_3+\emptyset_2 + m_{eff}\,g(t+\tau_1+\tau_2)]\big)\Big\}$$

where $\mathcal{F}^{-1}\{.\}$ stands for the inverse Fourier transform, $\theta_1 = \omega_c\tau_1 + \emptyset_1$ and $\theta_3 = \omega_c\tau_2 + \emptyset_3$.

# 3. GENERATION OF THE UWB SIGNALS

## 3.1. Gaussian Monocycle

To generate an UWB Gaussian monocycle, the polarization direction of the modulated light wave should be aligned with one principal axis of PMF1 (i.e., $\alpha_1 = 90^\circ$) and let $\theta_3 = 90^\circ$, $\emptyset_2 = 0$ [31]. Equation 8 is simplified to

$$e_{POL}(t) = -\exp\ (j\omega_c t)\{\cos\alpha_2\ sin\alpha_3 \exp\left(jm_{eff}g(t)\right) + \cos\alpha_3\ sin\alpha_2 \exp\left(j\left[\pi/2 + m_{eff}g(t+\tau_2)\right]\right)\} \quad (9)$$

Let $A = \cos\alpha_2\ sin\alpha_3$, $B = \cos\alpha_3\ sin\alpha_2$ and using the identities $\cos\left(x + \frac{\pi}{2}\right) = -\sin(x)$ and $\sin\left(x + \frac{\pi}{2}\right) = \cos(x)$ yields

$$\begin{aligned} e_{POL}(t) &= -\exp\ (j\omega_c t)\{A[\cos\left(m_{eff}g(t)\right) + jsin\left(m_{eff}g(t)\right) \\ &\quad + B[-\sin\left(m_{eff}g(t+\tau_2)\right) + jcos\left(m_{eff}g(t+\tau_2)\right)]]\} \end{aligned} \quad (10)$$

For a small-signal modulation, $|m_{eff}|$ is very small, $\sin(m_{eff}g(t)) \approx m_{eff}g(t)$ and $\cos(m_{eff}g(t)) \approx 1$, then

$$e_{POL}(t) = -\exp\ (j\omega_c t)\{[A - B\ m_{eff}g(t+\tau_2)] + j[A\ m_{eff}g(t) + B]\} \quad (11)$$

If the optical signal expressed in eqn. 11 is sent to a PD for square-law detection, the photocurrent is

$$\begin{aligned} i_{ph}(t) &= |e_{POL}(t)|^2 = \left|-\exp\ (j\omega_c t)\{[A - B\ m_{eff}g(t-\tau_2)] + j[A\ m_{eff}g(t) + B]\}\right|^2 \\ &= A^2 - 2AB\ m_{eff}g(t+\tau_2) + B^2\left(m_{eff}g(t+\tau_2)\right)^2 + B^2 + 2AB\ m_{eff}g(t) + A^2(m_{eff}g(t))^2 \\ i_{ph}(t) &= [A^2 + B^2] + 2AB\ m_{eff}[g(t) - g(t+\tau_2) \end{aligned} \quad (12)$$

Investigation eqn.12 reveals that the photocurrent corresponds to a first-order difference equation for the Gaussian pulse with a gain $G_{DE} = AB$ in addition to a DC term $I_{DC} = A^2 + B^2$. Note that both $G_{DE}$ and $I_{DC}$ depend on the PCs rotation angles $\alpha_2$ and $\alpha_3$

$$G_{DE} = \frac{1}{4}\sin(2\alpha_2)\sin(2\alpha_3) \quad (13a)$$

$$I_{DC} = (cos\alpha_2\ sin\alpha_3)^2 + (cos\alpha_3\ sin\alpha_2)^2 \quad (13b)$$

For the special case of $\alpha_2 = \alpha_3 = 45^\circ$ which is adopted in [31], eqn. 12 reduces to that reported in [31].

## 3.2. Gaussian Doublet

To generate an UWB Gaussian doublet, the polarization direction of the modulated lightwave should be oriented to have $45^\circ$ with principal axis of PMF1 (i.e., $\alpha_1 = 45^\circ$ ) and let $\theta_1 = \theta_3 = 90^\circ$, $\varnothing_2 = 0$ [31]. Equation 8 is simplified to

$$
\begin{aligned}
e_{POL}(t) = -1/\sqrt{2}\exp\ (j\omega_c t)\{&\cos\alpha_2\ sin\alpha_3 \exp\left(j\ m_{eff} g(t)\right) \\
&+ sin\alpha_2 sin\alpha_3 \exp\left(j\left[\frac{\pi}{2} + m_{eff} g(t+\tau_1)\right]\right) \\
&+ \cos\alpha_3\ sin\alpha_2 \exp\left(j\left[\frac{\pi}{2} + m_{eff} g(t+\tau_2)\right]\right) \\
&- \cos\alpha_2\ cos\alpha_3 \exp\left(j[\pi + m_{eff} g(t+\tau_1+\tau_2)]\right)\}
\end{aligned} \tag{14}
$$

Let $X = \sin\alpha_2\ \sin\alpha_3$ and $Y = \cos\alpha_2\cos\alpha_3$, then using the two trigonometric identities adopted in section 3.1 yields

$$
\begin{aligned}
e_{POL}(t) = -\frac{1}{\sqrt{2}}\exp(j\omega_c t).\{&A\left[\cos\left(m_{eff} g(t)\right) + jsin\left(m_{eff} g(t)\right)\right] \\
&+ X\left[-\sin\left(m_{eff} g(t+\tau_1)\right) + jcos\left(m_{eff} g(t+\tau_1)\right)\right] \\
&+ B\left[-\sin\left(m_{eff} g(t+\tau_2)\right) + jcos\left(m_{eff} g(t+\tau_2)\right)\right] \\
&+ Y\left[\cos\left(\left(m_{eff} g(t+\tau_1+\tau_2\right)\right) + jsin\left(\left(m_{eff} g(t+\tau_1+\tau_2\right)\right)\right]\}
\end{aligned} \tag{15}
$$

For a small-signal modulation, one can show that

$$
\begin{aligned}
e_{POL}(t) = -\frac{1}{\sqrt{2}}\exp(j\omega_c t).\{&\left((A+Y) - m_{eff}[Xg(t+\tau_1) + Bg(t+\tau_2)]\right) \\
&+ j\left(m_{eff}[Ag(t) + Yg(t+\tau_1+\tau_2)] + (X+B)\right)\}
\end{aligned} \tag{16}
$$

The photocurrent generated from the field described in eqn. 16 is given by

$$
\begin{aligned}
i_{ph}(t) &= \left|-\frac{1}{\sqrt{2}}\exp(j\omega_c t).\{\left((A+Y) - m_{eff}[Xg(t+\tau_1) + Bg(t+\tau_2)]\right) + j\left(m_{eff}[Ag(t) + Yg(t+\tau_1+\tau_2)] + (X+B)\right)\}\right|^2 \\
&= [(A+Y)^2 + (X+B)^2 + 2m_{eff}\{(X+B)[Ag(t) + Yg(t+\tau_1+\tau_2)] - (A+Y)[Xg(t+\tau_1) + Bg(t+\tau_2)]\}]
\end{aligned} \tag{17}
$$

Equation 17 contains a DC term $[(A+Y)^2+(X+B)^2]$ in addition to an ac term that reflects the second-order difference components. For special case of $\alpha_2=\alpha_3=45^\circ$ which is adopted in [31], $A=B=X=Y=\frac{1}{2}$. This yields an ac term that is proportional to $g(t)+g(t+\tau_1+\tau_2)-g(t+\tau_1)-g(t+\tau_2)$. This result is in accord with that in [31].

# 4. SIMULATION RESULTS

This section presents simulation results describing signal generation and transmission performance for both Gaussian monocycle and doublet UWBoF systems. The results are reported for both OOK and BPM modulation formats with 625 Mb/s data rate and 1550 nm operating wavelength. Time delays of PMF1 and PMF2 ($\tau_1$ and $\tau_2$) are 20.1 and 40.1 ps, respectively. Other parameter values used in the simulation are as given in Table 1.

Two types of transmission links are considered in the simulation. The first link consists of a single-mode fiber (SMF). The second link consists of a SMF and dispersion compensating fiber (DCF) along with their optical amplifiers. In this link, the average losses and dispersion is zero. The parameter values of SMF and DCF are listed in Table 2. This section contains two parts; the first part considers the case of $\alpha_1=\alpha_3=45^\circ$ while the second part addresses the effect of deviations of these angles from 45° on system performance.

## 4.1. Case of $\alpha_1 = \alpha_3 = 45^\circ$

Figure 3 shows the generated Gaussian monocycle and doublet waveform along with the ideal waveforms which are included here for comparison purpose.

The FWHM of the Gaussian pulse which is used to obtain all the waveforms is 85 ps. Note that the time-domain representation highlights that the signals generated in this work are in good agreement with the ideal waveforms. Received PSDs are simulated when photonically generated pulses are adopted in back-to-back transmission. A bandpass filter (3.1-10.6 GHz) is inserted after the PD to reconfigure the PSD according to FCC limit. The results are presented in

Fig. 4 for both OOK and BPM formats and show that the PSDs of the generated signals are within FCC mask.

**Table 1.** Parameter values used in the simulation

| Parameter | Value |
|---|---|
| Bit rate | 625 Mb/s |
| Wavelength | 1550 nm |
| FWHM of the Gaussian pulses | 85 ps |
| CW laser power | 0 dBm |
| MZM extinction ration | 20 dB |
| MZM insertion loss | 5 dB |
| MZM bias voltage | 2 V |
| Optical amplifier noise figure | 4 dB |
| Optical bandpass filter order | 5 |
| Optical bandpass filter depth | 100 dB |
| Electrical bandpass filter order | 5 |
| Electrical bandpass filter depth | 100 dB |
| Photodiodes dark current | 10 nA |
| Photodiodes responsivity | 1 A/W |
| Photodiodes thermal noise | $1\times10^{-22}$ W/Hz |
| Electrical amplifier noise | -60 dBm |

**Table 2.** Optical fiber parameter values used in the simulation

| Parameter | Value | |
|---|---|---|
| | SMF | DCF |
| Attenuation, $\alpha$ | 0.2 dB/km | 0.5 dB/km |
| Group velocity dispersion, D | 17 ps/(ns.nm) | -85 ps/(ns.nm) |
| Dispersion slop, S | 0.075 ps/nm$^2$/km | -0.3 ps/nm$^2$/km |
| Differential group delay | 0.2 ps/km | 0.2 ps/km |
| Effective area, $A_{eff}$ | 80 μm$^2$ | 22 μm$^2$ |
| Nonlinear refractive index, $n_2$ | 26×10$^{-21}$ m$^2$/W | 26×10$^{-21}$ m$^2$/W |

Figure 5 depicts the received eye diagrams when the generated signals are adopted in back-to-back transmission and used later to address the effect of fiber transmission link.

The simulation is carried further to address system performance when a SMF isused as the transmission link. No DCF or optical amplifiers are used to compensate the dispersion and losses of the fiber, respectively. Table 3 lists the lengths of the transmission fiber required to obtain a BER of $10^{-9}$ for different signals and modulation formats. The corresponding received eye diagrams and PSDs are displayed in Figs. 6 and 7, respectively.

The next step is to investigate system performance when the transmission link consists of SMF and DCF along with their loss-compensation amplifiers. The results corresponding to a BER of $10^{-9}$ are listed in Table 4 and Figs. 8 and 9.

## 4.2. Effect of Polarization Controllers Rotation Angles

This subsection addresses the effect of deviations of the angles $\alpha_2$ and $\alpha_3$ on system performance. The results are presented for four systems, monocycle- and doublet-based OOK and BPM systems. The length of the transmission link is set equal to the length that yields a BER of $10^{-9}$ for each of the four systems. It is worth to mention here that the total photocurrent is applied to decision circuit to estimate the BER characteristics. This is in contrast to the technique adopted in [31] which uses only the ac components of the photocurrent to estimate system BER. The receiver adopted here uses DC coupled front-end amplifiers to maintain the DC component of the photocurrent. Figure 10 shows the variation of the BER with $\alpha_2$ deviation and taking $\alpha_3$ as independent parameter for the four systems under investigation. The transmission link is assumed to be consisted of a SMF only. Results related to a transmission link consisting of SMF, DCF and optical amplifiers are given in Fig. 11.

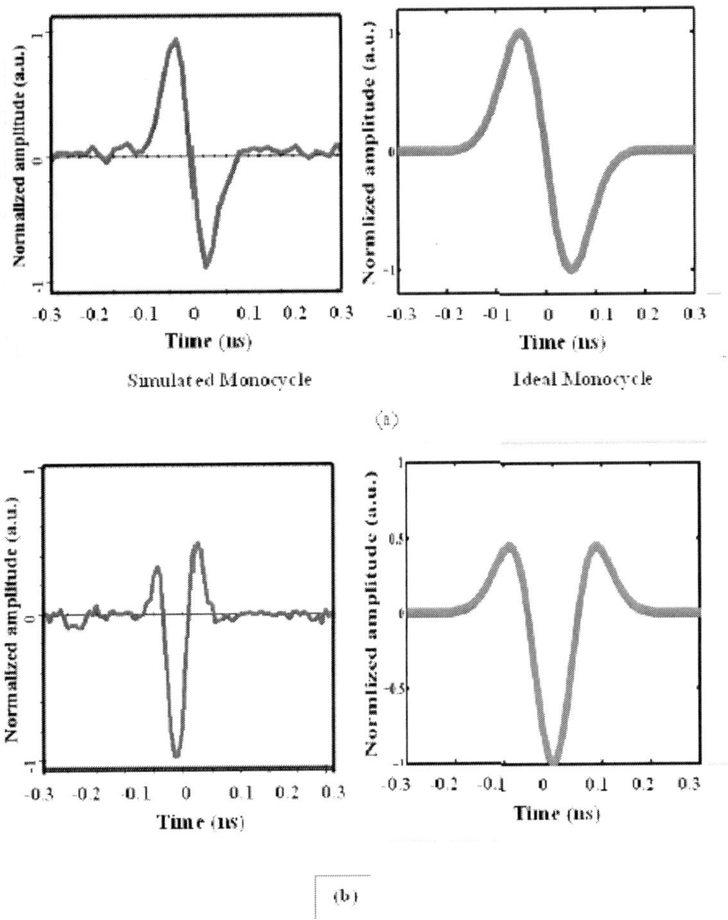


**Figure 3.** Simulated and ideal UWB pulses (a) Gaussian monocycle (b) doublet

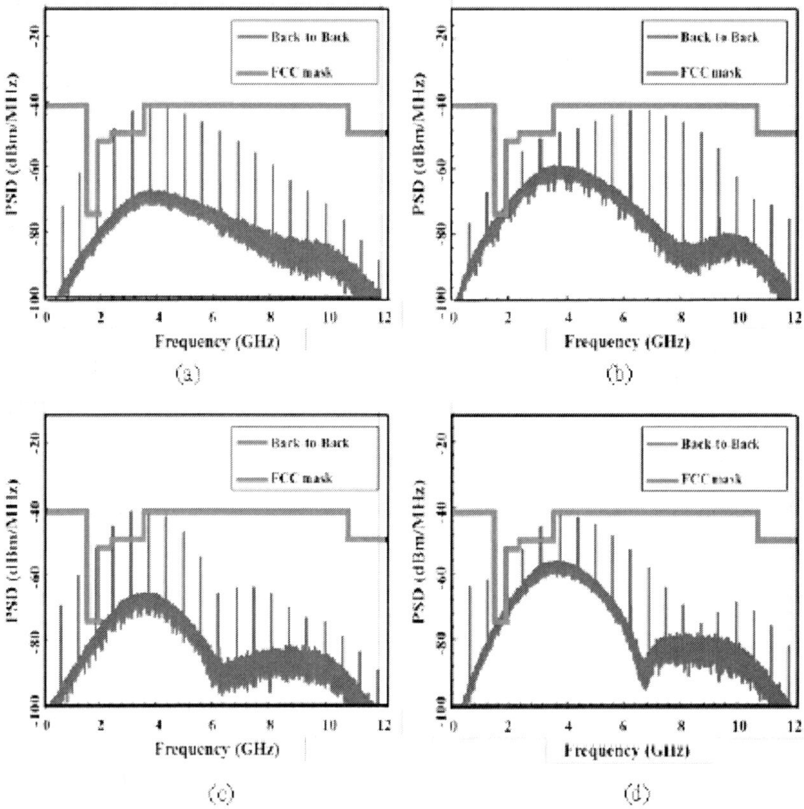


**Figure 4.** Power spectrum densities for back-to-back of (a) Gaussian monocycle OOK (b) Gaussian monocycle BPM (c) doublet OOK (d) doublet BPM

The results in Figs. 10 and 11 reveal that the BER performance can be improved by deviating $\alpha_2$ from $45^\circ$ and this deviation depends on the value of $\alpha_3$. This conclusion is also drawn when one consider the deviation of BER with $\alpha_3$ for different values of $\alpha_2$ (see Fig. 12 for monocycle OOK system).

The above results lead to the main finding that angles $\alpha_2$ and $\alpha_3$ should be chosen carefully to minimize the BER for each of the four systems. This is a two-dimensional optimization problem and it is here performed for the four systems and for both types of the transmission links. Table 5 lists the optimum values of $\alpha_2$ and $\alpha_3$ that give minimum BER. These results correspond to two transmission links, one consists of

SMF while the other consists SMF and DCF with their optical amplifiers. The last column of this table lists the values of a performance measure parameter which is denoted by Figure of Merit (FoM)

$$\text{FoM} = \text{Log}\,\frac{\text{BER}(45^{\circ},45^{\circ})}{\text{BER}(\alpha_2,\alpha_3)} \tag{18}$$

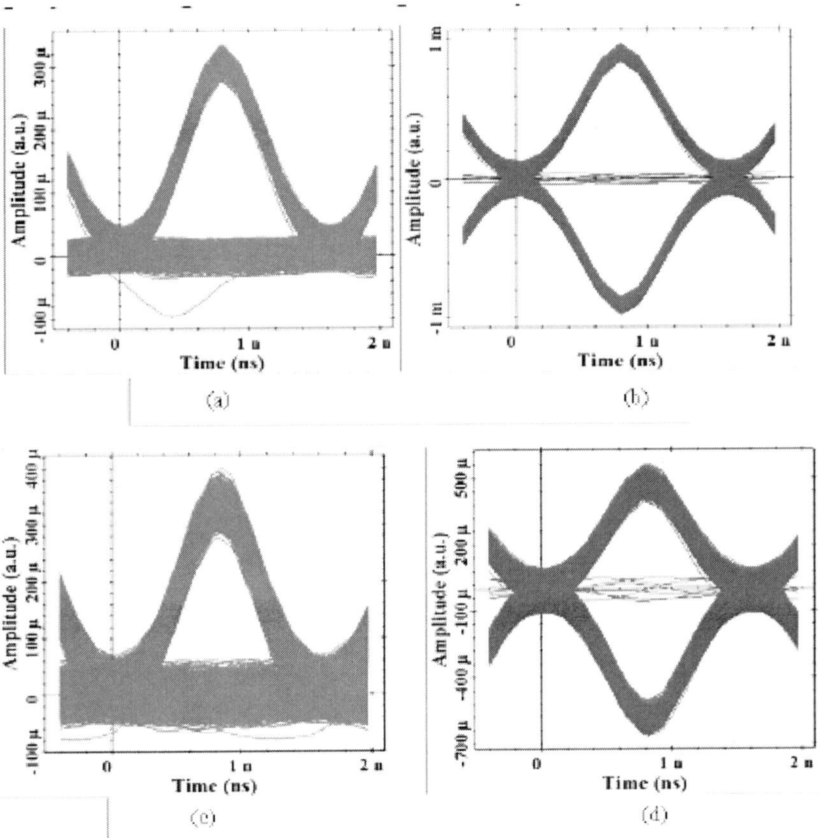


**Figure 5.** Eye diagrams of the received signals for back-to-back of (a) Gaussian monocycle OOK (b) Gaussian monocycle BPM (c) doublet OOK (d) doublet BPM

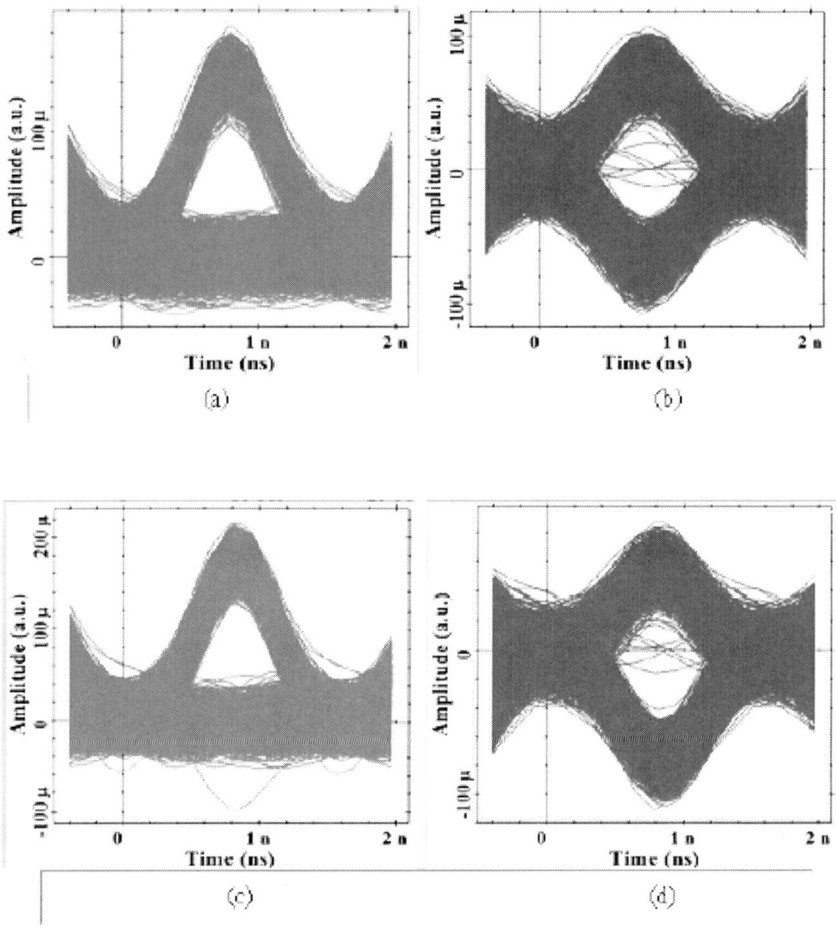


**Figure 6.** Eye diagrams of the received signals after transmission over SMF for (a) Gaussian monocycle OOK (b) Gaussian monocycle BPM (c) doublet OOK (d) doublet BPM

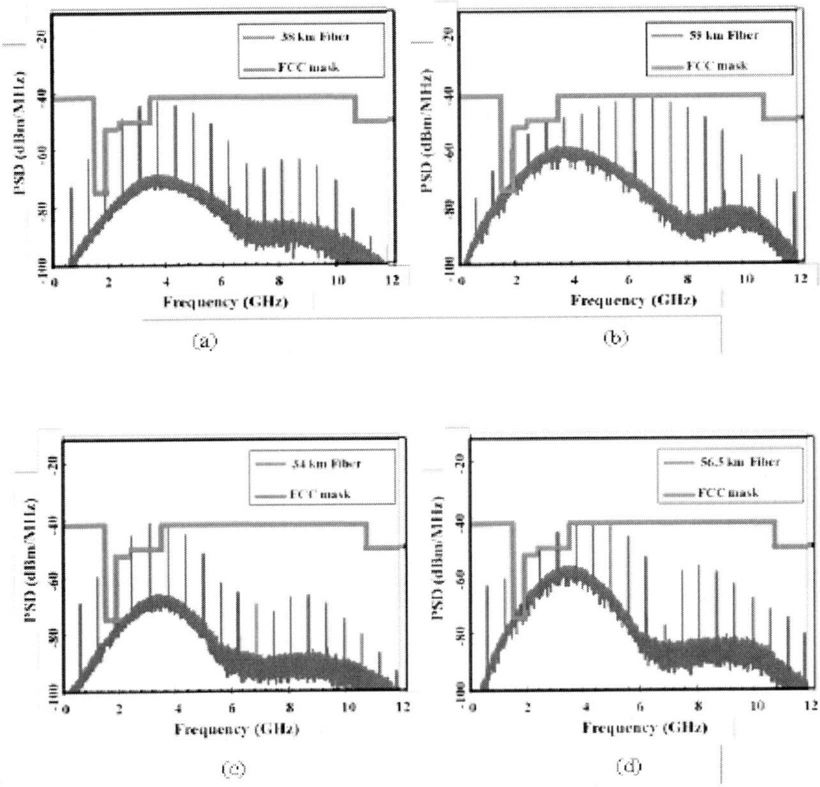


**Figure 7.** Power spectrum densities after transmission over SMF of (a) Gaussian monocycle OOK (b) Gaussian monocycle BPM (c) doublet OOK (d) doublet BPM

**Table 3.** SMF lengths at BER=$10^{-9}$ for monocycle and doublet signaling systems operating with OOK and BPM modulation formats

| Pulse Type | Modulation Format | SMF Length (km) |
|---|---|---|
| Gaussian Monocycle | OOK | 38 |
| | BPM | 59 |
| Doublet | OOK | 34 |
| | BPM | 56.5 |

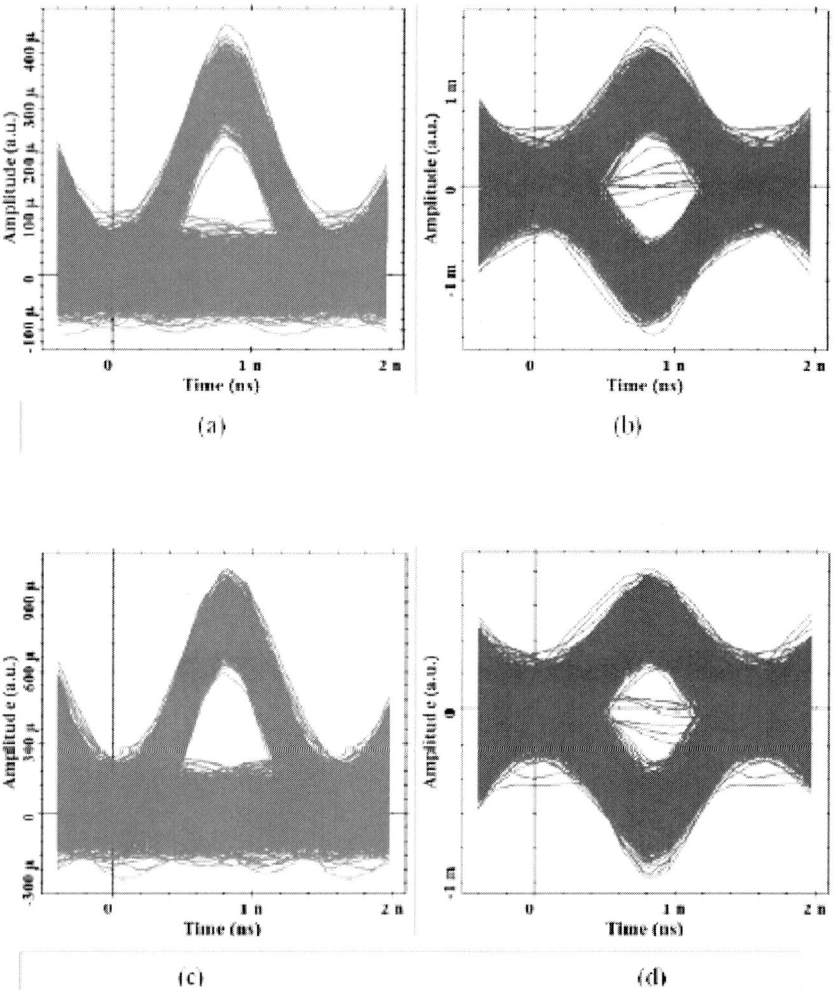


**Figure 8.** Eye diagrams of the received signals after transmission over optical link for (a) Gaussian monocycle OOK (b) Gaussian monocycle BPM (c) doublet OOK (d) doublet BPM

This parameter is introduced here to quantively measure the effect of deviations of rotation angles from $45^\circ$ on system performance. The results in Table 5 indicate clearly that BER performance can be enhanced by optimizing the rotation angles $\alpha_2$ and $\alpha_3$ and this effect is more pronounced when the transmission link consists of a SMF only. The monocycle BPM system offers the best performance (FoM=3.52) when operates under optimum conditions. In contrast, the optimum rotation angles do not yield notable BER performance enhancement when the monocycle OOK and BPM system redesigned with fully loss and dispersion compensation.

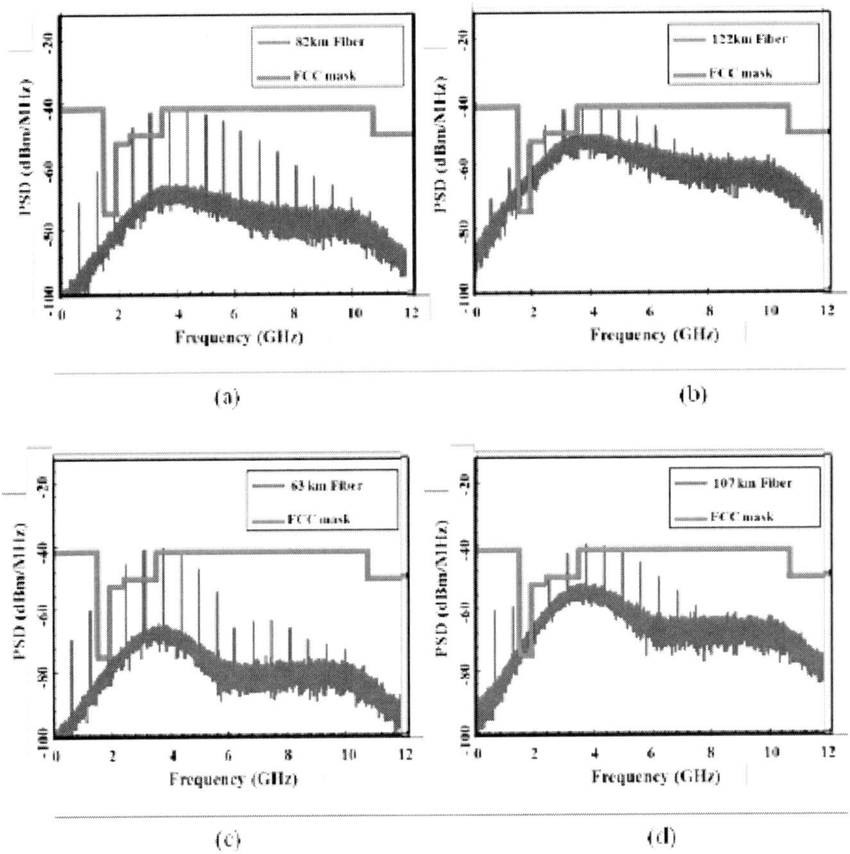


**Figure 9.** Power spectrum densities after transmission over optical link (SMF+DCF) for (a) Gaussian monocycle OOK (b) Gaussian monocycle BPM (c) doublet OOK (d) doublet BPM

The corresponding optimum PSDs and optimum received eye diagrams are given in Figs. 13-16.

**Table 4.** Transmission distances at BER=$10^{-9}$ for monocycle and doublet signaling systems operating with OOK and BPM modulation formats

| Pulse Type | Modulation Format | SMF Length (km) | Total Link Length (km) |
|---|---|---|---|
| Gaussian Monocycle | OOK | 82 | 98.5 |
| | BPM | 122 | 146.5 |
| Doublet | OOK | 63 | 75.5 |
| | BPM | 107 | 128.5 |

Note that applying doublet signal reduces the total length of the transmission link to 0.77 and 0.87 of that of monocycle signaling for OOK and BPM, respectively.

## 5. CONCLUSIONS

The analysis presented in [31] has been extended to take into account the effect of PCs rotation angles on the transmission performance of optically generated Gaussian monocycle and doublet UWB pulses. Simulation results have been reported for 625 Mb/s fiber communication systems operating with or without dispersion compensating fiber. The simulation results reveal that the BER performance can be enhanced by optimizing the rotation angles of the polarization controllers and this effect is more pronounced when the transmission link consists of a SMF only. The monocycle BPM system offers the best performance (FoM=3.52) when operates under optimum conditions. In contrast, the optimum rotation angles do not yield notable BER performance enhancement when the monocycle OOK and BPM systems are redesigned with fully loss and dispersion compensation.

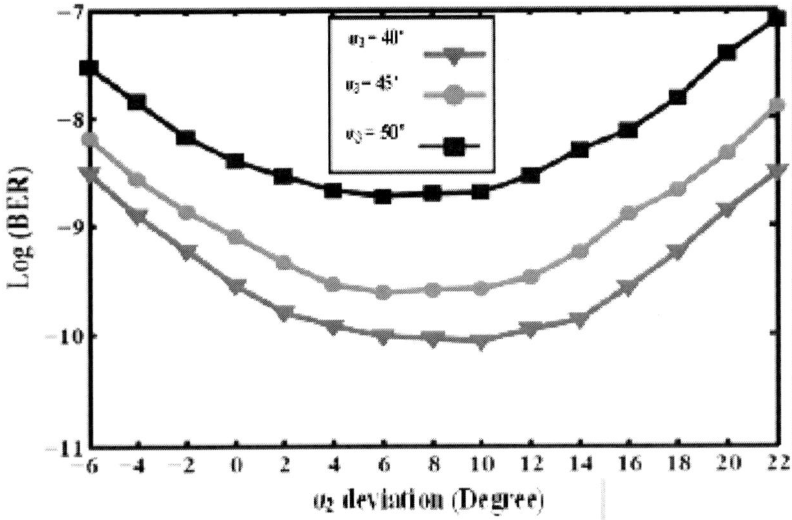
Log (BER)
θ2 deviation (Degree)
L1 = 59 km

(a)

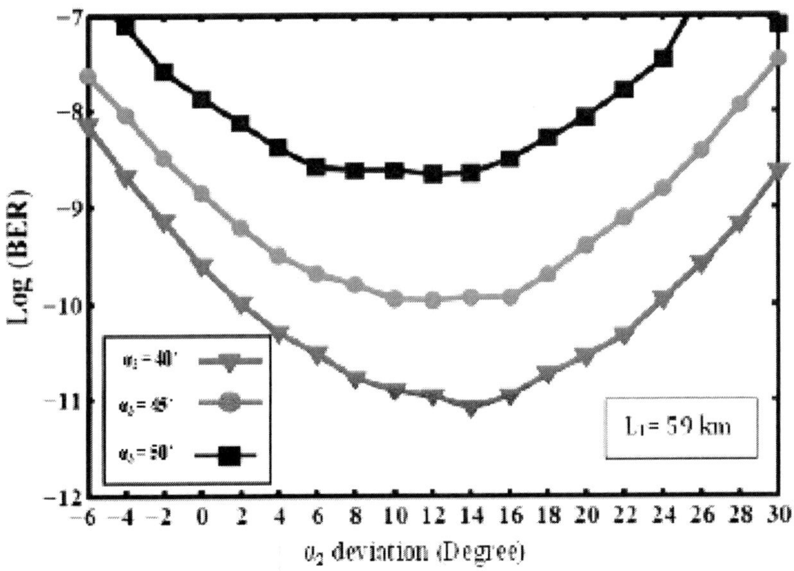
Log (BER)
θ2 deviation (Degree)
L1 = 59 km

(b)

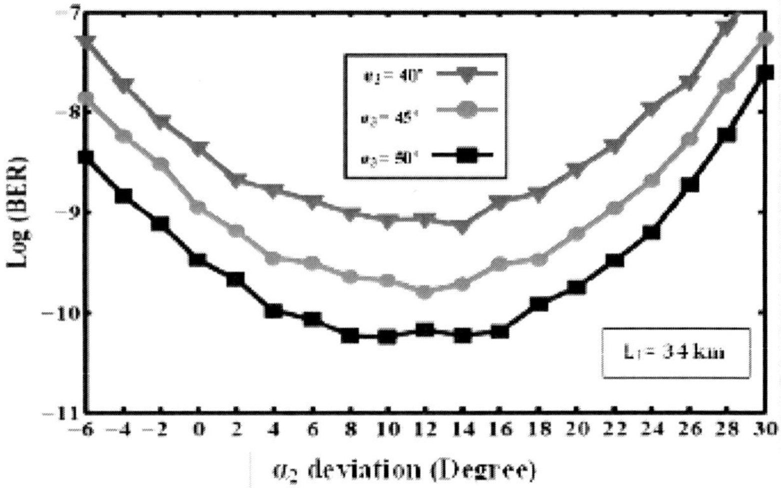


(c)

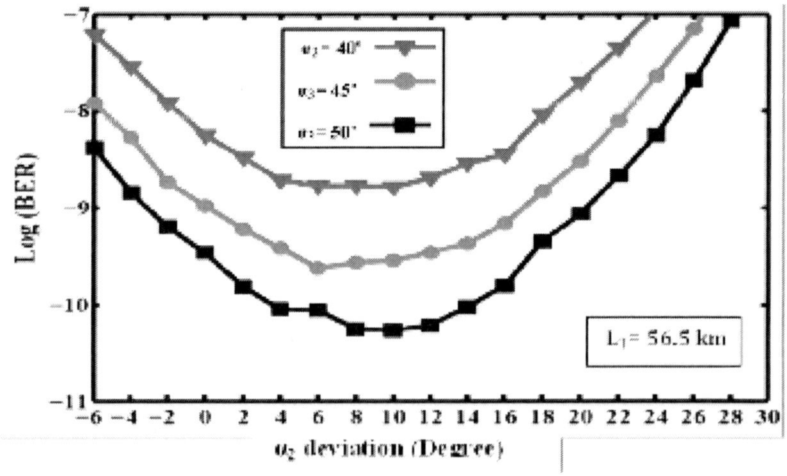


(d)

**Figure 10.** Variation of the BER with $\alpha_2$ deviation for different values of $\alpha_3$ after transmission over SMF for (a) Gaussian monocycle OOK (b) Gaussian monocycle BPM (c) doublet OOK (d) doublet BPM

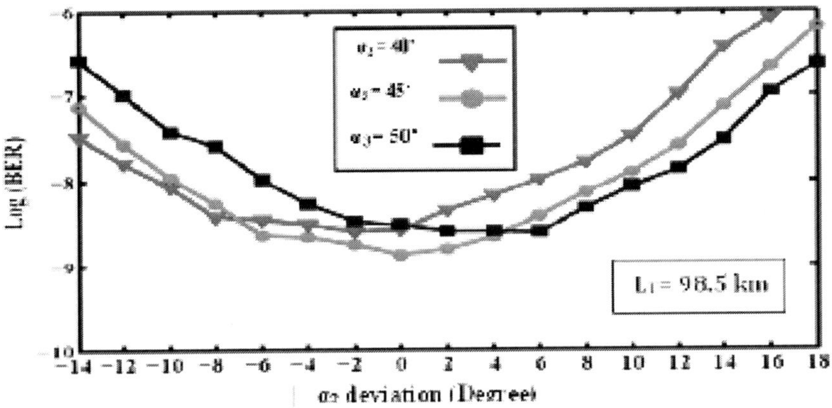
α3 = 40°
α3 = 45°
α3 = 50°
L1 = 98.5 km
Log (BER)
α2 deviation (Degree)

(a)

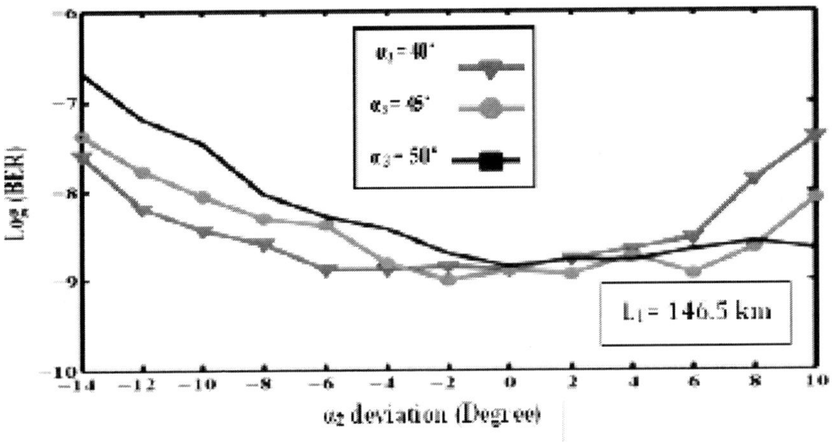
α3 = 40°
α3 = 45°
α3 = 50°
L1 = 146.5 km
Log (BER)
α2 deviation (Degree)

(b)

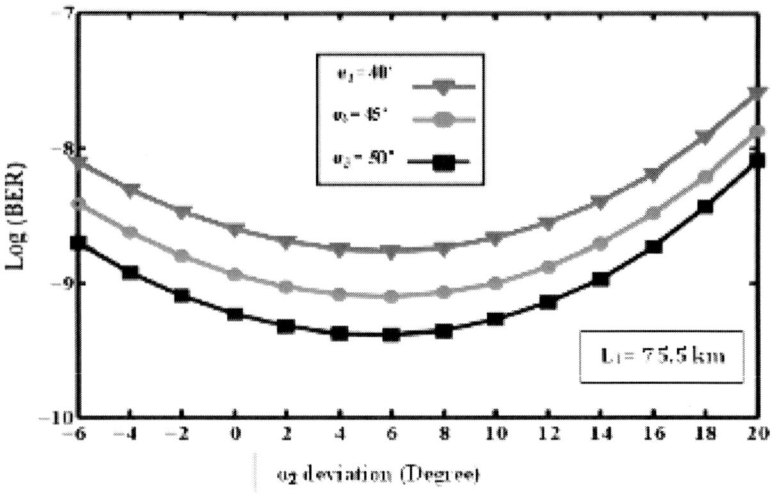


(c)

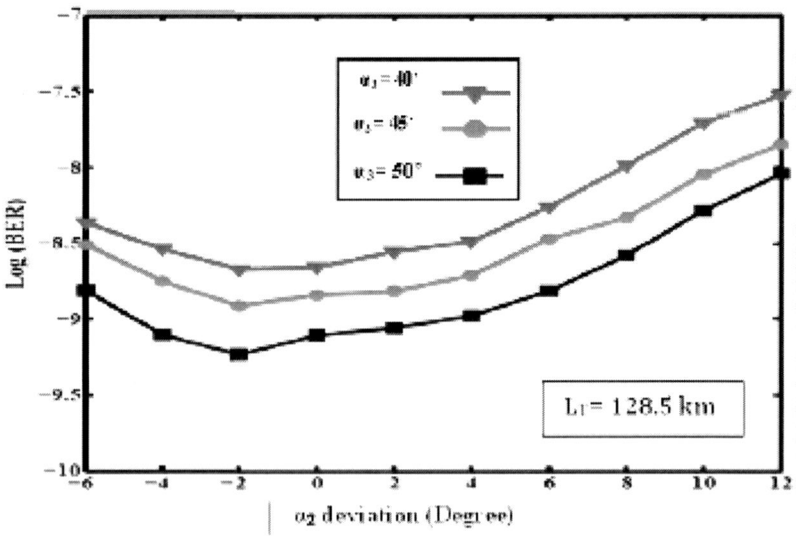


(d)

**Figure 11.** Variation of the BER with deviation for different values of after transmission over optical link (SMF+DCF) for (a) Gaussian monocycle OOK (b) Gaussian monocycle BPM (c) doublet OOK (d) doublet BPM

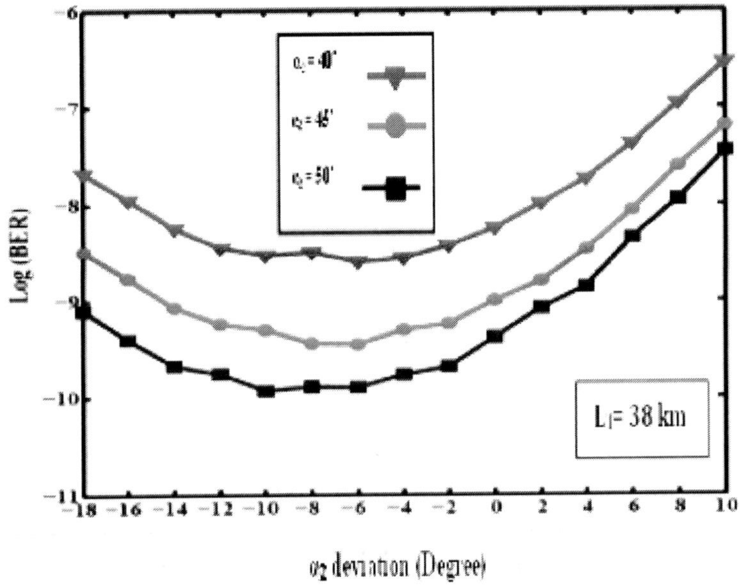


**Figure 12.** Variation of the BER with $\alpha_3$ deviation for different values of $\alpha_2$ for monocycle OOK signaling

**Table 5.** Optimum parameters of UWBoF systems using photonic generated Gaussian monocycle and doublet pulses

| Link Type | Pulse Type | Modulation Format | Link Length (km) | Optimum Values | | | FoM |
|---|---|---|---|---|---|---|---|
| | | | | Opt. $\alpha_2$ | Opt. $\alpha_3$ | Opt. BER | |
| Without Dispersion Compensation | Mono | OOK | 38 | 56° | 38° | $9.2\times10^{-11}$ | 1.04 |
| | | BPM | 59 | 58° | 30° | $3.0\times10^{-13}$ | 3.52 |
| | Doublet | OOK | 34 | 56° | 58° | $8.3\times10^{-12}$ | 2.08 |
| | | BPM | 56.5 | 54° | 58° | $1.6\times10^{-11}$ | 1.80 |
| With Dispersion Compensation | Mono | OOK | 98.5 | 48° | 48° | $8.3\times10^{-10}$ | 0.08 |
| | | BPM | 146.5 | 44° | 46° | $9.2\times10^{-10}$ | 0.04 |
| | Doublet | OOK | 75.5 | 50° | 58° | $1.8\times10^{-10}$ | 0.74 |
| | | BPM | 128.5 | 44° | 58° | $1.7\times10^{-10}$ | 0.77 |

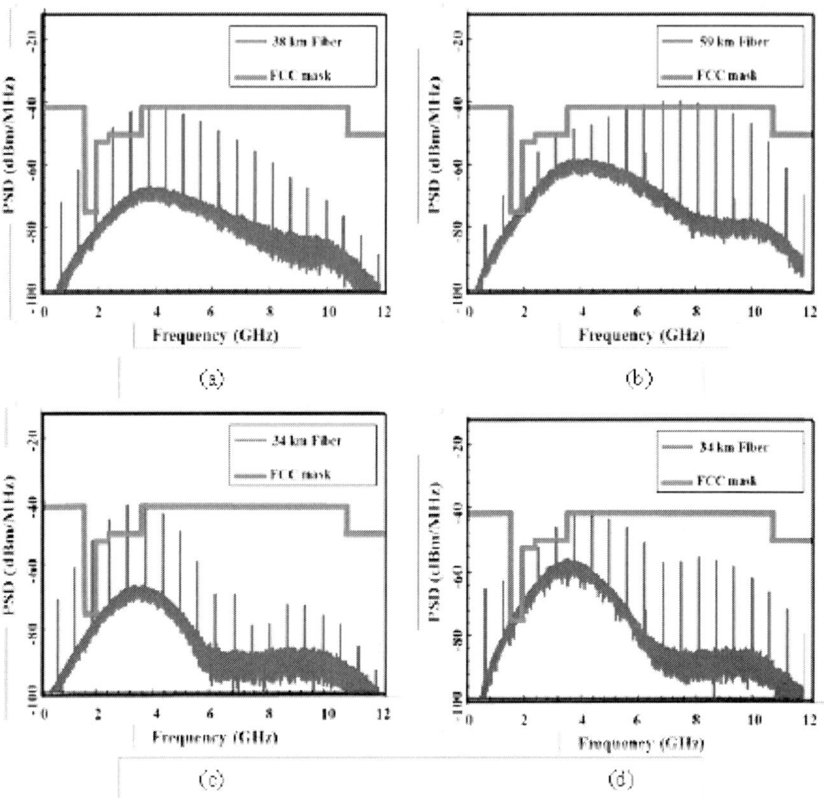


**Figure 13.** Optimum power spectrum densities after transmission over SMF of (a) Gaussian monocycle OOK (b) Gaussian monocycle BPM (c) doublet OOK (d) doublet BPM

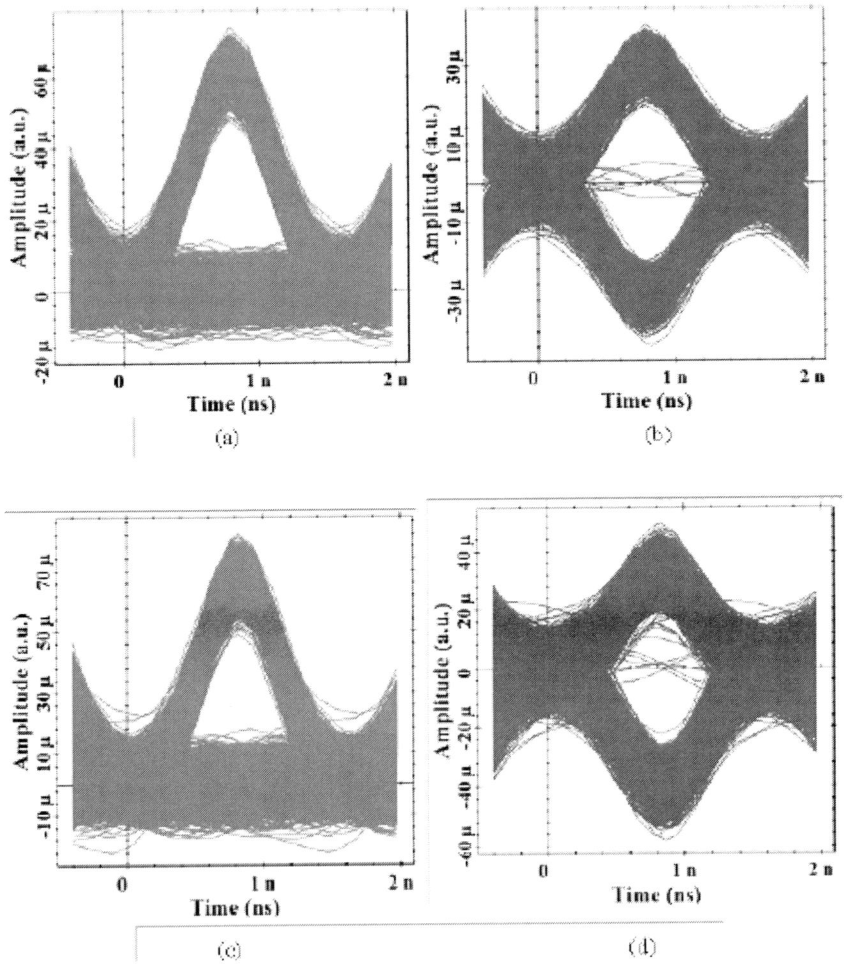


**Figure 14.** Optimum received eye diagrams after transmission over SMF of (a) Gaussian monocycle OOK (b) Gaussian monocycle BPM (c) doublet OOK (d) doublet BPM

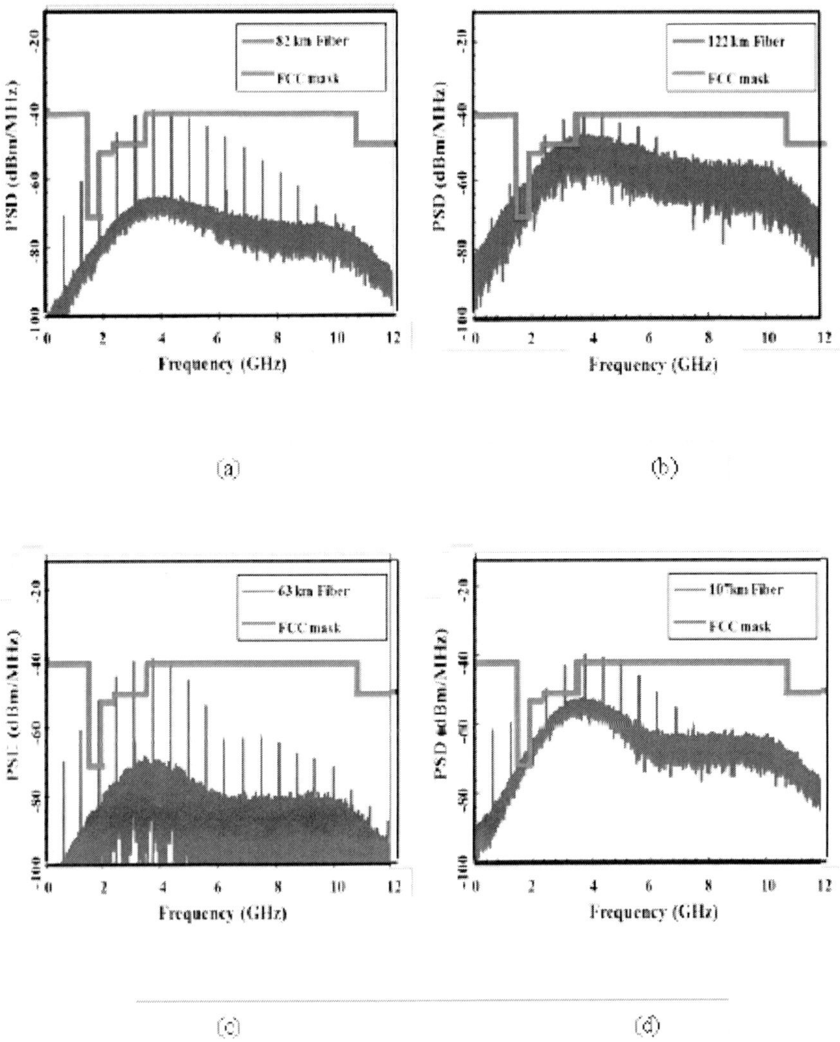


**Figure 15.** Optimum power spectrum densities after transmission over optical link (SMF+DCF) for (a) Gaussian monocycle OOK (b) Gaussian monocycle BPM (c) doublet OOK (d) doublet BPM

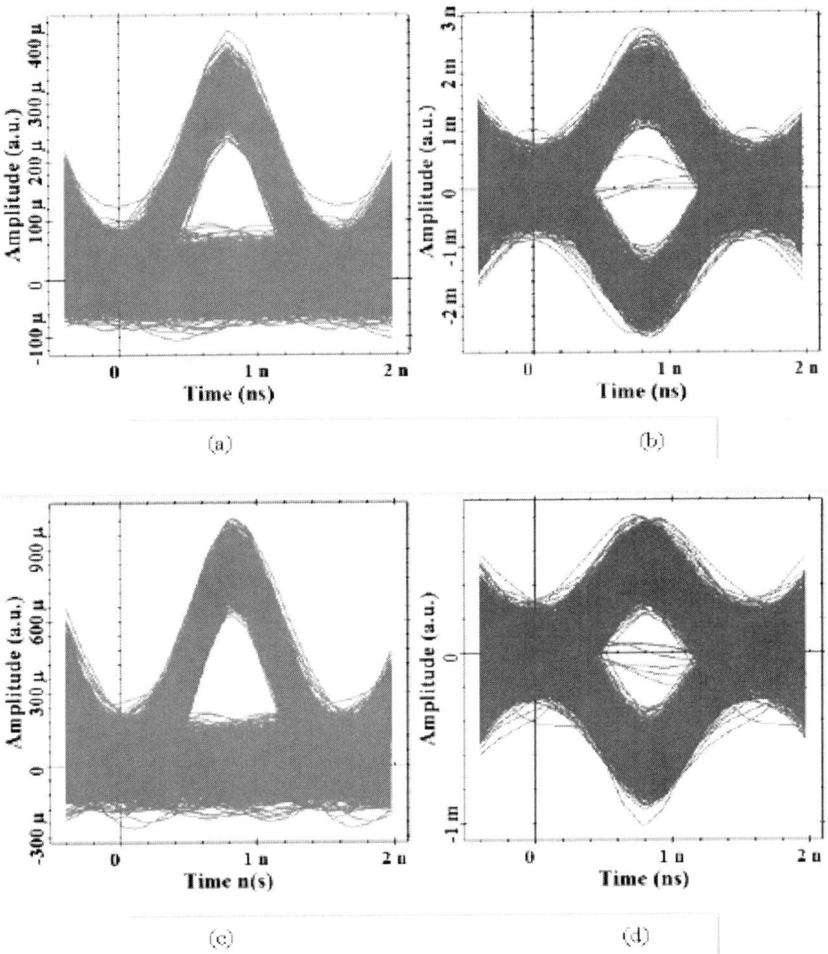


**Figure 16.** Optimum received eye diagrams after transmission over optical kink (SMF+DCF) for (a) Gaussian monocycle OOK (b) Gaussian monocycle BPM (c) doublet OOK (d) doublet BPM

# REFERENCES


1. J. Wang, and J. Sun, "All-optical ultrawideband monocycle generation using quadratic nonlinear interaction seeded by dark pulses," IEEE Photonics Technology Letters, Vol. 22, No. 3, PP. 140-142, FEBRUARY, 2010.

2. V. Moreno, M. Rius, J. Mora, M. A. Muriel, and J. Capmany, "UWB monocycle generator based on the non-linear effects of an SOA-integrated structure," IEEE Photonics Technology Letters, Vol. 26, No. 7, PP. 690-693, April, 2014.

3. H. Feng, M. P. Fok, S. Xiao, J. Ge, Q. Zhou, M. Locke, R. Toole, and W. Hu, "A Reconfigurable high-order UWB signal generation scheme using RSOA-MZI structure," IEEE Photonics Journal, Vol. 6, No. 2, PP. 7900307-7900307, April, 2014.

4. S. Pan, and J. Yao, "IR-UWB-over-fiber systems compatible with WDM-PON networks," Journal of Lightwave Technology, Vol. 29, No. 20, PP. 3025-3034, October, 2011.

5. Y. Yu, J. Dong, X. Li, and X. Zhang," UWB monocycle generation and bi-phase modulation based on Mach–Zehnder modulator and semiconductor optical amplifier," IEEE Photonics Journal, Vol. 4, No. 2, PP. 327-338, April, 2012.

6. P. Li, H. Chen, X. Wang, H. Yu, M. Chen, and S. Xie, "Photonic generation and transmission of 2-Gbit/s power-efficient IR-UWB signals employing an electro-optic phase modulator," IEEE Photonics Technology Letters, Vol. 25, No. 2, PP. 144-146, January, 2013.

7. F. Zhang, S. Fu, J. Wu, N. Q. Ngo, K. Xu, Y. Li, X. Hong, P. Shum, and J. Lin, "UWB impulse radio transmitter using an electrooptic phase modulator together with a delay interferometer," IEEE Photonics Technology Letters, Vol. 22, No. 20, PP. 1479-1481, October,2010.

8. X. Yu, T. B. Gibbon, R. Rodes, T. T. Pham, and I. T. Monroy, "System wide implementation of photonically generated impulse radio ultra-wideband for gigabit fiber-wireless access," Journal of Lightwave Technology, Vol. 31, No. 2, PP. 264-275, January, 2013.

9. M. Beltrán, and R. Llorente, "Dual photonic generation ultrawideband impulse radio by frequency shifting in remote-connectivity fiber," Journal of Lightwave Technology, Vol. 29, No. 24, PP. 3645-3653, December,2011.

10. P. Li, H. Chen, M. Chen, and S. Xie, "Gigabit/s photonic generation, modulation, and transmission for a reconfigurable impulse radio UWB over fiber system," IEEE Photonics Journal, Vol. 4, No. 3, PP. 805-816, June, 2012.

11. M. Mirshafiei, A. Ghazisaeidi, D. Lemus, S. LaRochelle, and L. A. Rusch, "Upconversion of gain-switched laser pulses for optical

generation of UWB signals," Journal of Lightwave Technology, Vol. 30, No. 2, PP. 207-214, January, 2012.

12. Z. Hu, J. Sun, J. Shao, and X. Zhang, "Filter-free optically switchable and tunable ultrawideband monocycle generation based on wavelength conversion and fiber dispersion," IEEE Photonics Technology Letters, Vol. 22, No. 1, PP. 42-44, January, 2010.

13. V. Moreno, M. Rius, J. Mora, M. A. Muriel, and J. Capmany, "UWB doublet generation employing cross-phase modulation in a semiconductor optical amplifier mach–zehnder interferometer," IEEE Photonics Journal, Vol. 5, No. 6, PP. 7101106- 7101106, December, 2013.

14. W. Li, L. X. Wang, J. Y. Zheng, M. Li, and N. H. Zhu, "Photonic generation of ultrawideband signals with large carrier frequency tunability based on an optical carrier phase-shifting method," IEEE Photonics Journal, Vol. 5, No. 6, PP. 5502007-5502007, October, 2013.

15. M. Abtahi, and L. A. Rusch, "RoF delivery over PONs of optically shaped UWB signals for Gigabit/s wireless distribution in the home," IEEE Journal On Selected Areas In Communications, Vol. 29, No. 6, PP. 1304-1310, June, 2011.

16. Y. M. Chang, J. Lee, D. Koh, H. Chung, and J. H. Lee, "Ultrawideband doublet pulse generation based on a semiconductor electroabsorption modulator and Its distribution over a fiber/wireless link," Journal Of Optical Communications Networks, Vol. 2, No. 8, PP. 600-608, August, 2010.

17. T. Huang, J. Li, J. Sun, and L. R. Chen, "Photonic generation of UWB pulses using a nonlinear optical loop mirror and Its distribution over a fiber link," IEEE Photonics Technology Letters, Vol. 23, No. 17, PP. 1255-1257, September, 2011.

18. B. Luo, J. Dong, Y. Yu, and X. Zhang, "Bandwidth-tunable single-carrier UWB monocycle generation using a nonlinear optical loop mirror," IEEE Photonics Technology Letters, Vol. 24, No. 18, PP. 1646-1649, September, 2012.

19. Y. Yu, J. Dong, X. Li, and X. Zhang, "Ultra-wideband generation based on cascaded Mach–Zehnder modulators," IEEE Photonics Technology Letters, Vol. 23, No. 23, PP. 1754-1756, December, 2012.

20. P. Cao, X. Hu, J. Wu, L. Zhang, X. Jiang, and Y. Su, "Reconfigurable UWB pulse generation based on a dual-drive Mach–Zehnder

modulator," IEEE Photonics Journal, Vol. 6, No. 5, PP. 7903206-7903206, October,2014.

21. W. Zhihu, W. Rong, P. Tao, S. Guodan, F. Tao, "Switchable optical ultra-wideband pulse generator based on polarization modulator," China Communications, PP. 113-117, July, 2013.

22. Y. M. Chang, J. Lee, H. S Lee, L. Yan, and J. H. Lee, "Generation and distribution of 1.25 Gb/s Ultrawideband doublet pulses based on the combination of nonlinear polarization rotation and parametric amplification," Journal Of Lightwave Technology, Vol. 29, No. 6, PP. 931-938, March, 2011.

23. Y. Min Chang, J. Lee, and J. H. Lee, "Bismuth nonlinear optical fiber for photonic ultrawideband radio-signal processing," IEEE Journal Of Selected Topics In Quantum Electronics, Vol. 18, No. 2, PP. 891-898, April, 2012.

24. Y. Peled, M. Tur, and A. Zadok, "Generation and detection of ultra-wideband waveforms using stimulated brillouin scattering amplified spontaneous emission," IEEE Photonics Technology Letters, Vol. 22, No. 22, PP. 1692-1694, November, 2010.

25. E. Zhou, X. Xu, K. S. Lui, and K. K. Y. Wong, "Photonic ultrawideband pulse generation with HNL-DSF-based phase and intensity modulator, " IEEE Photonics Technology Letters, Vol. 23, No. 7, PP. 396-398, April, 2011.

26. A. Zadok, Xi. Wu, J. Sendowski, A. Yariv, and A. E. Willner, "Photonic generation of ultra-wideband signals via pulse compression in a highly nonlinear fiber," IEEE Photonics Technology Letters, Vol. 22, No. 4, PP. 239-241, February, 2010.

27. J. Zheng, N. Zhu, L. Wang, J. Liu, and H. Liang, "Photonic generation of ultrawideband (UWB) pulse with tunable notch-band behavior," Vol. 4, No. 3, PP. 657-662, June, 2012.

28. C. Yang, L. Xia, S. Fu, and D. Liu, "Reconfigurable UWB pulse generation based on multi-taps and a programmable filter," IEEE Photonics Technology Letters, Vol. 26, No. 14, PP. 1395-1398, July, 2014.

29. E. Zhou, X. Xu, K. S. Lui, and K. K. Y. Wong, "A power-efficient ultra-wideband pulse generator based on multiple PM-IM conversions," IEEE Photonics Technology Letters, Vol. 22, No. 14, PP. 1063-1065, July, 2010.

30. H. Feng, S. Xiao, L. Yi, and W. Hu, "Photonic generation of reconfigurable orders ultrawideband signals by using cascaded RSOAs," IEEE Photonics Technology Letters, Vol. 26, No. 9, PP. 908-910, May, 2014.

31. S. Pan, and J. Yao, "UWB-over-fiber communications: modulation and transmission," Journal of Lightwave Technology, Vol. 28, No. 16, PP. 2445-2455, August, 2010.

3

# OPTICAL SYSTEMS FOR ULTRA-HIGH-SPEED TDM NETWORKING

**Michael Galili *, Hao Hu, Hans Christian Hansen Mulvad, Ashenafi Kiros Medhin, Anders Clausen and Leif Katsuo Oxenløwe**

DTU Fotonik, Department of Photonics Engineering, Technical University of Denmark, Ørsteds Plads 343, Lyngby DK-2800 Kgs., Denmark

## ABSTRACT

This paper discusses key results in the field of high speed optical networking with particular focus on packet-based systems. Schemes for optical packet labeling, packet switching and packet synchronization will be discussed, along with schemes for optical clock recovery, channel identification and detection of ultra-high-speed optical signals.

## KEYWORDS

Optical communication systems; Optical packet switching; Optical signal processing

## 1. INTRODUCTION

The ever-present demand for increased data bandwidth is pushing research and development in the field of optical communication in many different directions. One of the prevailing trends in the evolution of optical communication so far has been the gradual increase of the single-carrier symbol rate with is used to transport data. One of the main attractions of optical communication is that many optical components are practically agnostic to the bandwidth or speed of the optical signal. Huge amounts of optical data can be transported over long distances and distributed in complex reconfigurable circuits.

Ultra-high-speed interfaces that can aggregate traffic from a number of low-speed sources and transport it in a single channel will be desirable for a number of future ultra-fast communication networks. In data centers and super computer networks efficient aggregation and transport of large data0 volumes is essential. The use of optical technologies for this functionality could provide reduction in energy consumption [1, 2] as well as unprecedented performance in terms of data speeds and network flexibility. Many network functionalities can be achieved at very high speeds using optical means. These functionalities include: wavelength conversion and wavelength multicasting, and switching including both on-off switching and directional switching; all essential functionalities for data networking.

In this paper we will focus on aspects of optical networking more specific to running a network at ultra-high data rates. In this case conversion between optical (o) and electrical (e) signals, *i.e.*, e/o and o/e conversion, becomes non-trivial and should be avoided in the network except at the data origin and the data destination. This means that a number of tasks which are typically carried out by electronic circuits in more conventional optical transmission will be implemented optically. The key functionalities for high-speed optical networking which will be discussed in this paper are listed below:

- Optical labeling of data packets
- Optical labeling of TDM data channels
- Optical switching of data packets
- Optical synchronization of data packets
- Ultra-high-speed clock recovery
- Efficient detection of ultra-high-speed data signals

These six topics will be discussed in the following sections and the paper will be concluded with a general discussion of the achieved results for ultra-high-speed optical networking.

# 2. OPTICAL LABELING

In optical circuits where o/e conversion of data is to be reduced or avoided, optical labeling of data is an attractive alternative in combination with optical switching. Depending on the network architecture labeling of data packets or data channels may be required. In this section we will discuss systems for labeling of high speed data packets as well as labeling to identify individual high-speed time division multiplexed (TDM) data channels.

## 2.1. Principle of in-Band Optical Packet Labeling

There are several attractive features of in-band optical labeling of high speed data packets. In this context in-band labeling describes the situation where the label is transmitted within the optical bandwidth of the data payload. The ability to route packets with imbedded labels in a single operation is attractive for keeping network complexity low. Imbedding the labels into the optical packet serves the purpose of reducing the bandwidth used for network signaling and eliminating the risk of label and data packet becoming de-correlated. In-band optical labeling in the schemes outlined here relies on modifying the spectrum of the optical data packet to accommodate the optical labels within the bandwidth otherwise occupied by the signal. As such, this kind of labeling is best suited for high-speed packet data where the perturbation of the optical signal spectrum caused by labeling will be small relative to the total extent of the signal spectrum. The minimum width of the labels (L1–L4 in Figure 1a) is limited by practical filtering schemes.

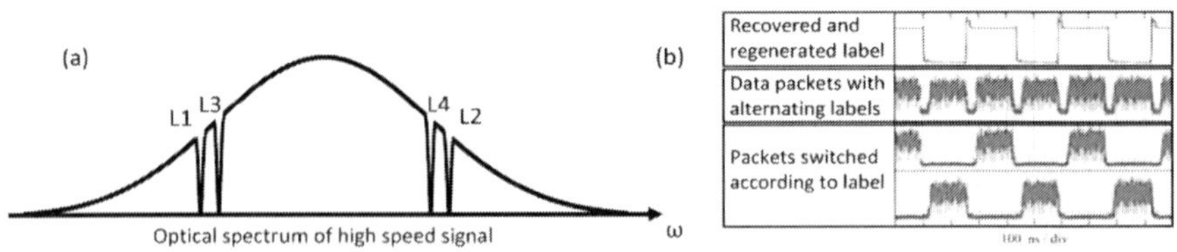


**Figure 1.** (**a**) Labeling of high speed data signals by spectral carving. L1–L4 indicate labeling positions where the presence or absence of power is interpreted as a digital label; (**b**) Results of labeling and switching 640 Gbit/s data packets.

A simple scheme for in-band optical labeling is illustrated in Figure 1a. The labeling relies on frequency sub-bands inside the signal spectrum, where optical power can be suppressed to indicate a label "mark" or passed to indicate a "space". In practice this is achieved by applying optical notch filters to shape the spectrum. As the labels encoded in this way rely on detecting small fractions of the actual signal spectrum, appropriate data scrambling is required to fulfill a mark-space ratio close to 50% in the data packet. This scheme has the clear advantage of keeping system complexity low since the labels are encoded by simple notch filtering and no extra signals are transmitted for labeling. In [3] it was verified that the scheme can be used to encode more than 60 logical labels while keeping the eye opening penalty less than 1 dB compared to the un-labeled data packet. The scheme has been shown in [4] to work for 640 Gbit/s data packets, where a single label was used to control a switch, separating labeled and unlabeled packets as shown in Figure 1b. As the scheme relies on detecting a fraction of the signal spectrum as a logical label, some regeneration of the amplitude of the detected label is required to extract a control signal suitable to drive an optical packet switch. This has been implemented using a simple electronic comparator circuit which regenerates the amplitude of the detected label and adds a fixed length to the label duration to ensure that stable and high-quality switching can be performed for every data bit in the packet. This fixed label padding is a few ns in this demonstration but can in principle be chosen freely according to the requirements of the label regenerator and switch. A packet guard interval of ~20 ns was used. In this demonstration the whole scheme could be implemented with a penalty of only 4 dB in receiver sensitivity of the labeled and switched data packet. Except for requiring slightly higher power into the receiver the signal could be detected with similar distribution in the performance of the 64 OTDM (optical TDM) channels compared with the un-labeled and un-switched signal.

It is expected that this labeling scheme could be particularly relevant in environments with high network complexity and relatively short transmission distances as is the case in e.g., data center networks.

A variation of the scheme mentioned above is demonstrated in [5] where narrow spectral bands in the broad data spectrum are allocated for transmission of optical labels. In this scheme optical labels are inserted into the assigned portions of the signal spectrum as low rate binary signals. Additionally, this scheme also allows for the insertion of a clock pilot signal which can be used for synchronizing the receiver to the incoming data packet relaxing the requirements for high-speed clock recovery circuits. High quality clock transmission does, however, require the optical spectrum of the data payload to be suppressed in the narrow spectral range where the clock is inserted. This scheme offers more freedom to adjust the duration and power of the optical labels. In this way the use of a regenerating circuit for the received labels can be avoided and the quality of the transmitted labels does not depend on scrambling of the data payload. This scheme does however require an optical transmitter for each optical label to be added. Consequently, complexity is moved from the receiver to the transmitter. One critical point in this labeling scheme is the relative stabilization of the filters used to extract the optical labels and the optical transmitters generating the labels. The wavelength stability of the label transmitters has to be carefully controlled.

This scheme has been demonstrated at 640 Gbit/s packet data rate with 50 km transmission to verify the performance of the transmitted pilot clock. Packet labeling and packet clock transmission was implemented in this demonstration. Only marginal phase noise was added by transmitting the clock as an in-band pilot clock increasing the total timing jitter (integrating the phase noise from 100 Hz to 10 MHz) from 80 fs to 90 fs. The quality of the transmitted labels allowed for switching the data packets with a penalty of ~1.5 dB in receiver sensitivity for achieving BER of $10^{-9}$, compared to the performance of unlabeled data packets.

Both packet labeling schemes mentioned above employ similar simple electronic circuits for detecting and processing the labels, imposing only a few ns of latency (<10 ns) when combined with the packet switching. Regarding dynamic allocation of labels to individual packets, both schemes require fast optical switching, similar to what is used for switching packets. This is used either to generate labels to be added to the signal in the case of the latter scheme or to switch the notch filters which the packet is subjected to in the case of the first scheme. In both cases the latency caused by label creation is expected to be even less than for label detection and processing.

## 2.2. Optical Labeling of TDM Data Channels

In very high speed TDM transmission it may be attractive to label the TDM channels in such a way that an optical receiver will be able to selectively receive only specific channels without having to have the capacity to receive the entire TDM data stream. In [6] a scheme for labeling of a 650 Gbit/s OTDM data stream is proposed and demonstrated. In this scheme the optical spectrum of one of the OTDM tributary channels is slightly altered compared to the remaining channels, allowing a receiver to identify that channel and subsequently enumerate the remaining channels based on this reference channel. Also, detecting a single channel allows for a low rate (in this case 10 GHz) clock recovery circuit to be used to extract the base rate clock, which is required for extracting the tributary channels. This should be seen in contrast to the pre-scaled clock recovery discussed in Section 4. In this scheme no dedicated clock signal is transmitted alongside the data signal, eliminating the potential signal or clock degradation due to cross talk between several co-propagating signals. The clock signal, which is also used for channel identification, is extracted simply by tuning a band pass filter to one edge of the optical data spectrum as illustrated in Figure 2a.

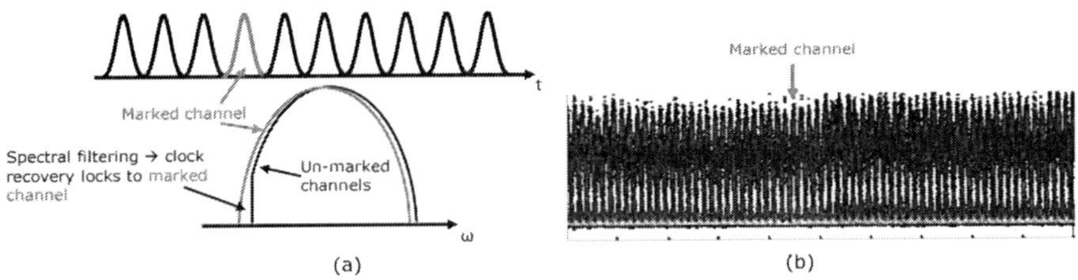


**Figure 2.** (**a**) Principle of channel labeling by spectral shaping; (**b**) 650 Gbit/s data signal after 80 km transmission with marked channel indicated.

The scheme is shown to work in an 80 km transmission demonstration of 650 Gbit/s data incorporating a fully polarization insensitive receiver. The transmitted 650 Gbit/s signal is shown in Figure 2b where the marked channel is also indicated. A slight penalty in receiver sensitivity of ~2 dB for achieving a BER of $10^{-9}$ is observed for the labeled channel compared to the remaining channels in back-to-back configuration. After 80 km transmission the labeled channel has similar performance to the remaining channels as Figure 2b would also indicate.

A main attraction of this labeling scheme is the low complexity of the setup in both generating and detecting the labels. Shaping the spectrum does alter the transmission properties of the channel slightly but it has been shown that this effect can be kept small while still achieving a detectable spectral change to the labeled channel.

## 2.3. Optical Switching of Data Packets

An attractive feature which is facilitated by the availability of optical packet labeling is optical packet switching. In optical packet switching data packets are switched and routed through a network without the need for o/e conversion of the entire data packet at every switching node. This is particularly attractive for high speed or very large data packets where detection and buffering requires significant resources. Significant benefit is expected if the data payload can be switched to the correct outlet port without requiring the switch to be able to process the full data payload. Optical packet switching has been realized in a number of different schemes. Two schemes will be briefly introduced in this section, having different properties and benefits.

## 2.4. Optical Packet Switching by Wavelength Conversion in a Si Waveguide

Optical packet switching can be combined with wavelength conversion in the scheme demonstrated in [7]. In this scheme cross phase modulation (XPM) is used to transfer amplitude modulation between original data packets and a control signal. The presence of this control signal determines which data packets are switched to the wavelength of the control signal. It is important to note that the data packet is not erased at the original wavelength in this scheme. This adds a multicasting capability to the scheme, which may be desirable. Figure 3a shows the working principle of the packet switching by wavelength conversion. A control signal indicated at $\lambda_2$ is selectively launched into a nonlinear device, in this case a silicon (Si) waveguide, to temporally overlap with the packet to be switched. The packet temporally aligned with the control signal is then wavelength converted by XPM and thus extracted from the original stream of packets as shown in Figure 3b. $\lambda_1$ is the data packet stream before switching—acting as pump for the XPM. $\lambda_2$ is the control signal used to select the packet to be switched—acting as probe for the XPM. $\lambda_2 + \Delta$ is the filtered output copy of the selected data packet.

This scheme for optical packet switching has been shown to introduce ~2.5 dB of power penalty in the receiver sensitivity for a BER of $10^{-9}$ compared to the un-switched packets. The target wavelength of the switched packet can be chosen with some flexibility depending on the properties of the specific chosen nonlinear switching material.

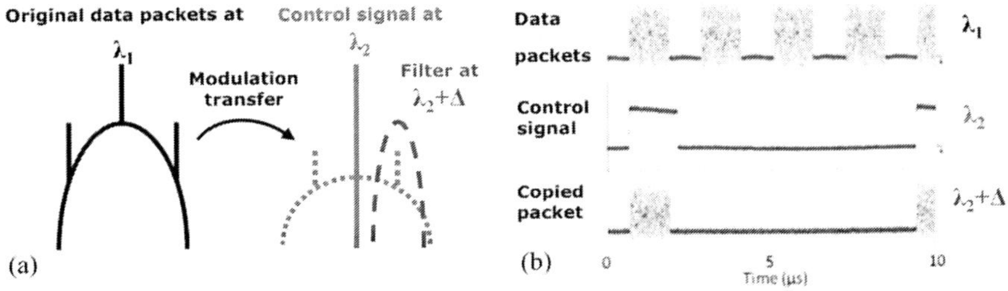


**Figure 3.** (**a**) Operating principle of packet switching by wavelength conversion and offset filtering; (**b**) Results of packet switching implementation in an 8 mm long Si nanowire.

Packet switching in this scheme is highly flexible. The nonlinear switching process is nearly instantaneous (based on Kerr nonlinearity) and the switching speed and the temporal resolution are limited only by the ability to generate appropriate optical control signals. One limitation of the packet switching scheme using wavelength conversion is the inability to remove the switched packet from the original packet flow. This prevents reusing time slots in the packet flow and may cause unnecessary traffic to be transmitted in parts of the network.

## 2.5. Optical Packet Switching in an Electro-Optic Switch

An alternate scheme for packet switching which has also been investigated and demonstrated, is switching of packets in an electro-optic switch [4]. In this scheme the packets remain at the same wavelength and switching is performed by transmitting or attenuating packets in the switch according to an applied electrical control signal. This scheme allows for reutilization of the cleared packet slots and is thus similar to packet routing based on o-e-o conversion. As electro-optic switches typically have limited modulation bandwidth, this scheme will require careful control of the guard interval between data packets and the electrical signal controlling the switch, as discussed below.

In [4], electro-optic packet switching is demonstrated in combination with the optical packet labeling scheme discussed inSection 2.1. The optical labels are detected for individual packets and used as control signal for the switch, which in this demonstration is a Lithium-Niobate (LN) Mach-Zehnder modulator (MZM) used as an on/off switch. The availability of high performance space switches would significantly improve the scaling potential of such optical switching schemes. Semiconductor-based circuits as e.g., [8] offer promising progress in this area in terms of switching speed and port count. In the current demonstration [4], scaling the scheme to a large number of outputs would incur a prohibitively large loss from splitting the signal to one switch for each output path. It is, however, clearly demonstrated that optical labels imbedded within data packets can be detected and used as control signals for an electro-optical switch achieving high-quality switching of the entire packet. This approach may offer attractive possibilities of combining control and transport layer infrastructure in optical networks.

## 3. OPTICAL SYNCHRONIZATION OF DATA PACKETS

Optical synchronization of data packets to a master clock addresses a key limitation regarding global synchronization in complex optical networks. The key functionality of optical data synchronization is demonstrated in [9,10]. The scheme for optical data synchronization relies on carefully applying a timing-dependent phase modulation to the individual data bits and subsequently adjusting their relative timing using a dispersive element. Figure 4a illustrates the basic principle of the scheme.

This scheme has been demonstrated to work on data packets up to 12,144 bits for a frequency offset of 200 kHz. Data synchronization according to this scheme can correct the clock offset with such precision that the synchronized data can be TDM multiplexed into a data stream of pulses running at the reference frequency $f$, achieving a total aggregated pulse rate of 1.29 Tbit/s while still allowing for error free (BER $< 10^{-9}$) detection of the synchronized data pulses. Figure 4b shows the multiplexing of the synchronized data packet into an empty time slot in a 1.29 Tbit/s OTDM signal. Precise synchronization and stable multiplexing is achieved.

It is important to notice that the data synchronization accomplished with this scheme is a localized phenomenon, meaning that the phase modulation transfers the frequency offset from the time domain to the frequency domain, allowing for synchronization to be achieved by adding a well-controlled amount of dispersion. Further dispersive propagation will cause the pulses to lose the synchronization to the clock *f*. In a larger optical network this synchronization scheme should ideally be combined with some means of canceling the spectral chirp caused by the phase modulation once synchronization is achieved. For amplitude modulated data this could e.g. be done by wavelength conversion by XPM as described in [11].

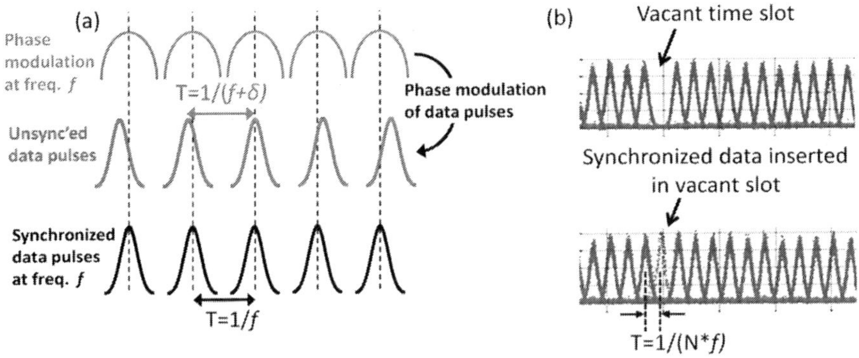


**Figure 4.** (**a**) Illustration of the operating principle of optical data synchronization. Phase modulation (marked in blue) is applied to the data signal (marked in red) which is at a pulse rate slightly offset from the target clock frequency *f*. After propagation through a dispersive medium the data pulses acquire the pulse rate *f* (marked in black); (**b**) Result of multiplexing a synchronized data packet into a vacant time slot in a high speed data stream.

# 4. ULTRA HIGH SPEED CLOCK RECOVERY

In network nodes where optical data aggregation and thus synchronization is not required a key functionality is the ability to recover a high fidelity clock, *i.e.*, a clock signal with ultra-low phase noise, from a high speed data signal, which can be used in the optical TDM demultiplexing (equivalent to electrical de-serializing). In Section 2.2 a labeling scheme was discussed which circumvents this issue by embedding a label from which a clock signal can be directly extracted. However, for unlabeled data channels pre-scaled optical clock recovery is required in order to extract a sub-rate clock from a high speed data signal at bit rates beyond the bandwidth of electronic receivers. A few schemes have been demonstrated for clock recovery at bit rates as high as 640 Gbit/s. In [12] an injection locked loop containing a semiconductor optical amplifier (SOA) was demonstrated for 640 Gbit/s clock recovery. In [13] pre-scaled clock recovery is demonstrated for a data signal at 640 Gbit/s using nonlinear mixing in a periodically poled lithium niobate (PPLN) waveguide. An opto-electronic phase locked loop (PLL) is used to lock a local oscillator (LO) radio frequency (RF) clock source to a tributary channel in the high speed data signal. The optical mixing in the PPLN waveguide is used to achieve very high resolution phase comparison between pulses derived from the LO clock and pulses in the data signal. The mixing output is then detected and used for feedback control of the LO clock. The operating principle of the pre-scaled electro-optic PLL clock recovery is sketched in Figure 5.

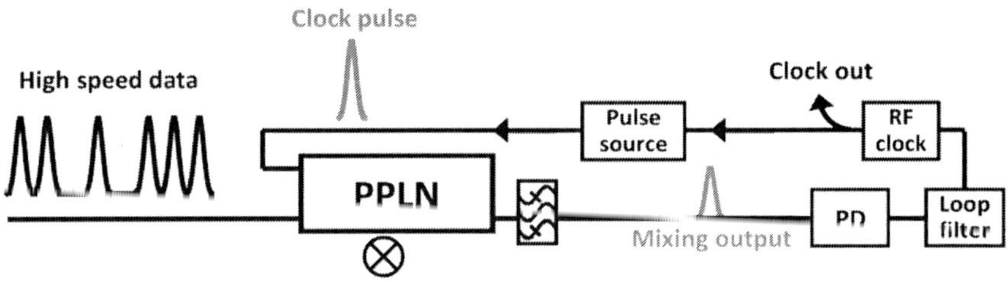


**Figure 5.** Working principle of pre-scaled electro-optic clock recovery.

This clock recovery scheme has been demonstrated in a transmission experiment in [14] to be able to recover a clock signal of sufficient high quality to enable detection of the transmitted 640 Gbit/s data signal with less than 1 dB penalty in receiver power to achieve a BER of $10^{-9}$. A recovered clock with timing jitter less than 100 fs could be derived from a transmitted 640 Gbit/s data signal using the clock recovery. This degree of synchronization is fully adequate even for bitrates as high as 640 Gbit/s.

# 5. EFFICIENT DETECTION OF ULTRA-HIGH-SPEED DATA SIGNALS

A final functionality of vital importance to high speed optical networking is the ability to efficiently receive very high speed optical data signals. As mentioned earlier high speed data signals must be optically demultiplexed (or parallelized) before they can be handled by high speed electronic receivers. This optical demultiplexing of time interleaved signals has typically been addressed in the time domain by applying very short temporal gates which sample out single tributary channels at a significantly reduced rate. This approach has been demonstrated for pulse rates as high as 1.28 Tbaud using various optical switches like a nonlinear optical loop mirror (NOLM) [15]. An alternate scheme which has recently been investigated extensively relies on optical Fourier transformation (OFT) to convert the distinct data channels from the time domain into the frequency domain. This allows for conventional wavelength division multiplexing (WDM) receivers to be used for detection of high speed data signals. In the OFT scheme dispersion is used to broaden the short optical pulses and a following parabolic phase modulation compresses the spectrum of the individual short pulses at different central wavelengths determined by the temporal alignment between the pulse and the applied phase modulation. The concept of the OFT-based receiver is outlined in Figure 6.

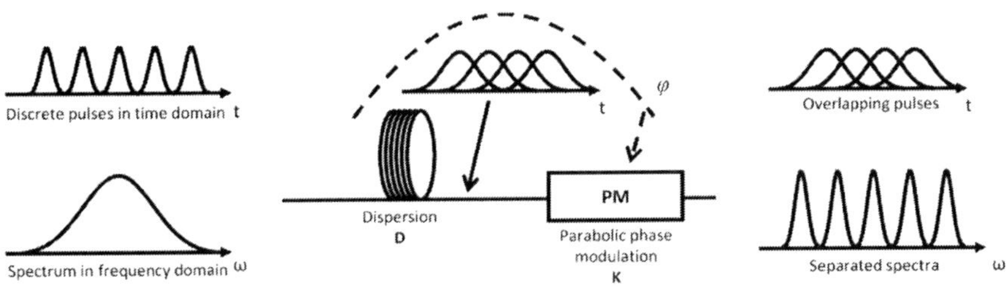


**Figure 6.** Principle of optical Fourier transformation (OFT) for transferring short pulses in the time domain into separated spectra in the frequency domain.

This scheme has been demonstrated mapping 10 Gbit/s tributary channels from a 640 Gbit/s data signal onto a 25 GHz WDM wavelength grid in [16]. A total of 40 channels could be converted simultaneously in a single OFT unit, however with significant performance penalty compared to conventional time gating of high speed signals. The OFT converted channels in this demonstration achieved a BER $\sim 10^{-5}$. The main limitation in converting such a large number of channels from a high speed data signal is the requirements to the optical mixing platform which performs the phase modulation (PM) shown inFigure 6. In this demonstration phase modulation was achieved by four wave mixing (FWM) of the data signal with a chirped pump signal. In this way phase modulation with large modulation index can be achieved, however efficient FWM over a large wavelength range is required. To achieve the large FWM bandwidth a dispersion-engineered Si waveguide was used in this demonstration. These devices can achieve very large FWM bandwidth but suffer from low FWM efficiency. This is the main cause for the severe signal degradation observed in the serial-to-parallel conversion. The OFT-based scheme for serial-to-parallel conversion holds great potential compared to the time-domain gating approach since a large number of channels can be processed simultaneously by the use of a single optical control pulse and subsequently demultiplexed using WDM de-multiplexers. It has been shown numerically in [17] that there is not an inherent performance penalty associated with the OFT scheme compared to the time gating, whereas the total power spent on switching the cannels is significantly lower when using serial-to-parallel conversion. The implementation penalty associated with the use of OFT for serial-to-parallel conversion is continuously being improved and recently error free signal detection was achieved for converting 10 Gbit/s tributary channels from a 1.28 Tbit/s serial data signal into WDM channels [18]. This was achieved in highly nonlinear fiber by reducing the number of converted channels. In [19] extension of the OFT scheme to convert all channels in a data stream simultaneously has been investigated. It is shown that this is possible at 160 Gbit/s at the expense of additional performance penalties on certain tributary channels.

## 6. DISCUSSION AND CONCLUSIONS

In the previous sections some key functionalities and demonstrations relevant for achieving ultra-high-speed optical networking have been described and discussed. Significant advances have been made in this field and many key system elements are already demonstrated. At the pulse repetition rates addressed here, literally hundreds of GHz, many system functionalities rely strongly on nonlinear optical signals processing. This is enabled by the availability of very high quality platforms for optical signal processing including nonlinear waveguides and nonlinear optical fiber. Nonlinear optical signal processing has yet to experience any substantial market penetration but a number of schemes for high speed signal processing are available and are continuously being improved and expanded.

The schemes and functionalities outlined in this paper already enable fairly sophisticated high speed optical networking. In a packet based network packet labeling, packet synchronization and packet switching have been demonstrated. In a TDM based transmission system channel identification and clock recovery have been demonstrated. Relevant for both cases is the demonstrated ability to perform highly efficient detection of these very high data rates by serial-to-parallel conversion.

## ACKNOWLEDGMENTS

OFS Denmark is acknowledged for providing the optical fiber used in the demonstrations discussed in this paper. This work was supported by the European Research Council (ERC) Starter Grant project SOCRATES (ref. no. 240467). This work was supported by the Danish Research Council grant for *Terabit Ethernet on Silicon Photonic Chips*.

## AUTHOR CONTRIBUTIONS

Writing of paper: Michael Galili. Comments and feedback: Anders Clausen. Experimental investigations: Michael Galili, Hao Hu, Hans Christian Hansen Mulvad, Ashenafi Kiros Medhin. Schemes and design of investigations: Michael Galili, Hao Hu, Hans Christian Hansen Mulvad, Ashenafi Kiros Medhin, Leif Katsuo Oxenløwe.

## CONFLICTS OF INTEREST

The authors declare no conflict of interest.

# REFERENCES


1. Namiki, S.; Hasama, T.; Ishikawa, H. Optical signal processing for energy-efficient dynamic optical path networks. In Proceedings of the 36th European Conference and Exhibition on Optical Communication (ECOC), Torino, Italy, 19–23 September 2010; pp. 1–6.

2. Tucker, R.S. Optical packet-switched wdm networks—A cost and energy perspective. In Proceedings of the Optical Fiber Communication Conference (OFC), San Diego, CA, USA, 24–28 February 2008. paper OMG1.

3. Mehdin, A.K.; Galili, M.; Oxenløwe, L.K. Novel Optical Labeling Scheme for Ultra-High Bit Rate Data Packets. In Proceedings of the OptoElectronics and Communications Conference (OECC), Kyoto, Japan, 30 June 2013–4 July 2013; pp. 1–2.

4. Medhin, A.K.; Galili, M.; Oxenløwe, L.K. 640 Gbit/s optical packet switching using a novel in-band optical notch-filter labeling scheme. In Proceedings of the Optical Fiber Communication Conference (OFC), San Francisco, CA, USA, 9–13 March 2014.

5. Gomez-Agis, F.; Hu, H.; Luo, J.; Hansen Mulvad, H.C.; Galili, M.; Calabretta, N.; Oxenløwe, L.K.; Dorren, H.; Jeppesen, P. Optical switching and detection of 640 Gbits/s optical time-division multiplexed data packets transmitted over 50 km of fiber. *Opt. Lett.* **2011**, *36*, 3473–3475.

6. Galili, M.; Hansen Mulvad, H.C.; Hu, H.; Oxenløwe, L.K.; Gomez-Agis, F.; Ware, C.; Erasme, D.; Clausen, A.; Jeppesen, P. 650 Gbit/s OTDM Transmission over 80 km SSMF Incorporating Clock Recovery, Channel Identification and Demultiplexing in a Polarisation Insensitive Receiver. In Proceedings of the Optical Fiber Communication Conference (OFC), San Diego, CA, USA, 21–25 March 2010; pp. 1–3.

7. Hu, H.; Galili, M.; Pu, M.; Yvind, K.; Jeppesen, P.; Oxenløwe, L.K. 160 Gbit/s optical packet switching using a silicon chip. In Proceedings of the IEEE Photonics Conference, Burlingame, CA, USA, 23–27 September 2012; pp. 915–916.

8. Varrazza, R.; Djordjevic, I.B.; Yu, S. Active Vertical-Coupler-Based Optical Crosspoint Switch Matrix for Optical Packet-Switching Applications. *J. Lightw. Technol.* **2004**, *22*, 2034–2042.

9. Hu, H.; Areal, J.L.; Hansen Mulvad, H.C.; Galili, M.; Dalgaard, K.; Palushani, E.; Clausen, A.; Berger, M.S.; Jeppesen, P.; Oxenløwe, L.K. Synchronization, retiming and time-division multiplexing of an asynchronous 10 Gigabit NRZ Ethernet packet to terabit Ethernet. *Opt. Express* **2011**, *19*, B931–B937.

10. Hu, H.; Areal, J.L.; Palushani, E.; Oxenløwe, L.K.; Clausen, A.; Berger, M.S.; Jeppesen, P. Optical Synchronization of a 10-G Ethernet Packet and Time-Division Multiplexing to a 50-Gb/s Signal Using an Optical Time Lens. *IEEE Photonics Technol. Lett.* **2010**, *22*, 1583–1585.

11. Galili, M.; Oxenløwe, L.K.; Hansen Mulvad, H.C.; Clausen, A.T.; Jeppesen, P. Optical wavelength conversion by cross-phase modulation of data signals up to 640 Gb/s. *J. Sel. Top. Quantum Electron.* **2008**, *14*, 573–578.

12. Hansen Mulvad, H.C.; Tangdiongga, E.; Waardt, H.d.; Dorren, H. 40 GHz clock recovery from 640 Gbit/s OTDM signal using SOA-based phase comparator. *Electron. Lett.* **2008**, *44*, 146–148.

13. Oxenløwe, L.K.; Gomez Agis, F.; Ware, C.; Kurimura, S.; Hansen Mulvad, H.C.; Galili, M.; Kitamura, K.; Nakajima, H.; Ichikawa, J.; Erasme, D.; *et al.* 640 Gbit/s clock recovery using periodically poled lithium niobate. *Electron. Lett.* **2008**,*44*, 370–372.

14. Oxenløwe, L.K.; Gómez-Agis, F.; Ware, C.; Kurimura, S.; Hansen Mulvad, H.C.; Galili, M.; Nakajima, H.; Ichikawa, J.; Erasme, D.; Clausen, A.T.; *et al.* 640-Gbit/s Data Transmission and Clock Recovery Using an Ultrafast Periodically Poled Lithium Niobate Device. *J. Lightw. Technol.* **2009**, *27*, 205–213.

15. Hansen Mulvad, H.C.; Oxenløwe, L.K.; Galili, M.; Grüner-Nielsen, L.; Jeppesen, P. 1.28 Tbit/s single-polarisation serial OOK optical data generation and demultiplexing. *Electron. Lett.* **2009**, *45*, 280–281.

16. Hansen Mulvad, H.C.; Palushani, E.; Hu, H.; Ji, H.; Lillieholm, M.; Galili, M.; Clausen, A.T.; Pu, M.; Yvind, K.; Hvam, J.M.; *et al.* Ultra-high-speed optical serial-to-parallel data conversion by time-domain optical Fourier transformation in a silicon nanowire. *Opt. Express* **2011**, *19*, B825–B835.

17. Lillieholm, M.; Hansen Mulvad, H.C.; Palushani, E.; Peucheret, C.; Jeppesen, P.; Oxenløwe, L.K. Numerical investigation of power requirements for ultra-high-speed serial-to-parallel conversion. In

Proceedings of the IEEE Photonics Conference (IPC), Burlingame, CA, USA, 23–27 September 2012; pp. 110–111.

18. Hu, H.; Kong, D.; Palushani, E.; Andersen, J.D.; Rasmussen, A.; Sørensen, B.M.; Galili, M.; Hansen Mulvad, H.C.; Larsen, K.J.; Forchhammer, S.; *et al.* 1.28 Tbaud Nyquist Signal Transmission using Time-Domain Optical Fourier Transformation based Receiver. In Proceedings of the Conference on Lasers and Electro-Optics (CLEO), San Jose, CA, USA, 9–14 June 2013.

19. Petrillo, K.G.; Foster, M.A. Full 160-Gb/s OTDM to 16x10-Gb/s WDM conversion with a single nonlinear interaction.*Opt. Express* **2013**, *21*, 508–518.

4

# SOFTWARE DEFINED ELASTIC OPTICAL NETWORKING IN TEMPORAL, SPECTRAL, AND SPATIAL DOMAINS

**S. J. Ben Yoo , Lei Liu, Roberto Proietti, Ryan P. Scott**

Department of Electrical and Computer Engineering,
University of California, Davis, CA 95616, USA

## ABSTRACT

This paper discusses architecture, protocol, technologies, systems, and networking testbed for software-defined elastic optical networking in temporal, spectral, and spatial domains. By exploiting the progress in elastic optical networking (EON) in temporal and spectral domains utilizing dynamic optical arbitrary waveform generation and measurement technologies, and by extending the EON concept into the spatial domain through the new orbital angular momentum-based spatial division multiplexing, we realize EON exploiting elasticity in temporal, spectral, and spatial domains (3D-EON). Routing, spectral, spatial mode, and modulation format assignment with fragmentation awareness as well as hitless defragmentation for high capacity, high quality of service, and resource-efficient networking will be pursued. OpenFlow-based 3D-EON testbed at UC Davis includes optical supervisory channel with optical performance monitoring for software-defined networking with an adaptive observe-analyze-act cycle.

## KEYWORDS

Elastic optical networking, Flexible grid networking, Orbital angular momentum, Arbitrary waveform measurement, Arbitrary waveform generation

# 1. INTRODUCTION

## 1.1 Motivation

The sustained pattern of exponential data traffic growths has completely transformed the landscape of modern telecommunication networks. Initially, the increases in capacity demands were successfully met by deployment of wavelength-division multiplexing (WDM) technology, which offered orders of magnitude increases in communication capacity from 10 Gb/s to multiple Tb/s. More recently, rapidly emerging new services and data centers are driving the peak link capacity demands beyond 10 Tb/s, hence, recent optical networking advances have addressed coherent optical communications with high spectral efficiency beyond 1–10 b/s-Hz employing advanced modulation formats to achieve link capacity as high as 100 Tb/s. However, it is extremely difficult to support such high spectral efficiency and high capacity under dynamically changing traffic conditions especially due to their sensitivity to physical layer impairments such as fiber amplifier noise, chromatic and polarization dispersion, and optical nonlinearity [1] even for moderate (~~500 km) transmission distances. Hence, capacity beyond 100 Tb/s must seek a new dimension other than temporal and spectral domains for growth. The spatial domain is the final unexplored frontier of capacity growth. In the past, space-division multiplexing (SDM) has constantly been an alternative to time-division multiplexing (TDM) and WDM, but now, we must look into a new coherent optical communication method exploiting temporal, spectral, and spatial domains. Figure 1 [2] shows steady increases in spectral efficiency reported for high-capacity transmission systems, which reached 100 Tb/s capacity and 10 b/s/Hz spectral efficiency on a single-mode fiber [3], and branching off to SDM for >>100 Tb/s capacity toward 10 Pb/s. SDM also helps to overcome the nonlinear capacity limit on a single-mode fiber [4].

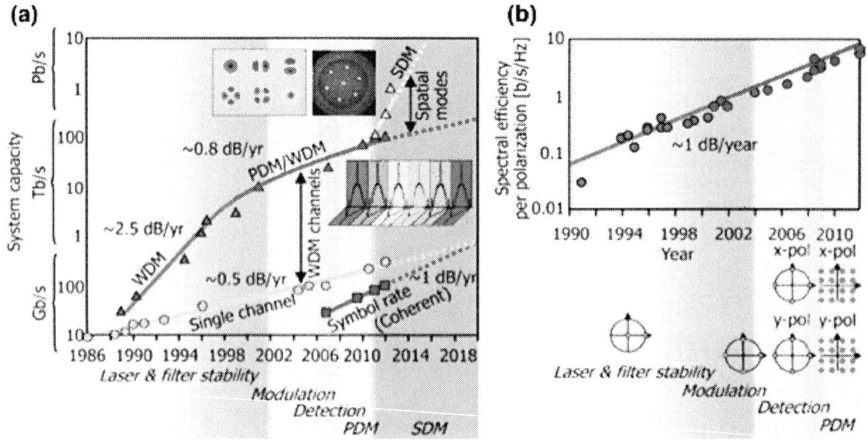


**Figure. 1 a** Evolution of experimentally achieved single-channel bit rates (single-carrier, single-polarization, electronically multiplexed; *green circles*), symbol rates in digital coherent detection (*purple squares*), and aggregate per-fiber capacities (*triangles*) using wavelength-division multiplexing (WDM; *red*), polarization-division multiplexing (PDM; *blue*), and space-division multiplexing (SDM; *yellow*) [2]. **b** Evolution of experimentally achieved per-polarization spectral efficiencies in single-(*red*) and dual-polarization (*blue*) experiments [2] (Color figure online)

In addition to the increased and flexible capacity use, future optical networks emphasize manageability and controllability of the networks. The networking community has so far adopted two types of network control frameworks: centralized and distributed control systems. Telecom networks have traditionally used centralized control planes for more accurate accountability and security while datacom networks have emerged from simple networking with rich end-system functionality based on the decentralized multi-administrative structure. As the datacom and telecom networks started to converge, telecom networks have considered the distributed Generalized Multi-Protocol Label Switching (GMPLS) architecture and its associated protocols [5] for IP-over-optical networking. While GMPLS provided flexibility for edge devices and functionalities, signaling schemes are still very complicated and it lacked a direct control and management of network resource. More recently, software-defined networking (SDN)-based on a centralized controller (e.g., NOX [6]) emerged. While the main concept of SDN is surprisingly similar to reconfigurable Multi-wavelength Optical NETworking pursued by the MONET consortium involving telecom companies in 1993–1999 [7] in separating the control plane from the data plane, and in offering hardware-independent software-defined programmable network elements, the OpenFlow-based SDN benefits from open source codes, in contrast to the fact that GMPLS is usually deployed in closed systems. SDN requires virtualization of networking resources, and it can benefit greatly from elasticity in networking resources in temporal, spectral, and spatial domains.

This paper investigates software-defined elastic optical networking in temporal, spectral, and spatial domains for future high performance cyberinfrastructures. The remainder of the paper is organized as follows. Section 1.2 summarizes the main concepts and feature of EON. Sections 2.1 and 2.2 introduce the two main enabling technologies used to exploit elasticity in the temporal/spectral and spatial domain, namely optical arbitrary waveform generation and measurement (OAWG-OAWM) and orbital angular momentum (OAM), respectively. Section 2.3 discusses how the two technologies above can be combined to form a 3D-EON communication system. Section 3 discusses the routing, spectral, spatial mode, and modulation format assignment (RSSMA) algorithm, which is an evolution of the RSMA considering the possibility to use the additional space domain for routing. Section 4 discusses the defragmentation problem augmented by the new spatial domain. Section 5 discusses some aspects of the SDN control plane design for the proposed 3D-EON. Section 6 discusses how the supervisory channel and performance monitoring technique demonstrated in [8] for a 2D EON scenario can be applied also in the new 3D elastic optical networking. Section 7 summarizes our 2D/3D-EON testbed. Finally, Sect. 8 concludes the paper.

## 1.2 Elastic optical networking (EON)

Recently, a new optical networking technology called *elastic optical networking* (EON) has been proposed [9,10]. As Fig. 2 illustrates, EONs utilize flexible (or elastic) spectral bandwidths for each data link without using fixed wavelength grids. The flexibility in spectrum allocation brings many appealing features to network operations. Current networks are designed for the worst case impairments in transmission performance and the assigned spectrum is over-provisioned. In contrast, the flexible-bandwidth networks can operate with the highest spectral efficiency and minimum bandwidth for the given traffic demand. In the case of a link failure in the network, flexible-bandwidth networks are more adaptive and likely to have spare spectrum to allocate the re-routed signal, ensuring a high survivable restoration compared to conventional optical networks. EONs employ coherent optical orthogonal frequency division multiplexing (CO-OFDM) [11, 12], coherent optical WDM (CO-WDM) [13], or Nyquist-WDM technologies [14], and adopt various modulation formats depending on the reach. More recently, UC Davis team has demonstrated dynamic optical arbitrary waveform generation (OAWG) and optical arbitrary waveform measurement (OAWM) [15, 16]-based EON, which have overcome a number of limitations of CO-OFDM, CO-WDM, and Nyquist-WDM technologies, providing opportunities for new architectures to be implemented [9, 15]. EON in temporal and spectral domain promises (a) to achieve high spectral efficiency by eliminating stranded spectrum between the fixed grid bandwidths, (b) to support both subchannel and superchannel traffic, (c) to provide multiple data rate and modulation formats. However, while EON can achieve 1.2×~×~1.6×× improvement in the network capacity, utilization, and availability, it cannot achieve 10×~×~100×× enhancements unless the spatial domain is explored. We will investigate EON in the temporal, spectral, and spatial domains.

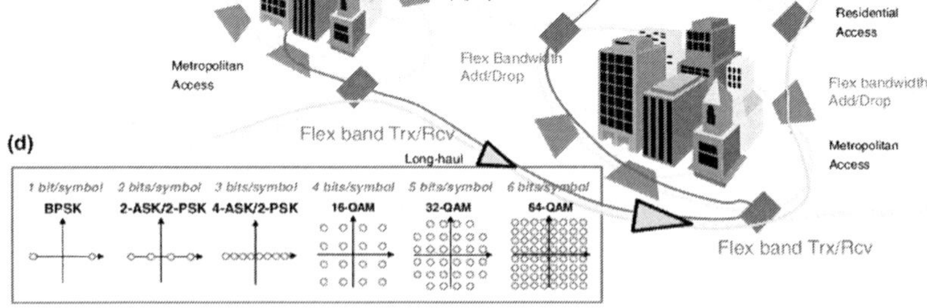


**Figure. 2** **a** WDM networks with five channels on a fixed grid and with fixed modulation format (e.g., QPSK). **b** using adaptive modulation optimized for the required bitrate and reach, the same demand can be satisfied with considerable spectrum savings [9]; **c** overview of metropolitan and access telecom network with flexible bandwidth TRXs and flexible add/drop multiplexer; **d** different modulation formats that can be used in EON

# 2. KEY ENABLING TECHNOLOGIES FOR EON

## 2.1 Dynamic OAWG and OAWM technologies for EON in spectral–temporal domains

We will use dynamic OAWG for EON transmitters and dynamic OAWM for EON receivers to exploit elasticity in temporal and spectral domains. As Fig. 3a illustrates, dynamic OAWG [15, 16] begins with a coherent optical frequency comb (OFC), which is spectrally demultiplexed with narrow passbands placing each comb line at a separate spatial location. A set of in-phase and quadrature-phase modulators (I/Q modulators), each with a bandwidth of ΔfG, apply temporal I/Q modulations to broaden the comb lines to create the spectral slices. Coherently combining the spectral slices using a gapless spectral multiplexer with broad overlapping passbands ensures a continuous bandwidth output waveform [17]. In this manner, arbitrary waveform (in amplitude and phase) across >>THz bandwidth can be generated using standard CMOS electronics (<<20 GHz). Preliminary experimental results achieved 240, 360 Gb/s, and 1.2 Tb/s data generation with BPSK, QPSK, and 16QAM modulation formats [18, 19].

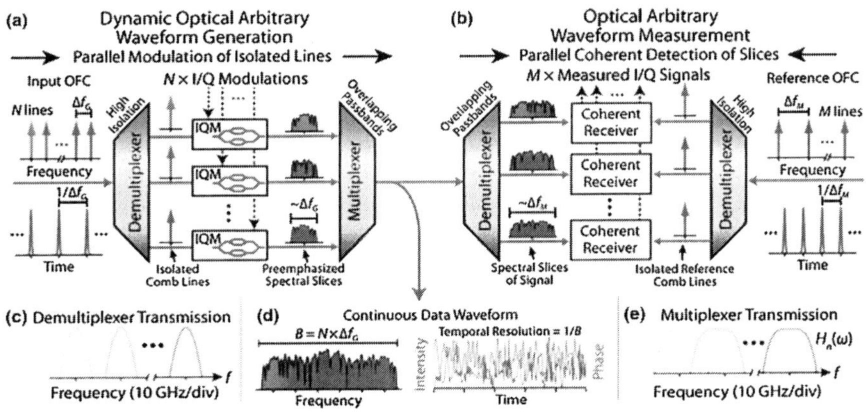


**Figure. 3 a** Dynamic OAWG transmitter and **b** OAWM receiver principle. Spectral slice **a** OAWG and **b** OAWM techniques for bandwidth scalable generation and detection of arbitrary modulation format data transmissions. **c** Demultiplexer and **e** multiplexer transmission functions. **d** Example continuous bandwidth, and complex data waveform [20]

On the other hand, the working principle of optical arbitrary waveform measurement (OAWM) is quite analogous to that of OAWG working in reverse, except for the fact that it will do coherent detection instead of coherent generation at each spectral slice. The optical comb and arbitrary optical waveform will propagate in an opposite direction in OAWM compared to OAWG. Figure 3b illustrates how an OAWM receiver characterizes waveforms through the coherent detection of MM spectral slices, each with bandwidth ΔfMΔfM. For the receiver, a reference OFC with MM-lines spaced at ΔfMΔfM provides a reference tone for the detection of each spectral slice [21]. The reference comb lines are isolated using a spectral demultiplexer with narrow and discrete passbands, and the signal is divided into spectral slices using a separate gapless spectral demultiplexer that has strongly overlapping passbands. Each reference comb line is then used to detect the corresponding spectral slice using a standard digital coherent receiver [22]. At this point, digital signal processing (DSP), as shown in Fig. 4 in case of two spectral slices, enables recombination of the spectral slices after electronic detection. In this transmission system, ΔfGΔfG can be different from ΔfMΔfM as long as the total measurement bandwidth (M×ΔfM)M×ΔfM) is greater than the generated waveform's bandwidth (N×ΔfG)N×ΔfG). The use of spectral slices enables independent optimization at the transmitter and receiver for the exact bandwidth of available electronics, and also allows utilization of the transmitters and receivers across heterogeneous network domains.

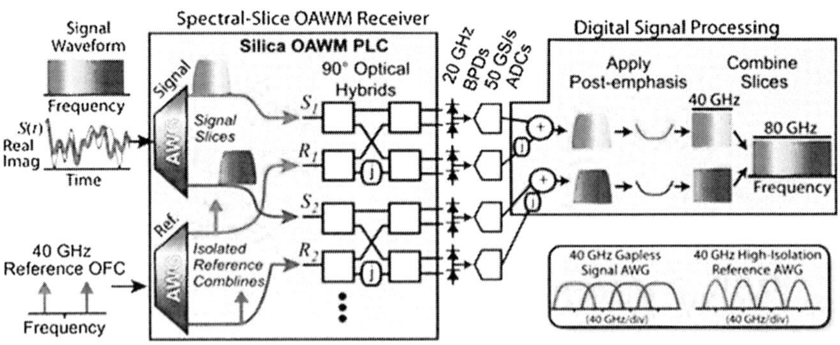


**Figure. 4** Example of 80 GHz silica PLC two-spectral-slice OAWM receiver with signal recovery DSP

The OAWG and OAWM can all-optically pre-compensate or post-compensate for chromatic dispersion [18,19], and they can also incorporate all-optical passband pre-emphasis or post-compensation filtering to equalize the non-uniform RF response of the modulators and detectors to ensure high-fidelity waveform generation after the multiplexer [17]. In addition, all-optical pre-compensation and post-compensation can shape the waveform to maintain the lowest peak-to-average power ratio (PAPR) across the transmission link to suppress nonlinear impairments. While recent flexible bandwidth networking and EON research activities have extensively used CO-OFDM, CO-WDM, or Nyquist-WDM technologies, the dynamic OAWG/OAWM technology has the following key advantages over the aforementioned counterparts for EON to be pursued here [9].

1) The dynamic OAWG/OAWM technology can generate and detect signals of any modulation format across the flexible bandwidth. Hence, OAWG/OAWM is backward compatible with other technologies.

2) While CO-OFDM and CO-WDM technologies can generate flexible bandwidth signals of multiple modulation formats, each group of the band must be parked on fixed frequency grids. On the other hand, the dynamic OAWG/OAWM technology can utilize any frequency spacing. Since CO-OFDM and CO-WDM rely on frequency comb with rigid frequency grid (due to the fixed channel spacing of the MUX and DEMUX used in the TX), the spectral efficiency decreases if the line rate per comb line decreases. Co-OFDM and Co-WDM systems can use a flexible grid only when wavelength-selective switches with flexgrid technology are used as MUX and DEMUX in the TRX architecture. On the other hand, dynamic OAWG/OAWM technique allows generation of arbitrary signal channels on arbitrary spacing as desired even when using a fixed frequency comb and MUX and DEMUX with fixed channel spacing.

3) While CO-OFDM, CO-WDM, and Nyquist-WDM technologies can generate coherent optical spikes and very high PAPR [23], the OAWG/OAWM technology can shape the total waveform to suppress PAPR and nonlinear impairments. In fact, since the DSP in OAWG can shape the total waveform at its output, rather than the waveform of each single subcarrier, this can potentially be used to control the PAPR evolution along the link by shaping the waveform in such a way that minimizes the nonlinear effects and maximizes the quality of the received signals.

## 2.2 Spatial domain enabling technology: orbital angular momentum (OAM)

An additional degree of freedom in the spatial domain can exploit orbital angular momentum (OAM) [24], so that together with dynamic OAWG and OAWM, it will enable 3D EON in the temporal, spectral, and spatial domains. Among many spatial domain methods, OAM [24] is attractive because it supports well-defined orthogonality and cylindrical symmetry. As Fig. 5a illustrates, the light beam carrying orbital angular momentum exhibits an azimuthal phase variation of $\varphi(r,\phi)=\exp(i\ell\phi)\varphi(r,\phi)=\exp(i\ell\phi)$, where $\phi\phi$ is the azimuthal angle and $\ell\ell$, a positive or negative integer, is the charge number of the OAM state. The OAM states of differing charge numbers ($\ell\ell$) are orthogonal to each other and the handedness of this helical phase front is determined by the sign of the charge number $\ell\ell$. For a nonzero charge number, the OAM beam leads to a phase singularity on the beam axis, and the amplitude along the beam axis vanishes. Compared to a spin mode (i.e., polarization), the light beam can in principle support an infinite number of OAM states at the same time (subject to the Shannon limitation imposed by the signal to noise ratio). Therefore, OAM has the potential to significantly improve the spectral efficiency or photon efficiency of free-space and fiber optical communications [25]. In addition, the inherent orthogonality of the various OAM states may reduce crosstalk, resulting in reduced reliance on digital signal processing (DSP) compared to cases employing standard multiple-input and multiple-output (MIMO) methods [25].

As illustrated in Fig. 5b, the UC Davis team has demonstrated [26] OAM multiplexing and demultiplexing devices based on photonic integrated circuits (PICs) containing a free propagation region (FPR) employing a Rowland circle. At the Roland circle, the wavefronts acquire a linear phase tilt based on the input waveguide position, thus providing progressive phase evolution in the azimuthal angle as required for the OAM state. The superposition principle of waves applied to Fig. 5b indicates that multiplexing of multiple OAM states is possible, and propagating the waves in reverse will achieve demultiplexing of multiple OAM states. Figure 5c illustrates a recent realization including a 3D photonic integrated circuit [27, 28] interfacing with the 2D PIC containing the FPR with a Roland circle. The 3D PIC is written by 3D laser inscription [29, 30] that works similarly as a 3D printer but uses a tightly focused ultrafast laser beam. The fabricated 3D/2D PIC supports Mux/Demux of up to 15 OAM states, with both TE and TM polarizations and relatively low loss performance at $1.55\mu m 1.55\mu m$. The elasticity in the spatial domain can be achieved by including all or part of the available 15 OAM states in the network, similarly to the way modulation formats are assigned. The compact design and single-mode interface easily connect with other high-speed optical components for future on-chip integration. The recent successful OAM transmission and switching experiments [26, 27, 31] exploited these 3D/2D PICs. We report below an example of OAM free space transmission exploiting the OAM PIC device described above.

In this experiment, we demonstrate a polarization-diversified free-space OAM link with simultaneous transmission of three OAM states and 14 wavelength channels using two OAM hybrid devices as a OAM mux/demux pair. Each wavelength channel has 10 GBd, dual-polarization quadrature-phase shift keying (DP-QPSK) data for an aggregate capacity of 1.68 Tb/s and spectral efficiency of 9.6 bits/s/Hz.

Figure 6a shows a conceptual diagram of a photonic integrated circuit (PIC) for OAM state multiplexing/demultiplexing. The free-propagation region (FPR) applies a linear phase tilt on the input light. The phase-matched waveguides sample the output of the FPR and transform the linear phase tilt to azimuthal phase variations at the output apertures. Illuminating multiple inputs simultaneously generates OAM states that spatially overlap at the output apertures, providing true OAM state multiplexing. Alternatively, when operated in reverse, the circular-patterned apertures sample and demultiplex spatially overlapped OAM states. Figure 6b shows a drawing of the 2D-3D hybrid integrated device where the 2D silica PLC couples with a 3D waveguide circuit fabricated by direct laser inscribing in borosilicate glass. Figure 6c shows the normalized transmission from OAM mux input (A) to OAM demux output (B) for both polarizations. The total loss for OAM state 0 is 19.4 dB (device TE) and 19.3 dB (device TM). Figure 6d shows the experimental arrangement where a 10 GBd QPSK signal (215−1PRBS215−1PRBS) drives all 14 WDM channels (12.5 GHz spacing). The modulated signal is polarization multiplexed with decorrelation of 500 ns between the two polarization states. After multiplexing up to three OAM states through the first hybrid device, the spatially overlapped beams are transmitted over a 81.8 cm free space link, demultiplexed by the second hybrid device and sent to the polarization-diversified coherent receiver with offline DSP. The DSP uses an adaptive equalizer, which includes a 17-tap finite impulse response (FIR) filter, power-of-four methods for phase and frequency recovery, and direct-decision-based bit error counting.

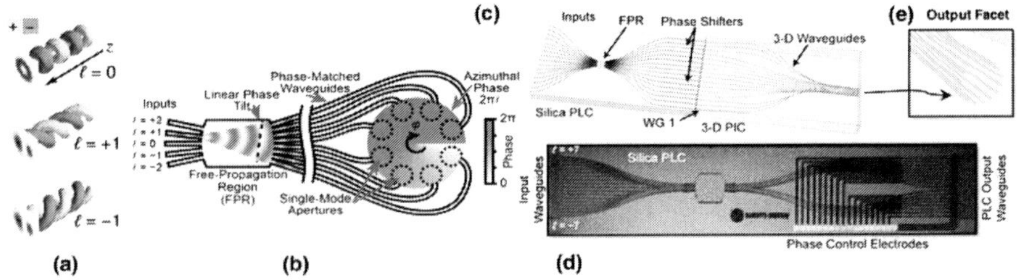


**Figure. 5 a** 3D illustration of OAM mode propagation for $\ell=0,+1,-1\ell=0,+1,-1$. **b** Planar waveguide configuration of OAM multiplexer/demultiplexer which includes the Rowland circle free-propagation region (FPR) and the phase-matched waveguides interfacing antenna apertures placed on a perimeter of the circle. Thus, the OAM multiplexer converts the linear phase front tilt to the azimuthal varying phase. Input OAM states at $\ell=+2,+1,0,-1,-2\ell=+2,+1,0,-1,-2$ will each get projected to an array of phase-matched waveguides to excite apertures with azimuthally varying phase dependent on the OAM state number $\ell\ell$. This way, multiple OAM states can be multiplexed spatially (SDM). **c** Construction of an OAM mux/demux based on a 2D silica planar lightwave circuit (PLC) containing the FPR and phase shifters and 3D waveguides leading to an output facet with apertures along the ring. **d** a photograph of the 2D silica PLC

The averaged bit error rate (BER) of two polarization states for a multichannel free space transmission link was measured. We used OAM states −6, −3 and 0 for the transmission link experiment. The measured crosstalk values at all three OAM states outputs are between −10.5 and −11.6 dB for TE, between −10.2 and −14.1 dB for TM. Figure 7a shows the measured BER for the following cases: (i) back to back transmission without WDM or OAM states crosstalk (XT); (ii) back to back transmission with all the wavelength channels; (iii) single OAM state and single wavelength channel transmission without crosstalk from other OAM states; (iv) all three OAM states transmission with single wavelength channel; (v) single OAM state with all the wavelength channels but without crosstalk from other OAM state; (vi) all three OAM states transmission with all the wavelength channels. There is ~ 3dB OSNR penalty for OAM state 0 and −3 transmission with OAM states crosstalk. However, the OSNR penalty for OAM state −6 is much higher due to a joint effect of crosstalk and polarization dependent loss (PDL). In the future, the better phase error correction can further reduce the crosstalk. PDL is mainly from the misalignment in the device setup. Figure 7b shows that the measured averaged BER values with all three

OAM states at all wavelength channels are below HD-FEC limit of 3.8×10−33.8×10−3 (BCH(1020,988) super FEC code, 7 % overhead).

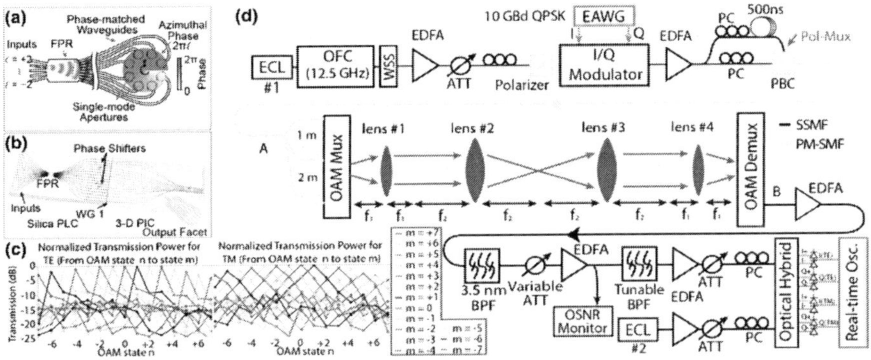


**Figure. 6 a** Concept of the OAM photonic integrated circuit. **b** Conceptual view of the 2D-3D hybrid integrated device for OAM mux/demux. **c** Normalized transmission of the OAM mux/demux pair. **d** Experimental arrangement of the multichannel OAM polarization-diversified free-space communication link testbed. *FPR* free-propagation region; *ECL* external cavity laser; *OFC* optical frequency comb; *WSS* wavelength-selective switch; *EDFA* erbium-doped fiber amplifier; *ATT* attenuator; *PC* polarization controller; *PBC* polarization beam combiner; *BPF* bandpass filter. *EAWG* electrical arbitrary waveform generator

## 2.3 System architecture for EON in spatial–temporal–spectral domains

The experimental example shown at the end of the above section is a practical simple example of how temporal and spectral domain can be combined with spatial domain. In fact, multiple WDM-QPSK signals illuminating multiple inputs of the OAM MUX simultaneously generate OAM states that spatially overlap at the output apertures, providing true OAM state multiplexing. At the RX, the same component used in the opposite direction acts as DEMUX to spatially separate the different OAM states.

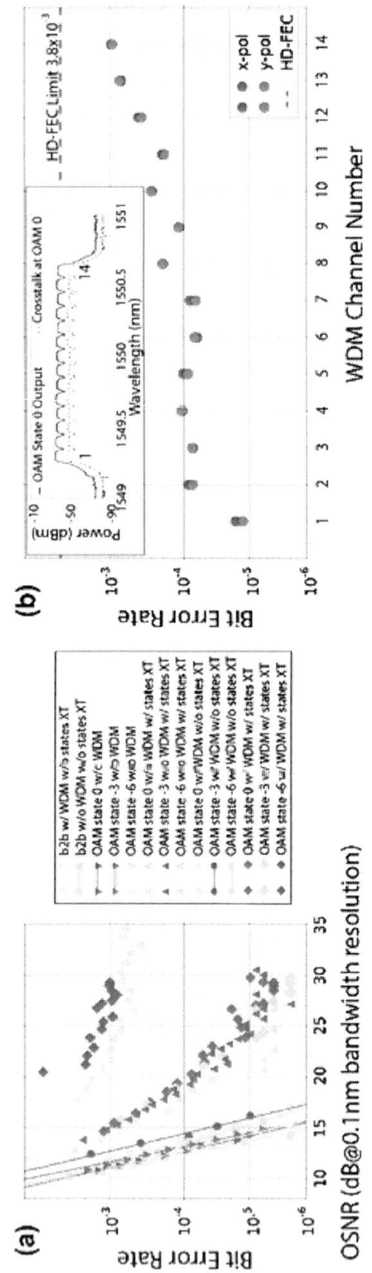


**Figure. 7 a** BER performance for the multichannel OAM link with 14×10GBd14×10GBd, DP-QPSK WDM channels. **b**Averaged BER for three OAM states (0, −−3 and −−6) for two received orthogonal polarizations (x-pol and y-pol not necessarily the same as device TE and TM)

More in general, as Fig. 8a shows, the 3D EON transmitter in the spatial, temporal, and spectral domains can be constructed by combining multiple parallel temporal and spectral OAWG transmitters feeding signals into the OAM Mux through a unitary matrix. The OAM Mux has been already introduced as a 2D/3D PIC (Fig. 5c), where circular-patterned apertures at the output facet will emit the superposition of OAM states weighted by the coefficients $a_{-m}, a_{-(m-1)}, ... a_m$. The unitary matrix provides the proper (arbitrary) weighting of $a_{-m}, a_{-(m-1)}, ... a_m$ contributed by each OAWG transmitter. Similarly, Fig. 8b shows the 3D EON receiver in spatial, temporal, and spectral domains constructed by combining multiple parallel OAWM receivers receiving signals from the OAM Demux through a unitary matrix. The unitary matrix provides the necessary transformation to map the received (arbitrary) superposition of multiple OAM states for each OAWM Rx.

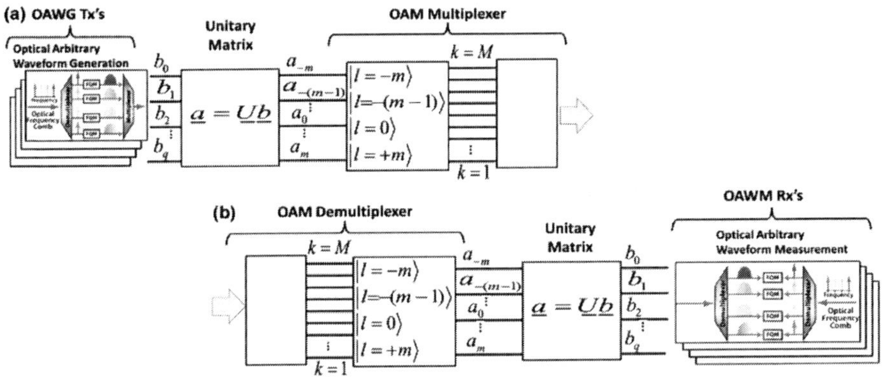


**Figure. 8 a** 3D EON transmitter in spatial, temporal, and spectral domains constructed by combining multiple parallel OAWG transmitters feeding signals into the OAM Mux through a unitary matrix. The unitary matrix provides the transformation necessary to create a desired (arbitrary) superposition of multiple OAM states for each OAWG Tx. **b** 3D EON receiver in spatial, temporal, and spectral domains constructed by combining multiple parallel OAWM receivers receiving signals from the OAM Demux through a unitary matrix. The unitary matrix provides the necessary transformation to map the received (arbitrary) superposition of multiple OAM states for each OAWM Rx

# 3. RSSMA IN TEMPORAL–SPECTRAL–SPATIAL DOMAINS

In standard EONs, we investigated routing, spectral, and modulation format assignment (RSMA) [32–37] along with defragmentation in the spectral domain. In [37], the proposing team investigated fragmentation-aware and alignment-aware RSMA algorithms so that the need for future defragmentation becomes minimal. In the proposed 3D EONs, we add the additional degree of freedom in the spatial mode (OAM states) and investigate routing, spectral, spatial mode, and modulation format assignment (RSSMA) with similar fragmentation awareness and alignment-awareness.

The fragmentation problem in the proposed EONs has three dimensions, namely those of spectral, spatial, and temporal fragmentation. Taking an example network as shown in Fig. 9a, the vertical dashed rectangular indicates the fragmented spectrum resources in link AB, which is the fragmentation in the spectral dimension. On the other hand, Fig. 9b depicts the spatial dimension fragmentation when multiple OAM states are introduced. We should consider the spectral, spatial and temporal dimensions when designing solutions to the fragmentation problem. Also, considering the fact that spectrum misalignment between neighboring links will most probably increase the end-to-end blocking probability, as shown in the horizontal dashed rectangular in Fig. 9b, therefore, an optimized RSSMA algorithm should assign a new connection in such a way that it fragments the least number of continuous spectral blocks on candidate links, while it fills up as many misaligned spectral slots as possible on neighboring links, for a given OAM state.

Figure 10 shows a more detailed example of the effect on the spectral fragments before and after a lightpath provisioning process for a given OAM state. Suppose that we have a 6-node, 8-link mesh network and that the spectral resources are distributed as shown in Fig. 10a. For simplicity, in this example, we assume only 12 spectrum slots on each fiber link. When a new request "AA to EE" arrives with a bandwidth requirement of 1 slot, the routing algorithm first calculates all possible routes, resulting in the 5 shortest paths as shown in Fig. 10b. There are 6 paths from AA to EE in total, but path *ADBCFE* is omitted since we consider only "kk" (k=5k=5) shortest paths for each source-destination pair. All arrows in Fig. 10b indicate the different possible routes and spectral positions to provision the requested circuit "AA to EE". Note that, when a spectral block with more than one slot is available for this one-slot request, the first-fit rule applies and only the slot assignment in the bottom (the lowest spectrum frequency) of the block is taken. In Fig. 5b, one "cut" indicates that the candidate RSSMA solution will break the contiguousness of a spectral block on one of the links on the current path. The "cuts" are considered to be the costs of the candidate solutions, since more "cuts" create more fragments on the candidate links of the routes. For example, the provisioning of the request on path *ABCE* with slot 10 will "cut" two spectrum blocks on links *BC* and *CE*, namely the contiguous spectral slots 9–12 on link *BC* and spectral slots 2–12 on link *CE*. Likewise, the other provisioning choices "cut" different numbers of spectrum blocks on their corresponding routes. The spectrum blocks become more fragmented as they lose continuousness in the spectral domain. In the example, in Fig. 10b, the provisioning choices with solid arrows on the path *ADE* and *ABDE* both give zero "cuts"; therefore, they are the most preferred solutions in terms of spectral fragmentation awareness. On the other hand, the provisioning of a request can also increase the misalignment of the available spectral blocks between the candidate links and their neighboring links. The optimized spectrum assignments are keeping the unused spectrum on neighboring links aligned for future requests. For example, the candidate provisioned on slot 8 (the top slash blocks) will change the alignment of the neighboring links as shown in Fig. 10c. The misalignment for the link pair *BA* and *AD* will increase by one since the provisioning on link *AD*on slot 8 reduces the commonly available spectrum by one slot. Likewise, all the misalignment changes of neighboring link pairs along the candidate path *ADE* can be calculated. Note that, if the provisioning fills up the originally misaligned spectrum, as shown for link pair *DE* and *EF* in Fig. 10c, the misalignment decreases by one. If the requested bandwidth is more than one slot, the "misalignment increase" is counted accumulatively over all slots. The increased misalignment in a network is considered as an additional cost for future light-path provisioning and spectrum defragmentation. Therefore, among all candidate solutions, the algorithm should minimize the misalignment cost as well. For a given OAM state, the proposed fragmentation-aware RSSMA simply gives higher priority to the cost "cuts" and calculates the spectral fragmentation first. It tries to minimize the number of "cuts" on the candidate routes and spectrum slots. If there is more than one solution that achieve the identical minimum number of "cuts", then the algorithm starts to calculate the misalignment increase between the candidate links and their neighboring links. In the end, the algorithm provides the optimized RSSMA solution with both minimum "cuts" and minimum increase in misalignment. The shortest-path first-fit rule kicks in if more than one RSSMA solutions are found. If no RSSMA solution can be found for this OAM state, the algorithm will repeat the above procedure and check the feasible solutions for the next OAM state until a RSSMA solution is found.

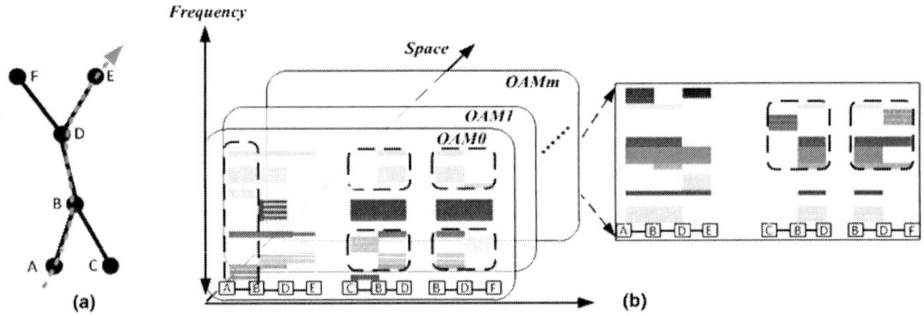


**Figure. 9 a** An example topology for elastic optical networking; **b** spectral fragmentation on spectral-spatial-temporal dimensions

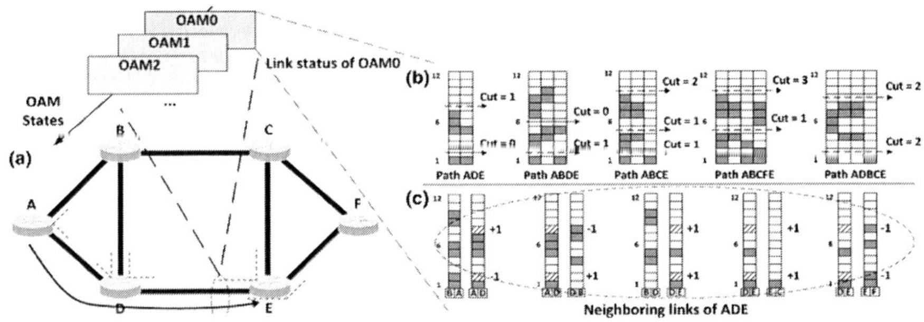


**Figure. 10 a** Example FISH network and the spectral assignment status on the links, **b** number of "Cuts" to all the candidate solution to connection request A–EA–E, and **c** misalignment increase if choosing path *ADE* with slot 8 or slot 2

To show the benefits of the 3D EON and fragmentation-aware RSSMA algorithm, we have conducted simulations on sample network topologies to compare the blocking probability of newly proposed fragmentation-aware RSSMA algorithm with our previously proposed 2D fragmentation-aware RSMA algorithm [37] for traditional EON. Dynamic connection arrival and departure events are simulated on a 14-node NSFNET network and on a 24-node USBN network, respectively. Each spectral slot is set to be 12.5 GHz, and each fiber link has 400 slots. In the simulation, each source-destination pair generates connection requests randomly according to a Poisson process. The offered load is controlled by changing the average of the Poisson process in the range over [0.8, 10] arrivals per time unit. The holding time of each connection follows a negative exponential distribution averaging 5 time units. The connection bandwidth is randomly distributed in the range of LL = [1-slot, 10-slot]. The number of OAM states is set to 8 for the 3D-EON. Figure 11 compares the blocking probability of the different provisioning algorithms in the 14-node NSFNET and the 24-node USBN network. The results show that fragmentation-aware RSSMA algorithm for 3D EON can greatly reduce the blocking probability compared with our previously proposed 2D fragmentation-aware RSMA algorithm for traditional EON.

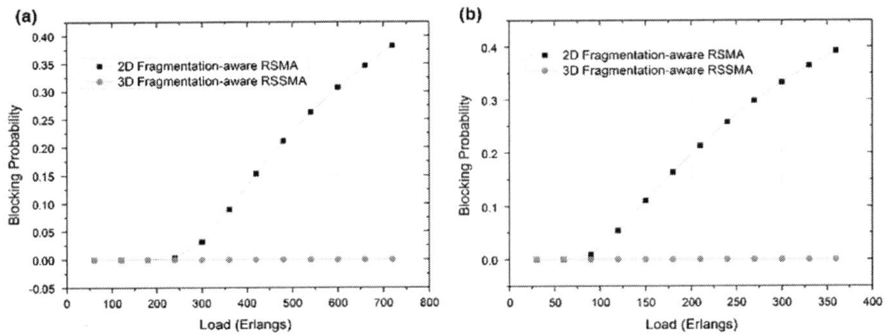


**Figure. 11** Blocking probability comparison of different algorithms in the **a** 14-node NSFNET network, **b** 28-node USBN network

# 4. DEFRAGMENTATION IN SPATIAL–SPECTRAL DOMAIN

As mentioned above, the fragmentation problem in the 3D EON is not simply related to the spectral domain, but also to the new spatial domain. Therefore, in 3D EONs, it becomes necessary to consider spectral-spatial domain defragmentation. In the standard EON, defragmentation required either wavelength conversion or reassignment of spectrum or routes. Interestingly, in the 3D EON, the OAM-state conversion [38] is far easier than wavelength conversion or time-slot interchange.

The first step in 3D defragmentation is to reconfigure the network so that the fragments can be consolidated into contiguous blocks. While the fragmentation-aware RSSMA should minimize the need for defragmentation, the reconfigurations for defragmentation can be done either periodically or on-demand in the 3D space. Periodical defragmentation usually operates with the goal of confining the spectral usage to one side of the spectrum and requires the entire network to be considered for defragmentation simultaneously. On the contrary, dynamic on-demand defragmentation can be adaptively tailored to the specific demand of the important and urgent incoming connections. With either approach, the existing connections need to be reconfigured either by changing routes, assigning different spectrum and/or OAM states at the transceivers, or converting wavelength and/or states in the intermediate nodes. It is important to avoid or minimize disruption of live connections. UC Davis team has demonstrated hitless defragmentation [39] in EON by introducing a tunable laser. For 3D EONs, we can also realize hitless defragmentation in the temporal, spectral, and spatial domains by utilizing tunable components.

# 5. SDN CONTROL PLANE FOR 3D EON

Control plane is a key element to enable intelligent optical networks, especially for the dynamic and cost-efficient provisioning and restoration of end-to-end optical connections. For the 3D EON, two different control planes can be deployed. One is the distributed GMPLS with an optional, dedicated path computation element (PCE). The other one is the centralized SDN architecture, which offers hardware-independent software-defined network programming capability. The advantages and disadvantages of GMPLS/PCE and SDN, as well as their comparison have been widely discussed in previous studies, such as [40–42]. In this work, we focus on the SDN-based solution, in particular, the OpenFlow-based architecture and its associated protocol. The reason is that OpenFlow-based SDN provides satisfactory flexibility for the operators to control a network given its open interfaces as well as the open nature of its source code, in contrast to the fact that GMPLS is usually deployed in closed systems. In addition, OpenFlow-based SDN requires virtualization of networking resources, and it can benefit greatly from elasticity in networking resources in temporal, spectral, and spatial domains.

Figure 12 shows the OpenFlow-based SDN architecture for the 3D EON. An intelligent OpenFlow-based controller can be deployed to support multi-thread processing and can dynamically perform RSMMA and defragmentation through its PCE module. The PCE can get the network information from the traffic engineering database (TED) for the path computation, and after a successful path computation, the PCE can notify the OpenFlow engine to send the extended OpenFlow messages to the corresponding OpenFlow agents for path provisioning.

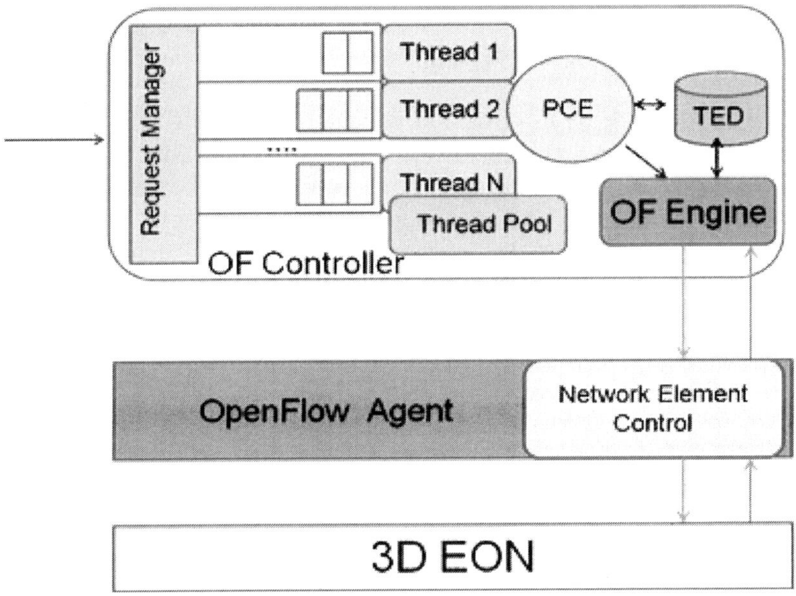


**Figure. 12** OpenFlow-based SDN architecture for 3D EON

To support 3D EON using an OpenFlow-based control plane, all the network elements (NEs) in the 3D EON are required to be extended with the OpenFlow capability. This can be achieved by introducing an OpenFlow agent on top of each NE, as shown in Fig. 12. A centralized OpenFlow controller (e.g., NOX) can communicate with all the OpenFlow agents through the extended OpenFlow protocol. The OpenFlow protocol extensions are summarized below:

1) The *Feature Reply* message is extended to report the new features of a 3D EON (e.g., switching capability, OAM states, available spectrum ranges, etc.) to the NOX controller;

2) The *Packet In* message is extended to carry the bit rate of each incoming flow

3) The *Flow Mod* message is extended to carry the RSMMA results from the NOX, including input/output ports, OAM state, central frequency, slot width, and modulation format, etc. for each OpenFlow agent to control the underlying hardware.

# 6. SUPERVISORY CHANNEL AND PERFORMANCE MONITORING FOR 3D EON

3D EON can benefit from optical supervisory channels that can achieve optical performance monitoring so that pertinent link states can be measured across the entire spectrum and spatial modes at all times. This will be a critically important functionality to realize impairment-aware networking [8, 43]. The spread-spectrum optical supervisory channel [8] used for the EON to monitor the entire spectrum at all times can be also used in 3D EON by introducing it on each OAM states. At each monitoring point, it will be necessary to use an OAM DEMUX, as the one shown above, to separate the different OAM states so that each state can be monitored independently. Performance monitoring is of particular importance in a SDM system. In fact, while different OAM states can be orthogonal in free space, there will be crosstalk once the signal will be launched and transmitted over multimode optical fibers, thus necessitating sophisticated MIMO DSP.

As we demonstrated in [8], using a real-time performance monitoring method that spans a broad and elastic spectral width, the network can dynamically and adaptively adjust the modulation format to maximize spectral efficiency while maintaining the required quality of service (QoS) and bit error rate (BER) performance even for signals that experience time-varying physical layer impairments (QoS-aware and impairment-responsive networking). We can encode supervisory channel information at low speed (e.g., 1.25 Gb/s) modulation with low modulation index (e.g., ~~0.1) on top of the data spectrum. Since the data and the supervisory channel signal follow that same path, the BER of the data is strongly correlated with that of the supervisory channel. Figure 13a shows the correlation between the measured BERs for the supervisory channel versus the data channel for three data modulation formats: BPSK, QPSK, and 8PSK. The strong correlation indicates that the BER of the supervisory channel can be used to estimate the BER of the data for different modulation formats, or to choose the modulation format that maximizes the spectral bandwidth while meeting the QoS requirement (or, data BER<1E−4BER<1E−4 in this case). The same spread-spectrum supervisory channel can be applied to each OAM state in the OAM-based SDM system to monitor the QoT of each single OAM state [31, 44].

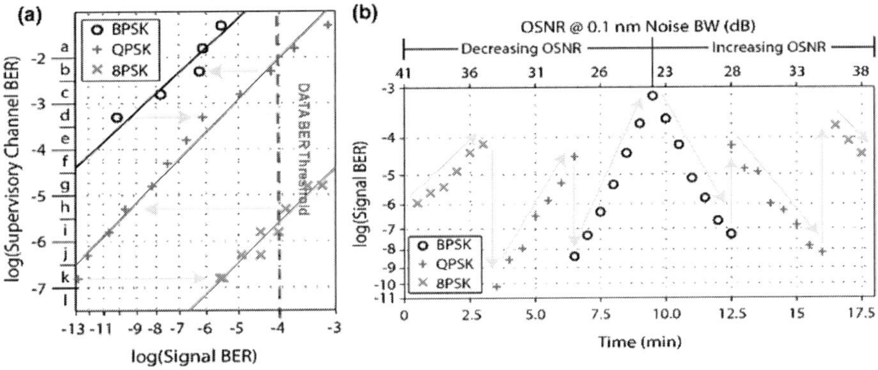


**Figure. 13 a** Correlation between the BER of Supervisory Channel and that of data, **b** autonomous adaptation of the QoT in the testbed shown in Fig. 14a

# 7. TESTBED INTEGRATION

Figure 14 illustrates 2D-EON testbed (temporal and spectral) at UC Davis [45, 46] with OAWG and OAWM EON transceivers with optical supervisory channel-based optical performance monitoring. The testbed has been upgraded to include OpenFlow-based NOX with RSSMA-based Path Computation Element (PCE). In this 3-node 2D-SDN-EON testbed, we have shown autonomous adaptation to network impairments to maintain QoT by utilizing the observe-analyze-act cycle facilitated by optical supervisory channel. We already demonstrated transmission of multiple OAM states using the photonic integrated MUX/DEMUX device shown in Fig. 5 [31], and we have been currently working on fully integrating the OAM-based spatial dimension in the current testbed, so that it will be possible to run a full system 3D-EON demonstration with RSSMA and reconfigurable OAWG, OAWM, and OAM based on supervisory channel monitoring technique.

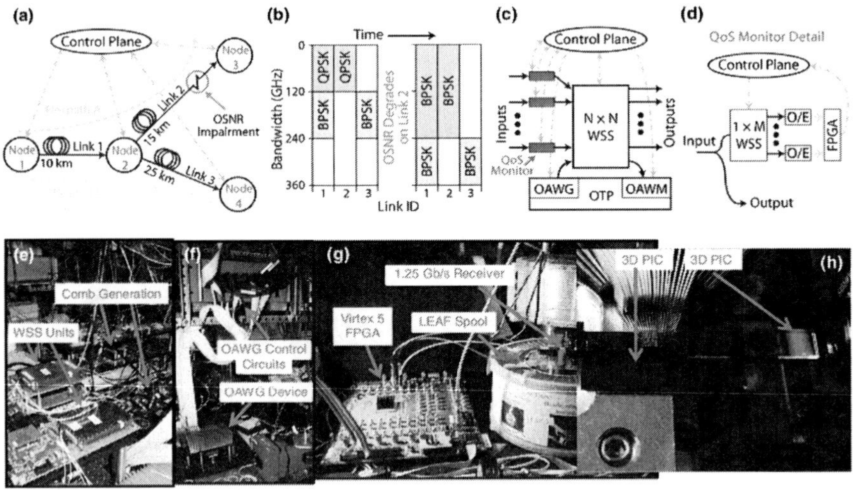


**Figure. 14** **a** Flexible bandwidth networking scenario including time-varying OSNR on Link 2. **b** Spectral usage for flexpaths (corresponding to color) as OSNR degrades over time. **c** Flexible bandwidth wavelength cross connect (WXC) node detail. **d** Quality of service (QoS) monitor detail. **e–h** hardware testbed with OpenFlow-based NOX with PCE for 3D-SDN. (*WSS* wavelength-selective switch. *OAWG* optical arbitrary waveform generation. *OAWM* optical arbitrary waveform measurement. *OTP* optical transponder. *FPGA* field programmable gate array)

# 8. CONCLUSION

This paper presented a new approach to temporal, spectral and spatial elastic optical networking (3D-EON) exploiting, at the physical layer, optical arbitrary waveform generation and measurement techniques (OAWG and OAWM) for elasticity in the temporal and spectral domain, and orbital angular momentum technology for elasticity in the spatial domain. The control plane part of the proposed 3D-EON relies on an OpenFlow-based SDN, which is preferred to a GMPLS approach. We described the routing and spectrum, spatial mode and modulation format assignment (RSMMA) algorithm that take advantage of the additional dimension to reduce the blocking probability, and we discussed how to apply the supervisory channel monitoring technique in a 3D-EON scenario. Finally, we introduced our current 2D-EON testbed and the work in progress to fully integrate the OAM dimension in the current network testbed.

## ACKNOWLEDGMENTS

This work was supported by DOE under the grant DE-FC02-13ER26154, by NSF under the EECS grant 1028729, and by DARPA DSO under the grants HR0011-11-1-0005 and W911NF-12-1-0311.


## REFERENCES


1. Essiambre, R., Kramer, G., Winzer, P.J., Foschini, G.J., Goebel, B.: Capacity limits of optical fiber networks. J. Lightw. Technol. **28**, 662–701 (2010)

2. Winzer, P.J.: High-spectral-efficiency optical modulation formats. J. Lightw. Technol. **30**, 3824–3835 (2012)

3. Qian, D., Huang, M.-F., Ip, E., Huang, Y.-K., Shao, Y., Hu, J., Wang, T.: 101.7-Tb/s (370××294-Gb/s) PDM-128QAM-OFDM transmission over 3××55-km SSMF using pilot-based phase noise mitigation. In: Optical Fiber Communication Conference and Exposition (OFC/NFOEC), 2011 and the National Fiber Optic Engineers Conference, pp. 1–3 (2011)

4. Essiambre, R.J., Kramer, G., Winzer, P.J., Foschini, G.J., Goebel, B.: Capacity limits of optical fiber networks. J. Lightw. Technol. **28**, 662–701 (2010)

5. Muñoz, R., Casellas, R., Martínez, Liu, L., Tsuritani, T., Morita, I.: Experimental evaluation of efficient routing and distributed spectrum allocation algorithms for GMPLS elastic networks. Presented at the 38th European Conference and Exhibition on Optical Communications (ECOC 2012) (2012)

6. Lowery, A.J., Wang, S., Premaratne, M.: Calculation of power limit due to fiber nonlinearity in optical OFDM systems. Opt. Express **15**, 13282–13287 (Oct 2007)

7. Yoo, S.J.B., Xin, W., Garrett, L.D., Young, J.C., Ellinas, G., Chiao, J.C., Rauch, M., Baran, J.E., Meagher, B., Leblanc, H., Chang, G.K.: Observation of prolonged power transients in a reconfigurable multiwavelength network and their suppression by gain-clamping of optical amplifiers. IEEE Photonics Technol. Lett. **10**, 1659–1661 (1998)

8. Geisler, D.J., Proietti, R., Yin, Y., Scott, R.P., Cai, X., Fontaine, N.K., Paraschis, L., Gerstel, O., Yoo, S.J.B.: Experimental demonstration of flexible bandwidth networking with real-time impairment awareness. Opt. Express **19**, B736–B745 (2011)

9. Gerstel, O., Jinno, M., Lord, A., Yoo, S.J.B.: Elastic optical networking: a new dawn for the optical layer? IEEE Commun. Mag. **50**, s12–s20 (2012)

10. Jinno, M., Takara, H., Kozicki, B., Tsukishima, Y., Sone, Y., Matsuoka, S.: Spectrum-efficient and scalable elastic optical path network: architecture, benefits, and enabling technologies. IEEE Commun. Mag.**47**, 66–73 (2009)

11. Jansen, S.L., Morita, I., Schenk, T.C.W., Takeda, N., Tanaka, H.: Coherent optical 25.8-Gb/s OFDM transmission over 4160-km SSMAF. J. Lightw. Technol. **26**, 6–15 (2008)

12. Lowery, A.J., Du, L.B., Armstrong, J.: Preformance of optical OFDM in ultralong-haul WDM lightwave systems. J. Lightw. Technol. **25**, 131–138 (2007)

13. Ellis, A.D., Gunning, F.C.G.: Spectral density enhancement using coherent WDM. IEEE Photonics Technol. Lett. **17**, 504–506 (2005)

14. Bosco, G., Curri, V., Carena, A., Poggiolini, P., Forghieri, F.: On the performance of Nyquist-WDM terabit superchannels based on PM-BPSK, PM-QPSK, PM-8QAM or PM-16QAM subcarriers. J. Lightw. Technol. **29**, 53–61 (2011)

15. Scott, R.P., Fontaine, N.K., Heritage, J.P., Yoo, S.J.B.: Dynamic optical arbitrary waveform generation and measurement. Opt. Express **18**, 18655–18670 (2010)

16. Yoo, S.J.B., Scott, R.P., Geisler, D.J., Fontaine, N.K., Soares, F.M.: Terahertz information and signal processing by RF-photonics. IEEE Trans. Terahertz Sci. Technol. **2**, 167–176 (2012)

17. Geisler, D.J., Fontaine, N.K., Scott, R.P., He, T., Paraschis, L., Gerstel, O., Heritage, J.P., Yoo, S.J.B.: Bandwidth scalable, coherent transmitter based on the parallel synthesis of multiple spectral slices using optical arbitrary waveform generation. Opt. Express **19**, 8242–8253 (2011)

18. Geisler, D.J., Fontaine, N.K., He, T.T., Scott, R.P., Paraschis, L., Heritage, J.P., Yoo, S.J.B.: Modulation-format agile, reconfigurable

Tb/s transmitter based on optical arbitrary waveform generation. Opt. Express **17**, 15911–15925 (2009)

19. Geisler, D.J., Fontaine, N.K., Scott, R.P., He, T., Heritage, J.P., Yoo, S.J.B.: Single channel, 200 Gb/s, chromatic dispersion precompensated 100 km transmission using an optical arbitrary waveform generation based optical transmitter. In: Proceedings of the OFC 2010 (2010)

20. Geisler, D.J., Fontaine, N.K., Scott, R.P., Yoo, S.J.B.: Demonstration of a flexible bandwidth optical transmitter/receiver system scalable to terahertz bandwidths. IEEE Photon. J. **3**, 1013–1022 (2011)

21. Fontaine, N.K., Scott, R.P., Zhou, L., Soares, F.M., Heritage, J.P., Yoo, S.J.B.: Real-time full-field arbitrary optical waveform measurement. Nat. Photon. **4**, 248–254 (2010)

22. Ip, E., Lau, A.P.T., Barros, D.J.F., Kahn, J.M.: Coherent detection in optical fiber systems. Opt. Express**16**, 753–791 (2008)

23. Schmogrow, R., Winter, M., Meyer, M., Hillerkuss, D., Wolf, S., Baeuerle, B., Ludwig, A., Nebendahl, B., Ben-Ezra, S., Meyer, J., Dreschmann, M., Huebner, M., Becker, J., Koos, C., Freude, W., Leuthold, J.: Real-time Nyquist pulse generation beyond 100 Gbit/s and its relation to OFDM. Opt. Express **20**, 317–337break (2012)

24. Allen, L., Beijersbergen, M.W., Spreeuw, R.J.C., Woerdman, J.P.: Orbital angular momentum of light and the transformation of Laguerre–Gaussian laser modes. Phys. Rev. A **45**, 8185–8189 (1992)

25. Bozinovic, N., Yue, Y., Ren, Y., Tur, M., Kristensen, P., Huang, H., Willner, A.E., Ramachandran, S.: Terabit-scale orbital angular momentum mode division multiplexing in fibers. Science **340**, 1545–1548 (2013)

26. Su, T., Scott, R.P., Djordjevic, S.S., Fontaine, N.K., Geisler, D.J., Cai, X., Yoo, S.J.B.: Demonstration of free space coherent optical communication using integrated silicon photonic orbital angular momentum devices. Opt. Express **20**, 9396–9402 (2012)

27. Guan, B. Scott, R.P., Fontaine, N.K., Su, T., Ferrari, C., Cappuzzo, M., Klemens, F., Keller, B., Earnshaw, M., Yoo, S.J.B.: Integrated optical orbital angular momentum multiplexing device using 3-D waveguides and a silica PLC. In: CLEO: 2013, San Jose, CA, p. CTu1L.4 (2013)

28. Scott, R.P., Guan, B., Qin, C., Fontaine, Su, N.K.T., Ferrari, C., Cappuzzo, M., Klemens, F., Keller, B., Earnshaw, M., Yoo, S.J.B.: Free-space coherent optical communication demonstration using a 3D photonic integrated circuit device for orbital angular momentum multiplexing/demultiplexing. In: European Conference and Exhibition on Optical Communication (ECOC), p. Tu.3.B.2 (2013)

29. Macdonald, J.R., Thomson, R.R., Beecher, S.J., Psaila, N.D., Bookey, H.T., Kar, A.K.: Ultrafast laser inscription of near-infrared waveguides in polycrystalline ZnSe. Opt. Lett. **35**, 4036–4038 (2010)

30. Thomson, R.R., Bookey, H.T., Psaila, N.D., Fender, A., Campbell, S., MacPherson, W.N., Barton, J.S., Reid, D.T., Kar, A.K.: Ultrafast-laser inscription of a three dimensional fan-out device for multicore fiber coupling applications. Opt. Express **15**, 11691–11697 (2007)

31. Guan, B., Scott, R.P., Qin, C., Fontaine, N.K., Su, T., Ferrari, C., Cappuzzo, M., Klemens, F., Keller, B., Earnshaw, M., Yoo, S.J.B.: Free-space coherent optical communication with orbital angular, momentum multiplexing/demultiplexing using a hybrid 3D photonic integrated circuit. Opt. Express **22**, 145–156 (2014)

32. Patel, A.N., Ji, P.N., Jue, J.P., Wang, T.: Routing, wavelength assignment, and spectrum allocation in transparent flexible optical WDM (FWDM) networks. In: Photonics in Switching, p. PDPWG1 (2010)

33. Christodoulopoulos, K., Tomkos, I., Varvarigos, E.A.: Elastic bandwidth allocation in flexible OFDM-based optical networks. J. Lightw. Technol. **29**, 1354–1366 (2011)

34. Jinno, M., Kozicki, B., Takara, H., Watanabe, A., Sone, Y., Tanaka, T., Hirano, A.: Distance-adaptive spectrum resource allocation in spectrum-sliced elastic optical path network [topics in optical communications]. IEEE Commun. Mag. **48**, 138–145 (2010)

35. Liu, L., Yin, Y., Xia, M., Shirazipour, M., Zhu, Z., Proietti, R., Xu, Q., Dahlfort, S., Yoo, S. B.: Software-defined fragmentation-aware elastic optical networks enabled by OpenFlow. In: 39th European Conference and Exhibition on Optical Communication (ECOC 2013), pp. 1–3 (2013)

36. Wen, K., Cai, X.R., Yin, Y.W., Geisler, D.J., Proietti, R., Scott, R.P., Fontaine, N.K., Yoo, S.J.B.: Adaptive spectrum control and management in elastic optical networks. IEEE J. Sel. Areas Commun. **31**, 39–48 (2013)

37. Yin, Y., Zhang, H., Zhang, M., Xia, M., Zhu, Z., Dahlfort, S., Yoo, S.J.B.: Spectral and spatial 2D fragmentation-aware routing and spectrum assignment algorithms in elastic optical networks [invited]. J. Opt. Commun. Netw. **5**, A100–A106 (2013)

38. Qin, C., Guan, B., Scott, R., Proietti, R., Fontaine, N., Su, T., Ferarri, C., Capuzzo, M., Clemens, F., Keller, B., Earnshaw, M., Yoo, S.J.B.: Demonstration of Orbital Angular Momentum State Conversion Using Two Hybrid 3D Photonic Integrated Circuits Technical Digest Optical Fiber Communication Conference, vol. Paper Th4A.1 (2014)

39. Proietti, R., Qin, C., Guan, B., Yin, Y., Scott, R.P., Yu, R., Yoo, S.J.B.: Rapid and complete hitless defragmentation method using a coherent RX LO with fast wavelength tracking in elastic optical networks. Opt. Express **20**, 26958–26968 (2012)

40. Liu, L., Tsuritani, T., Morita, I., Guo, H., Wu, J.: Experimental validation and performance evaluation of OpenFlow-based wavelength path control in transparent optical networks. Opt. Express **19**, 26578–26593 (2011)

41. Das, S., Parulkar, G., McKeown, N.: Why OpenFlow/SDN can succeed where GMPLS failed. Presented at the European Conference and Exhibition on Optical Communications (ECOC 2012), Amsterdam, Netherlands (2012)

42. Liu, L., Tsuritani, T., Morita, I., Casellas, R., Martínez, R., Muñoz, R.: Control plane techniques for elastic optical networks: GMPLS/PCE vs OpenFlow. In: International Workshop on Flexible Optical Networks, IEEE Global Communications Conference (GLOBECOM), pp. 352–357, Anaheim, CA, USA (2012)

43. Cai, X.R., Wen, K., Proietti, R., Yin, Y.W., Geisler, D.J., Scott, R.P., Qin, C., Paraschis, L., Gerstel, O., Yoo, S.J.B.: Experimental demonstration of adaptive combinational QoT degradation restoration in elastic optical networks. J. Lightw. Technol. **31**, 664–671 (2013)

44. Ren, Y., Zhang, Y., Yue, Y., Bozinovic, N., Xie, G., Huang, H., Tur, M., Kristensen, P., Djordjevic, I.B., Ramachandran, S., Willner, A.E.: Efficient crosstalk mitigation of OAM based 400-Gbit/s QPSK data transmission in 1.1-km vortex fiber by using soft-decision LDPC codes. In: CLEO, p. CM2G.5, San Jose, CA (2013)

45. Wen, K., Yin, Y., Geisler, D., Chang, S., Yoo, S.J.B.: Dynamic on-demand lightpath provisioning using spectral defragmentation in flexible bandwidth networks. In: ECOC 2011, p. Mo.2.K.4, Geneva, Switzerland (2011)

46. David, R.P., Geisler, J., Yin, Y., Scott, R.P., Cai, X., Fontaine, N.K., Paraschis, L., Gerstel, O., Yoo, S.J.B.: The first testbed demonstration of a flexible bandwidth network with a real-time adaptive control plane. In: ECOC 2011, p. Th.13.K.2, Geneva, Switzerland (2011)

5

# DESIGN AND PERFORMANCE EVALUATION OF NARROWBAND RECTANGULAR OPTICAL FILTER BASED ON STIMULATED BRILLOUIN SCATTERING IN FIBER

**Invited Paper**

Lilin Yi , Wei Wei, Mengyue Shi, Yves Jaouen, Weisheng Hu

## ABSTRACT

We propose a rectangular optical filter based on stimulated Brillouin scattering (SBS) in fiber with tunable bandwidth from 50 MHz to 4 GHz at 15-MHz tuning resolution. The steep-edged rectangular shape of the filter is precisely controlled utilizing digital feedback compensation of the multi-tone pump light. The passband ripple is ~~1 dB by nonlinearity management of the pump light and using the fiber with a single Brillouin peak. The filter selectivity is improved to more than 40 dB by using pump-splitting dual-stage configuration. We analyze the noise performance of the proposed SBS filter and demonstrate a sub-band extraction of a multi-band orthogonal frequency division multiplexing (OFDM) signal. Furthermore, we validate the amplification performance with different gains for OFDM signal, which shows the potential capability of the filter in the fields of optical signal processing.

## KEYWORDS

Rectangular optical filter, Stimulated Brillouin scattering, Optical signal processing

# 1. INTRODUCTION

A tunable rectangular filter with an ultra-flat passband and steep edges brings about many benefits. The flat passband will not distort the signal thus keeping high signal fidelity. The steep edges can suppress interference from adjacent bands at the extreme. And the tunability makes it more flexible to meet different requirements. Several methods have been proposed to implement such filters, including specially designed fiber Bragg gratings (FBG) [1], Fabry–Perot etalons [2] and stimulated Brillouin scattering (SBS) [3–6]. Among all the above methods, SBS is considered as one of the most promising techniques. The wavelength of the SBS-based filter can be tuned easily by changing the wavelength of the pump. The filter bandwidth and the shape can be also flexibly changed by controlling the pump spectrum using external phase modulation [3], direct current modulation [4] and cascaded phase and intensity modulation [5]. However, it is very difficult to control the pump spectrum precisely in the previous works; therefore, the exact flat top and steep edges as the properties of ideal rectangular filter are hardly achieved.

In this paper, we present a narrowband rectangular SBS filter with tunable bandwidth from 50 MHz to 4 GHz at 15-MHz tuning resolution. We modulate the pump using external intensity modulation by an electrical frequency comb whose amplitude and initial phase can be digitally controlled accurately. A feedback compensation process is proposed to revise the pump shape precisely. Some preliminary results have been presented in a previous paper [7] which shows the validity of the feedback method. On the foundation of the feedback process, we further propose the nonlinearity management to mitigate the four-wave mixing (FWM) effect and utilize a more suitable fiber with a single Brillouin gain peak to increase the filter flatness [8]. As a result, the in-band ripple is decreased to ~~1 dB for all bandwidth cases, while the unwanted out-of-band gain is further suppressed. The out-of-band SBS gain originated from the FWM effect of the multiple pump lines will degrade the filter selectivity. We propose a pump-splitting dual-stage confirmation to improve the filter selectivity to more than 40 dB. Based on this filter, we analyze the noise performance and demonstrate a sub-band extraction of a multi-band orthogonal frequency division multiplexing (OFDM) signal; furthermore, we validate the SBS amplification with different gains for OFDM signal and prove the flexibility and validity of the filter.

## 2. RECTANGULAR FILTER

The natural SBS gain is Lorentzian shape with bandwidth ranging from 10 to 50 MHz depending on different types of fibers [9]. Theoretically, in order to obtain the ideal rectangular SBS gain spectrum, a pump consisting of equal-amplitude spectral lines with interval equaling the natural SBS gain bandwidth is required. Instead of controlling the pump spectrum in the time domain, we firstly generate electrical frequency combs with equal amplitude at each spectral line using an arbitrary waveform generator (AWG). Then, we modulate them with optical carrier-suppressed single-sideband (OCS-SSB) modulation to generate optical frequency combs as the Brillouin pump. The SBS filter bandwidth adjustment is achieved simply by controlling the number of electrical spectral lines so as to the pump spectral lines. In addition, we can control the amplitude, the initial phase of each electrical spectral line and the frequency spacing of spectral lines digitally and precisely. It guarantees that there are no spectral components out of the comb region at all, resulting in a steep-edged pump light; thus, the SBS gain spectrum obtained has very steep edges as well, which is unachievable in the previously reported schemes.

Since the SBS gain increases with the pump power exponentially, even little pump power drifting results in great gain difference. However, on account of the nonlinear responses of electrical and optical components such as the optical modulator and its driver, the flat electrical spectral lines lead to uneven SBS gain inevitably as sketched in Fig. 1. Thus, we propose an effective feedback compensation scheme as follows: firstly generate the original pump lines by OCS-SSB modulation using flat electrical spectral lines. Then measure the SBS gain spectrum by an electrical vector network analyzer (EVNA) and calculate the new amplitude of each electrical spectral line. Taking the linear approximation error and the non-ideal responses of electrical and optical components into account, a rectangular gain spectrum can only be obtained by repeating the processes of generation, measurement and feedback process successively for several times. Once the feedback compensation is completed, the electrical spectral lines can be recorded for rectangular SBS filter generation.

In the feedback process, it is assumed that the SBS gain at a certain frequency is only related to the corresponding electrical spectral line. It works well when the filter bandwidth is narrow. However, with the increasing of the filter bandwidth, the increasing pump power and number of spectral lines cause remarkable FWM in the optical fiber which redistributes the pump power and generates new spectral lines at the stop band. Another factor not taking into account is the nonlinearity of the modulation such as third-order intermodulation that also introduces new spectral lines and cannot be neglected when the pump power is high. The nonlinearity-induced spectral components are sketched in Fig. 2a, b in the color of yellow. It results in a partial failure of the feedback process, leading to larger passband ripple and out-of-band gain. Therefore, an effective nonlinearity management is required.

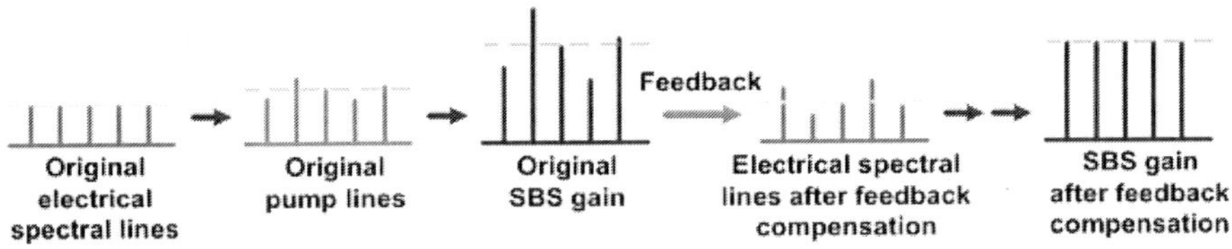


**Figure. 1** Feedback compensation scheme

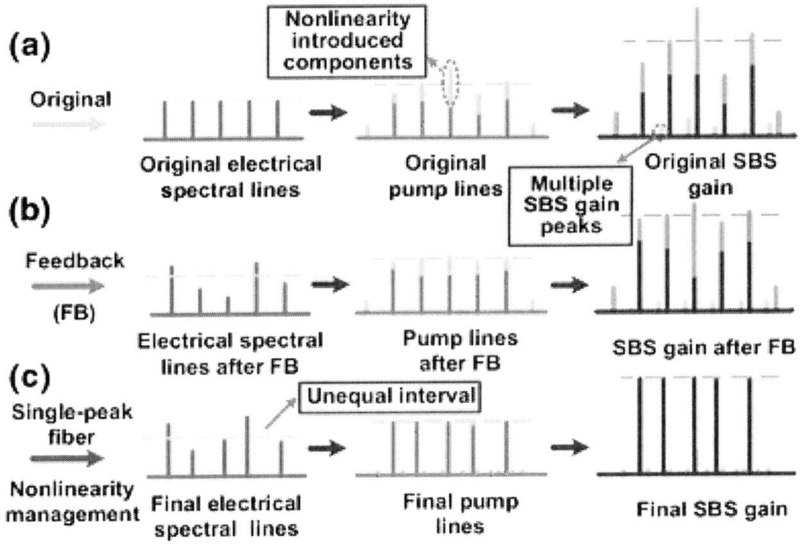


**Figure. 2** The relationship between electrical spectral lines, pump lines and the SBS gains in different cases

We propose a nonlinearity management method as shown in Fig. 2c. We set frequency interval of the electrical spectral lines randomly around the natural SBS gain bandwidth instead of the equal interval, i.e., 14, 15 and 16 MHz. In this case, the new spectral lines generated by the FWM and the nonlinear modulation are no longer superposing on the original lines. As a result, the power of these new spectral lines is very small compared with the original lines and can just induce tiny gain or even under the threshold of SBS effect. Therefore, the flatness of the passband gain can be greatly improved, and the unwanted out-of-band gain can be partly suppressed as well.

Moreover, the natural SBS gain spectrum is related to the material doping profile [10] and the structure of the fiber [11]. In the Germanium-doped G-652 fiber that we used in the previous work [7], there are 4 SBS peaks (i.e., four first Lom acoustic modes) as sketched in Fig. 2a, b in the color of green, while the measured result is shown in Fig. 3a. The multiple minor SBS gain peaks superpose with other main SBS gain peaks generated by different pump lines and will affect the feedback process. It contributes to the un-flat filter passband and induces out-of-band gain as well. Therefore, we employ a fiber with a single Brillouin peak (a standard single-mode fiber in our lab) as shown in Fig. 3b which is highly desired in our experiment. By employing both the nonlinear management and the new fiber with single Brillouin peak, the final SBS gain will be ultra-flat as sketched in Fig. 2c.

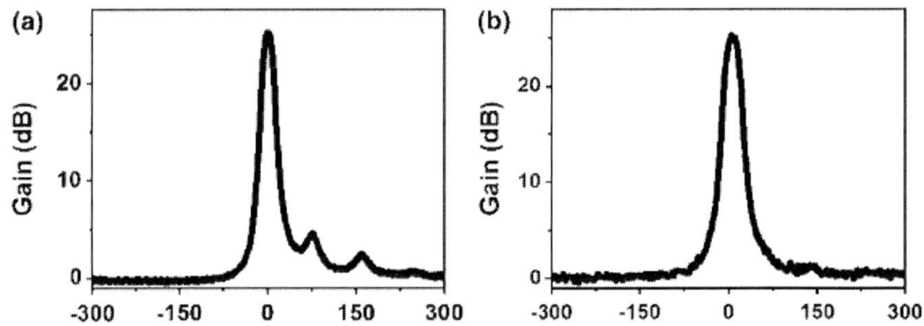


**Figure. 3** The natural SBS gain in **a** G-652 fiber, **b** fiber with a single Brillouin peak

The experimental setup is shown in Fig. 4. An external cavity laser (ECL) operating at 1550 nm is split into two branches. In the upper branch, an AWG is used to generate the electrical spectral lines with random frequency intervals within ±±1-MHz deviation from the natural SBS bandwidth of 15 MHz, i.e., 14, 15 and 16 MHz. Then it is modulated on the light to generate the SBS pump lines utilizing a dual-parallel Mach–Zehnder modulator (DPMZM). With proper phase control of the driving signals and bias control of the DPMZM, the OCS-SSB modulation is achieved. The OCS-SSB signal is then amplified by an erbium-doped fiber amplifier (EDFA) and launched into the 12.5 km single-peak fiber through an optical circulator. In the lower branch, a sweeping signal covering the whole SBS gain region from an EVNA is modulated on the light to generate the probe signal. An optical bandpass filter (BPF) removes the left sideband of the probe signal for stable SBS gain measurement. Then the probe light goes through the single-peak fiber and is amplified once it sweeps within the SBS gain region. A polarization controller (PC3) is used to achieve the maximum SBS gain. After the SBS process, the probe signal is detected by a photodiode (PD) and then sent into the EVNA.

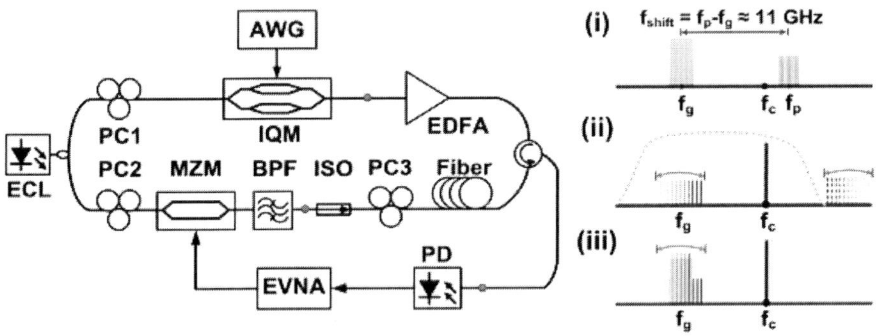


**Figure. 4** Experimental setup. *Inset* (*i*) single-sideband pump fpfp, SBS gain around fgfg, and ECL laser frequency fcfc, (*ii*) sweeping probe signal, (*iii*) sweeping probe signal amplified by the SBS gain

The measured SBS spectra with 3-GHz bandwidth in different cases are shown in Fig. 5. The original SBS gain generated by the electrical spectral lines with equal amplitude and equal frequency interval is seriously affected by the FWM and nonlinear response of the electrical and optical components. As shown in Fig. 5a, the SBS gain is far from the rectangular shape. When implementing the feedback compensation as shown in Fig. 5b, the ripple is still as high as 3.70 dB with large unwanted gain out of the passband. After utilizing the unequal interval spectral lines, the passband ripple in Fig. 5c is significantly reduced to 1.56 dB. By using the single-peak fiber, the ripple is further decreased to

1.00 dB as shown in Fig. 5d, and the gain at stop band is smaller than the previous cases.

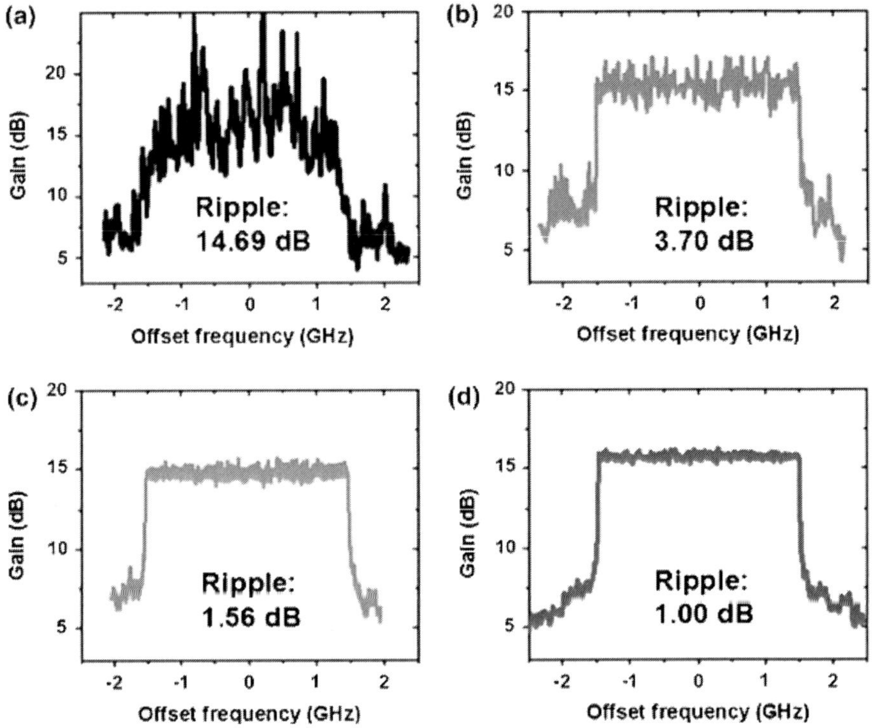


**Figure. 5** The measured SBS spectra at different cases

The bandwidth of the proposed filter can be changed flexibly with a resolution of 15 MHz by changing the number of the electrical spectral lines generated by the AWG. The maximal rectangular SBS filter bandwidth is 4 GHz with 12-dB gain limited by the pump power.

# 3. FILTER SELECTIVITY IMPROVEMENT

The filter selectivity cannot be increased by simply increasing pump power due to two reasons. The first reason is the competition between SBS and SRS effect. When the Brillouin pump power is higher than several hundreds of mW, it starts acting as Raman pump at the same time which results in SBS pump power consumption. The second reason for filter selectivity limitation is FWM effect induced by multi-frequency comb. It generates new components out of the filter passband and cannot be completely mitigated by the nonlinearity management method. When the pump power increases, FWM effect is getting much more severe, especially in broad bandwidth cases. The out-of-band gain will consume considerable amount of pump power and disrupt the rectangular shape generation. Thus when the pump power continues to increase, most of the power is wasteful and brings about hardly no benefit.

In order to increase the filter selectivity, we propose a pump-splitting dual-stage scheme. Instead of using a single pump with high power, we split it into two stages and amplify the signal twice successively. Note that the filter selectivity is only related to the on–off gain and has nothing to do with the transmission loss; extra attenuation can be induced between the two stages to prevent the signal saturation in the second stage. In this case, the pump power of each stage is not too high to stimulate high Raman gain nor to induce too much out-of-band FWM components. Moreover, the decrease in pump power for each stage will decrease the induced noise from the spontaneous Brillouin emission [12]. Thus the filter selectivity can be increased dramatically, and better noise performance can be achieved at the same time.

The experimental setup is shown in Fig. 6. The process of generating SBS pump lines, sweeping signal, feedback compensation, as well as nonlinearity management is described as above. The difference in the OCS-SSB signal of the upper branch is then split into two parts equally and send into two identical 25-km-long SMFs, which are under the same strain and temperature conditions to ensure the same Brillouin characteristics. At each stage, a polarization controller (PC) is used to maintain the SBS gain at the maximum value. An optical attenuator (ATT) between the two stages can prevent signal saturation in the second stage therefore effectively increase the filter selectivity. The amplitude response sketched in Fig.6(iv) is measured by the EVNA, and the SBS gain spectrum can be obtained by comparing the results between the SBS pump active or not.

The amplitude and phase responses of both single- and dual-stage filters are shown in Fig. 7. The pump power is set to 21, 24 and 27 dBm for filter bandwidth of 1, 2 and 3 GHz, respectively. As shown in Fig. 7a, for all bandwidth cases, the filter gain has been improved obviously with the same total pump power, while the passband flatness remains at the same level. For the 3-GHz filter, the maximum gain of single-stage filter is limited to ~18.5 dB due to SRS and the severe out-of-band FWM components. By using the dual-stage scheme, the gain has been increased to ~32 dB. The high selectivity makes the filter shape more like rectangular than flat-top trapezoid. For the 1-GHz case, by using dual-stage configuration, the gain is increased while the out-of-band FWM effect is slightly mitigated. As to the phase response shown in Fig. 7b, all the curves are very smooth due to the small passband ripple, and higher gain of the dual-stage filter leads to larger phase variation compared with the single-stage case. The results have clearly proved the validity of the dual-stage scheme to increase the filter selectivity.

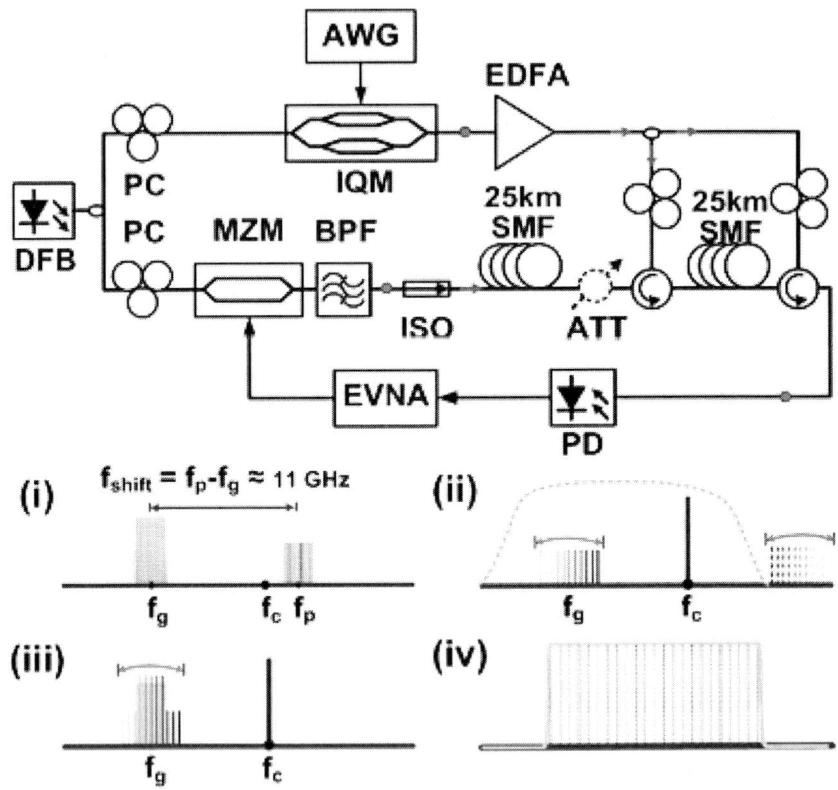


**Figure. 6** Experimental setup. *Inset* (*i*) single-sideband pump fpfp, SBS gain around fgfg, and ECL laser frequency fcfc, (*ii*) sweeping probe signal, (*iii*) sweeping probe signal amplified by the SBS gain, (*iv*) measured gain spectrum

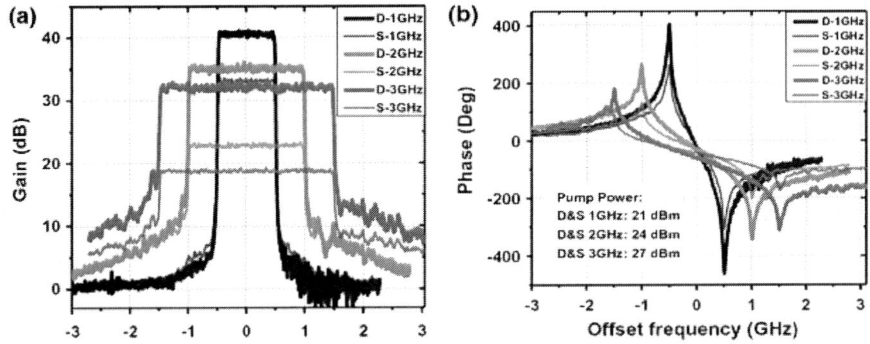


**Figure. 7** Filter comparison between single- and dual-stage approaches. **a** The amplitude and **b** phase response of the two types of filter. *D* dual stage, *S* single stage

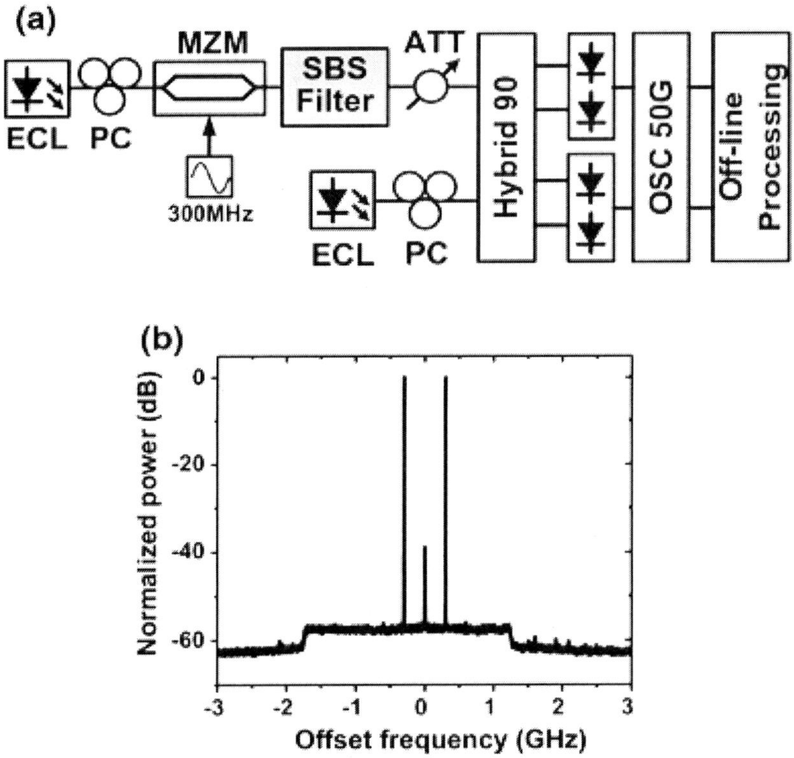


**Figure. 8 a** Noise measurement setup and **b** the electrical spectrum of the amplified signal by a 3-GHz SBS filter

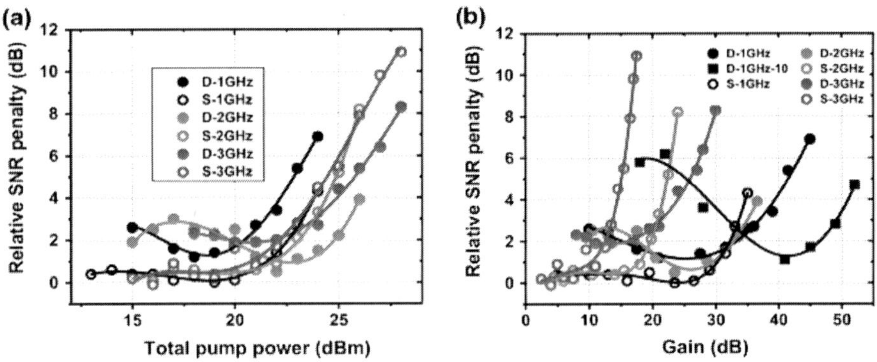


**Figure. 9** Noise comparison between the single and cascade filters, **a** the relation between SNR and pump power, **b**the relation between SNR and SBS gain. *D* dual stage, *S* single stage. "–10–10" means 10 dB optical attenuation between two stages

# 4. SYSTEM PERFORMANCE EVALUATION

It is well-known that the Brillouin amplification is accompanied with strong noise [13]. We analyze the noise performance of the proposed SBS filter with the experimental setup shown in Fig. 8a. A 300-MHz single-frequency signal modulates the CW light from an external cavity laser (ECL). After passing through the active SBS filter and an optical attenuator to maintain the same power level, the signal is detected by a coherent receiver. Figure 8b shows the spectrum of the SBS amplified signal which is calculated from the waveform obtained by a real-time oscilloscope from the receiver since we do not have an ultra-high-resolution optical spectrum analyzer for direct observation. The symmetric frequency line is due to the carrier-suppressed double-sideband modulation by the Mach–Zehnder modulator. We do not use the CW light as the signal since the DC noise from the oscilloscope itself is significant. The noise pedestal is linked to SBS gain which is shifted off-center due to slight frequency detuning between pump and signal waves.

Then the signal to noise ratio (SNR) is carefully measured. We use the amplified signal peak power and the average noise power in the SBS gain region to calculate the SNR. We define the SNR penalty at 1-GHz bandwidth with 19-dBm pump power as reference since the SNR penalty under this condition is not obvious compared with the back-to-back case. The relation between the pump power and the SNR is shown in Fig. 9a. We can conclude that the general trend of SNR penalty increases with the increase of the pump power, and the pump power level is the main factor to induce noise. For the single-stage filter, the turning point is around 20 dBm, and the three curves show strong consistency. As to the dual-stage filter, the turning points generally shift 2–3 dB higher in both the 2-GHz and 3-GHz cases, while in 1-GHz case, the turning point remains the same. So if we can improve the pump efficiency to reduce the required pump power for the same gain value, the SNR performance can be improved. From this regard, the pump-splitting dual-stage configuration is preferred. Figure 9b demonstrates the SNR variation with different SBS gains. We can observe steep rise in SNR penalty with the increase in SBS gain after a certain turning point for both single- and dual-stage cases. For the small gain cases, the dual-stage-induced SNR penalty is only 1–2 dB. Once the gain reaches 32, 19, and 13 dB for the bandwidth of 1, 2, and 3 GHz, respectively, the noise performance of the dual-stage scheme is always better than the single-stage method. If 10-dB attenuation between the two stages is used, the SNR will degrade when the gain is small, but when the gain reaches 34.5 dB, it induces smallest penalty and becomes the best choice. Note that for a SBS-based microwave photonic filter, high selectivity with low noise is preferred; thus, the pump-splitting dual-stage approach is a simple but effective method to reach this goal.

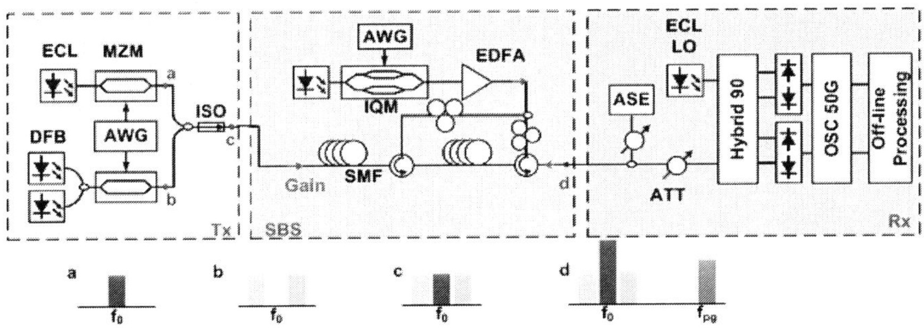


**Figure. 10** Experimental setup and the optical spectrum schemes at different points

After evaluating the noise performance of the SBS-based filter, we demonstrate a sub-band extraction of a multi-band orthogonal frequency division multiplexing (OFDM) signal using the pump-splitting dual-stage SBS filter. The experimental setup consists of three parts: the transmitter, the SBS-based filter and the coherent receiver as shown in

Fig. 10. In the transmitter part, the light from an external cavity laser (ECL) and two distributed feedback lasers (DFB) operating at ~~1543 nm are modulated by the electrical OFDM signal. As only a single output of the Tektronix7221B AWG is available for OFDM signal generation, an OFDM signal satisfying the Hermitian symmetry is generated. This constraint does not affect our analysis as all sub-carriers are treated independently at the receiver side. Given the instability of the three lasers, the minimum band gap of the three sub-bands is set to 300 MHz. For each sub-band, 128 sub-carriers are used in order to mitigate the effect of the ECL phase noise of ~~100 KHz. Both quadrature-phase-shift-keying (QPSK) and 16-quadrature-amplitude-modulation (16-QAM) formats are employed for each sub-carrier at the sampling rate of 2.5GSamples/s. The bandwidth is set to 2 GHz by adjusting the number of empty sub-carriers. After passing through an isolator (ISO) to block inverse pump light, the three OFDM sub-bands are launched into the pump-splitting dual-stage SBS filter with bandwidth of 2.2 GHz, where the extra 200 MHz bandwidth is dedicated to laser wavelength drift mitigation. The amplified central sub-band from the SBS filter is adjusted to the optimal power of the receiver, and a broadband ASE noise source is added for the SNR-BER measurement. Finally the OFDM signal is detected by an intradyne coherent receiver followed by a high-speed real-time oscilloscope. QPSK and 16-QAM constellations are then obtained by off line processing. A more precise description of OFDM signal generation and detection is in [14, 15].

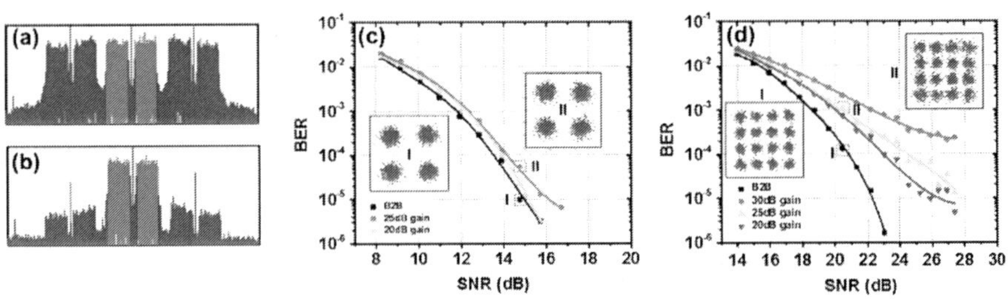


**Figure. 11** Amplified OFDM signals: **a** original 3 sub-bands OFDM signal, **b** amplified central band by using the SBS gain filter, **c** BER-SNR performance for QPSK, **d** BER-SNR performance for 16-QAM

The validity of the SBS amplification with different gains for OFDM signal has been assessed. Figure 11a shows the electrical spectrum of the received original 3-band OFDM signal, and Fig. 11b illustrates the spectrum when the central band is amplified by the SBS gain filter. The SNR-BER curves for the QPSK format are presented in Fig. 11c while the 16-QAM case in Fig. 3d. Different constellation diagrams are also given in the insets. After being amplified by the SBS gain filter with 25-dB gain, the SNR penalties are only ~~0.5 and ~~2 dB at a BER of 10–310–3 for QPSK and 16-QAM, respectively. For 16-QAM, larger penalty has been observed when the SNR is high in large gain cases because of the large SBS-ASE noise. The results have proved that the SBS gain induced penalty is not significant, especially when the signal format is QPSK. It also validates the feasibility of the proposed rectangular SBS gain filter in the OFDM system.

# 5. CONCLUSION

We have presented an ultra-flat rectangular optical filter based on stimulated Brillouin scattering and demonstrated a typical filter application of OFDM sub-band extraction and amplification. With flexible bandwidth from 50 MHz to 4 GHz and high tuning resolution of 15 MHz, only small penalties are induced by the SBS-ASE noise which prove the feasibility of the SBS amplification in OFDM systems. The proposed steep-edged rectangular filter can find versatile applications in optical and microwave signal processing.

# ACKNOWLEDGMENTS

This work is supported by 973 Program (2012CB315602), National Nature Science Foundation of China (61322507 and 61132004) and Program of Excellent Ph.D. in China (201155).

# REFERENCES

1. Zou, X.H., Li, M., Pan, W., Yan, L.S., Azana, J., Yao, J.P.: All-fiber optical filter with an ultranarrow and rectangular spectral response. Opt. Lett. **38**, 3096–3098 (2013)

2. Pruessner, M.W., Stievater, T.H., Goetz, P.G., Rabinovich, W.S., Urick, V.J.: Cascaded integrated waveguide linear microcavity filters. Appl. Phys. Lett. **103**, 11105 (2013)

3. Tanemura, T., Takushima, Y., Kikuchi, K.: Narrowband optical filter, with a variable transmission spectrum, using stimulated Brillouin scattering in optical fiber. Opt. Lett. **27**, 1552–1554 (2002)

4. Zadok, A., Eyal, A., Tur, M.: Gigahertz-wide optically reconfigurable filters using stimulated Brillouin scattering. J. Lightwave Technol. **25**, 2168–2174 (2007)

5. Sakamoto, T., Yamamoto, T., Shiraki, K., Kurashima, T.: Low distortion slow light in flat Brillouin gain spectrum by using optical frequency comb. Opt. Express **16**, 8026–8032 (2008)

6. Wise, A., Tur, M., Zadok, A.: Sharp tunable optical filters based on the polarization attributes of stimulated Brillouin scattering. Opt. Express **19**, 21945–21955 (2011)

7. Wei, W., Yi, L., Zhang, Y., Jaouen, Y., Song, Y., Dong, Y., Hu, W.: A bandwidth-tunable narrowband rectangular optical filter based on stimulated Brillouin scattering. In: Optical Fiber Communication Conference, OSA Technical Digest (online) (Optical Society of America, 2014), paper W4F.5

8. Wei, W., Yi, L., Jaouën, Yves: et al.: An Ultra-Flat Rectangular Optical Filter Based on Stimulated Brillouin Scattering in Fiber, ECOC'2014, paper P.1.7

9. Mamdem, Y., Burov, E., de Montmorillon, L., Jaouën, Y., Moreau, G., Gabet, R., Taillade, F.: Importance of residual stresses in the Brillouin gain spectrum of single mode optical fibers. Opt. Express **20**, 1790–1797 (2012)

10. Jaouën, Y., Canat, G., Sikali-Mamdem, Y., Gabet, R., Lombard, L., Burov, E.: Stimulated Brillouin scattering in specialty optical fibers: importance of material, structure and manufacturing parameters. In: Conference on Lasers and Electro-Optics 2012, OSA Technical Digest (online) (Optical Society of America, 2012), paper CF3N.1

11. Yeniay, A., Delavaux, J.M., Toulouse, J.: Spontaneous and stimulated Brillouin scattering gain spectra in optical fibers. J. Lightwave Technol. **20**, 1425–1432 (2002)

12. Olsson, N., Van Der Ziel, J.: Characteristics of a semiconductor laser pumped Brillouin amplifier with electronically controlled bandwidth. J. Lightwave Technol. **5**, 147–153 (1987)

13. Tkach, R.W., Chraplyvy, A.R., Derosier, R.M.: Performance of a WDM network based on stimulated Brillouin scattering. IEEE Photonics Technol. Lett. **1**, 111–113 (1989)

14. Pincemin, E., Song, M., Karaki, J., Zia-Chahabi, O., Guillossou, T., Grot, D., Thouenon, G., Betoule, C., Clavier, R., Poudoulec, A., Van der Keur, M., Jaouën, Y., Le Bidan, R., Le Gall, T., Gravey, P., Morvan, M., Dumas-Feris, B., Moulinard, M., Froc, G.: Multi-band OFDM transmission at 100 Gbps with sub-band optical switching. J. Lightwave Technol. **32**, 2202–2219 (2014)

15. Awwad, E., Jaouën, Y., Othman, G.R.-B., Pincemin, E.: Polarization-time coding for PDL mitigation long-haul PolMux OFDM systems. Opt. Express **21**, 22773–22790 (2013)

6

# OPTICAL BACKPLANE BASED ON RING-RESONATORS: SCALABILITY AND PERFORMANCE ANALYSIS FOR 10 GB/S OOK-NRZ

**Giuseppe Rizzelli [1,†], Domenico Siracusa [2,†], Guido Maier [3,†,*], Maurizio Magarini [3,†], Mehmood Alam [3,4,†] and Andrea Melloni [3,†]**

[1] Network Rail Telecom, The Quadrant: Elder Gate, Milton Keynes MK91EN, UK

[2] CREATE-NET, Via alla Cascata 56, 38123 Trento, Italy

[3] Politecnico di Milano, Piazza L. da Vinci 32, 20133 Milano, Italy

[4] National University of Sciences & Technology, Sector H-12, Islamabad 44000, Pakistan

## ABSTRACT

The use of architectures that implement optical switching without any need of optoelectronic conversion allows us to overcome the limits imposed by today's electronic backplane, such as power consumption and dissipation, as well as power supply and footprint requirements. We propose a ring-resonator based optical backplane for router line-card interconnection. In particular we investigate how the scalability of the architecture is affected by the following parameters: number of line cards, switching-element round-trip losses, frequency drifting due to thermal variations, and waveguide-crossing effects. Moreover, to quantify the signal distortions introduced by filtering operations, the bit error rate for the different parameter conditions are shown in case of an on-off keying non-return-to-zero (OOK-NRZ) input signal at 10 Gb/s.

## KEYWORDS

optical backplane; optical interconnections; network optimization; assignment and routing algorithms

## 1. INTRODUCTION

The deployment of High Performance Switches and Routers (HPSRs) capable of managing huge aggregated bandwidth is becoming mandatory to address traffic growth-rate projections [1]. An HPSR is composed of one or more racks, each one containing a set of Line Cards (LCs). Each LC hosts an HPSR bidirectional port that connects the node to an outer link towards another node of the network. The LCs also host the hardware implementing low-layer functions (transceiver, framer, *etc.*) and other network-processing functions (address lookup, packet classification, buffer management, scheduling,*etc.*). Internally to the HPSR, LCs are interconnected via a backplane [2]. The bandwidth of the internal LC-to-LC connections supported by the backplane should be roughly equal (except for some small extra bandwidth due to internal signaling) to the line speed of the LCs: in this paper we consider 10 Gbit/s line-speed and, thus, also backplane bandwidth. The backplane can be static, connecting LCs to a separated switching-fabric module, or it can be dynamic, performing the function of a switching fabric itself. We are interested, in this work in the dynamic switching-backplane case. (In some cases some LCs, instead of having a single 10 Gbit/s port, may be equipped by multiple smaller-bandwidth physical ports (e.g., 4 ° 2.5 or 10 ° 1 Gbit/s, *etc.*). This usually does not have an impact on the backplane, since traffic grooming of low-granularity tributaries into high-granularity flows is usually managed within the LCs.).

In order to overcome current electronic-backplane limitations, Optical Interconnections (OI) represents an attractive alternative: attenuation does not increase at high bit-rates; optical transmission lines do not suffer the crosstalk impairment,*etc.* Generally speaking, OI can be deployed between racks (rack-to-rack), between line-cards inside a rack (card-to-card or backplane), between chips of a line-card (chip-to-chip), or even between the different cores of a single chip (on-chip). In this context, we decided to investigate the card-to-card OI. Literature presents some proposals of all-optical backplane interconnection fabrics with Semiconductor Optical Amplifiers (SOAs) and Arrayed Waveguide Gratings (AWGs) as implementation technologies [3]. In this study, we deal with Ring-Resonators [4] (RRs), typically used for filtering and routing [5], as the basic switching element for LCs interconnection. In [6], authors propose different RR-based backplane fabrics focusing on a power-budget scalability analysis. In our work, (1) we propose a backplane fabric made of a cascaded RR; (2) we introduce a new methodology for scalability analysis based on accurate transfer function evaluation; and (3) we confirm the feasibility of the architecture via analysis of the bit error rate (BER) for transmission of an On-Off Keying (OOK) Non-Return-to-Zero (NRZ) signal at 10 Gb/s. This work is an extension of [7], in which BER analysis was not available. Our procedure goes beyond the simple power budget and Cross-Talk (XT) analyses since it accounts for waveguide dispersion effects, which may have a non-negligible impact in all-optical backplanes.

# 2. OPTICAL INTERCONNECTION ARCHITECTURE

In this section we briefly introduce a general multi-plane switching architecture, which is the basis, technology-independent version of the RR based switching backplane we have investigated in this study, followed by a description of the design procedure for routing assignment.

The RR-based switching backplane exploits transmitter wavelength tunability as switching mechanism: wavelength *h*selection by the source LC transmitter is then translated into selection of a target destination LC by means of a passive wavelength-routing structure composed of waveguides, star couplers, and RRs.

In Figure 1, *N* line cards, each one comprising a tunable-wavelength transmitter (TX) and a fixed-wavelength receiver (RX), are divided into *S* Input-Output planes (*h* from 0 to *S*-1); at each Input plane, *N*/*S* TXs are connected to the *plane selection circuit* (PSC). For instance, TX and RX of each LC are depicted as separated, though they are actually co-located on board of each LC. Routing properties of such a circuit let each TX select the output port corresponding to the plane a certain RX belongs to. Each selection-circuit output port is connected to one of the *S* couplers (SCs) ( *S*:1 ) that convey signals in order to set-up connections between TXs and RXs which lie into different planes. Signals coming from different transmission planes are coupled together and an Erbium Doped Fibre Amplifier (EDFA) is placed to restore signal optical power. A*receiver selection circuit* (RSC) separates channels directed to different

RXs of the same Output plane. The number of planes $S$is allowed to range from 1 to $N$. If $N/S \notin \mathbb{N}$, then $[N/(S\text{-}1)]$ cards will be connected to the first $S$-1 planes, while the remaining ones will occupy the first available positions within the last plane.

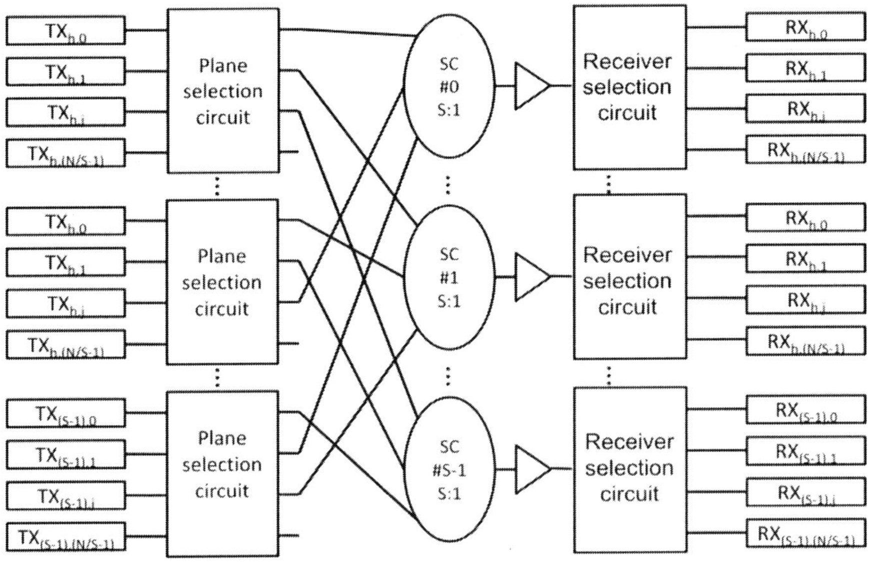


**Figure 1.** General multiplane switching architecture.

## 2.1. Ring Resonator Based Backplane

The switching backplane we aim at investigating in this work is shown in Figure 2. A Binary-Tree Ring Router (BTRR), used to select the output plane, acts as PSC whereas an RR array demultiplexer represents the RSC. We assume that only point-to-point connections are established between TXs and RXs, with no output contention, *i.e.*, each RX is to be connected to at most one TX using one single wavelength.

The principle of the PSC tree-structure is the same as that based on a cascade of Mach-Zehnder filters, the so-called interleaving. An interleaver is a multi-stage demultiplexer with T stages where at each stage an RR spatially discriminates the incoming wavelength-channels partitioning them into two alternate sub-groups, sending half of them towards the "through" port and half of them towards the "drop" port. Note that, compared to Figure 1, this is a completely passive switching architecture as we do not investigate the use of EDFA before the RSCs. The number of stages T has to be large enough to completely discriminate all the $S$ switching planes, thus T = (log2($S$)). The $j$-th stage will be composed of 2($j$ − 1) RRs, with $1 \leq j \leq$ T and after $j$ stages 2$j$ planes can be selected.

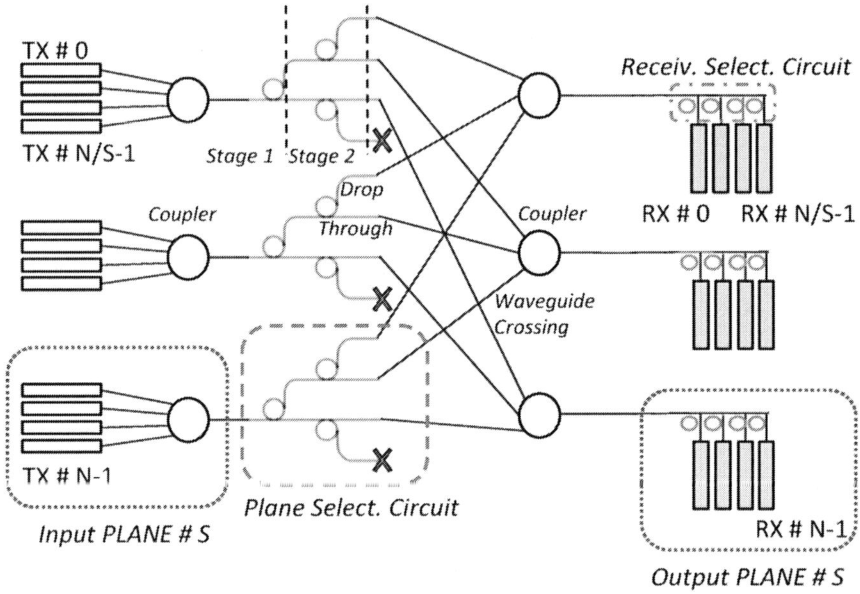


**Figure 2.** Ring Resonator-based switching architecture.

Ring resonators are infinite-impulse response filters with a periodic transfer function of the input frequency [4]. The family of resonance wavelengths $\lambda_{res}$ at which the transfer function has its peaks is $\beta n_{eff} L_r = 2\pi N$, where $\beta = 2\pi/\lambda_{res}$, $n_{eff}$ is the effective refractive index, $L_r$ is the geometrical ring length and $N$ is an integer number ($N \geq 1$). The period of the transfer function is knows as Free Spectral Range ($FSR$) and is given by $FSR = c/(2\pi n_g R)$, where $c$ is the speed of the light, $n_g$ is the group index and $R$ is the ring radius. The $FSR$ is inversely proportional to the ring radius $R$. The smallest achievable radius is constrained by the fabrication technology and optical losses inside the ring increase as well with bending [8].

Let us consider a tree-structure composed of three stages as in Figure 3a. If we assume that $\lambda_{(i,j)}$ is the resonance wavelength of the $i$-th ring of the *j*-th stage, the first ring of the first stage has to be dimensioned with a $FSR = 2\Delta\lambda$ and $\lambda^{res}_{11} = \lambda_0$, then both rings of the second stage have $FSR = 4\Delta\lambda$ but $\lambda^{res}_{12} = \lambda_0$, $\lambda^{res}_{22} = \lambda_1$. Each ring of the third stage is dimensioned with $FSR = 8\Delta\lambda$ where $\lambda^{res}_{14} = \lambda_0$, $\lambda^{res}_{24} = \lambda_2$, $\lambda^{res}_{34} = \lambda_3$, $\lambda^{res}_{44} = \lambda_1$ are the resonance wavelengths of ring resonators respectively. As an example, Figure 3b shows general ring drop port transfer function of the first, second, and third stage respectively. The −3 dB bandwidth of drop port transfer function can be very narrow, thus, showing higher filter selectivity than the through port [4].

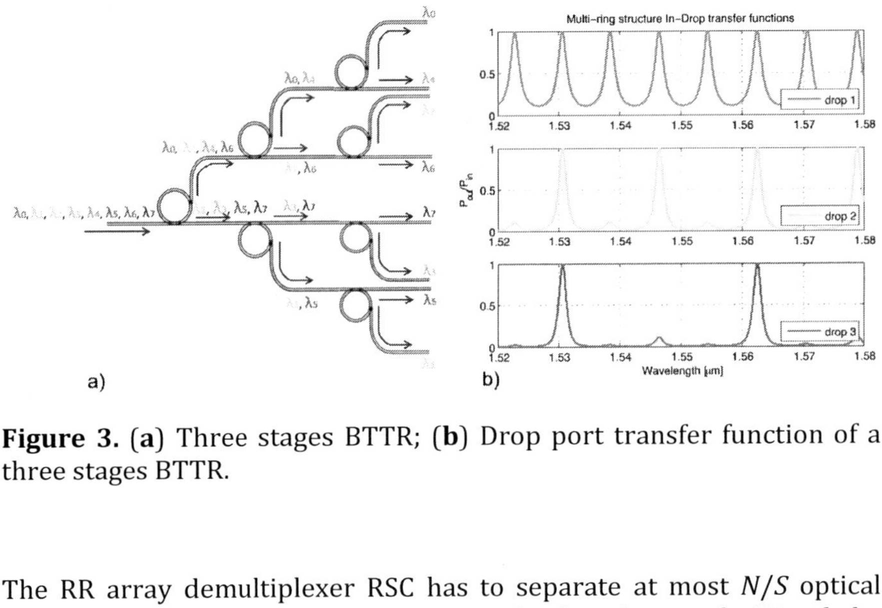


**Figure 3.** (**a**) Three stages BTTR; (**b**) Drop port transfer function of a three stages BTTR.

The RR array demultiplexer RSC has to separate at most $N/S$ optical channels (see Figure 2) spaced as $2^T\Delta\lambda$; therefore each RR of the considered array is designed with a $FSR_{RSC} = N/S\ 2^T\Delta\lambda$ centered at a fixed wavelength. In order to set up a connection between a TX and a given RX on a certain Output plane to which the $\lambda_x$ is assigned, the TX has to be tuned on the $\lambda_x$.

## 2.2. Backplane Transfer Function

Once each RR is designed (*i.e.*, setting its physical parameters on the basis of the target filter-function physic [8]), we have used the Advanced Simulator for Photonic Integrated Circuits (ASPIC) [9,10] to calculate the transfer function of the backplane. Figure 4 shows the transfer function of a certain TX-RX pair from the architecture shown in Figure 2 with $S$ = 4 planes and $N/S$ = 8 TX/RX per plane. The grey vertical bars represent optical channels at 10 Gb/s with 0.4 nm (50 GHz) spacing. Typical parameters used to assess the performance are: in-band insertion losses ($IL$), bandwidth at −3 dB from the resonance peak ($B_{3\ dB}$), adjacent channel cross-talk (XT) ($X_A$) and non-adjacent channel XT ($X_{NA}$). When considering the transfer function of all TX-RX pairs, other parameters are defined: $<IL>$ is the average insertion-loss value across filter resonance peaks; $IL_m$ and $IL_M$ are the best and worst insertion loss, respectively; $B_m$ and $B_M$ are the smallest and largest −3 dB bandwidth, respectively; and $X_{FL}$ represents the worst-case floor XT (*i.e.*, the gap between $IL_M$ and the ground XT).

Besides the amplitude response of the architecture, in this study we also take into account the phase response $\Phi(\omega)$. Actually, the most important parameter which is used to be investigated is the *group delay*, which is defined as: $\tau_g = -d\Phi(\omega)/d\omega = \lambda^2 d\Phi/(2\pi cd\cdot\lambda)$. The group delay represents the time that a signal takes to pass through a filter. Usually, frequency components of the signal spectrum emerge from the filter with a changed time relationship as they propagate with different velocities and it can cause signal distortion and pulse broadening [11]. It is also worth mentioning that the derivative of the group delay with $\omega$ is the second order dispersion which is expressed in (ps$^2$) and it can be calculated as $D= d\tau_g/d\omega$ or, equivalently, by $D = \lambda^2 d\cdot d\tau_g/(2\pi cd\lambda)$ expressed in (ps$^2$/nm).

Figure 5a shows a typical "*drop*" port transfer function of a ring resonator and its group delay. Optical −3 dB bandwidth is set to 20 GHz. Figure 5b shows the "*through*" port transfer function and its group delay. It is evident that for the latter the dispersion is negligible as the group delay is constant and all the frequency components near to the resonant frequency roughly experience the same "pass-through" time. On the other hand, drop port exhibits a group delay distortion due to a non-linear phase response of the filter. At the resonance, the group delay increase with the *finesse*, which is the ratio between the *FSR* and the bandwidth, whereas, for the through, port it becomes no longer significant. Second order dispersion is zero at the resonance peak, leading to a small contribution of the third order dispersion on the pulse shape. In our architectures, on the basis of the number of stages of the PSC, signals pass through a given number of ring resonators. Thus, transfer functions show different values of the performance parameters since they are related to the particular path through the architecture a certain signal is routed.

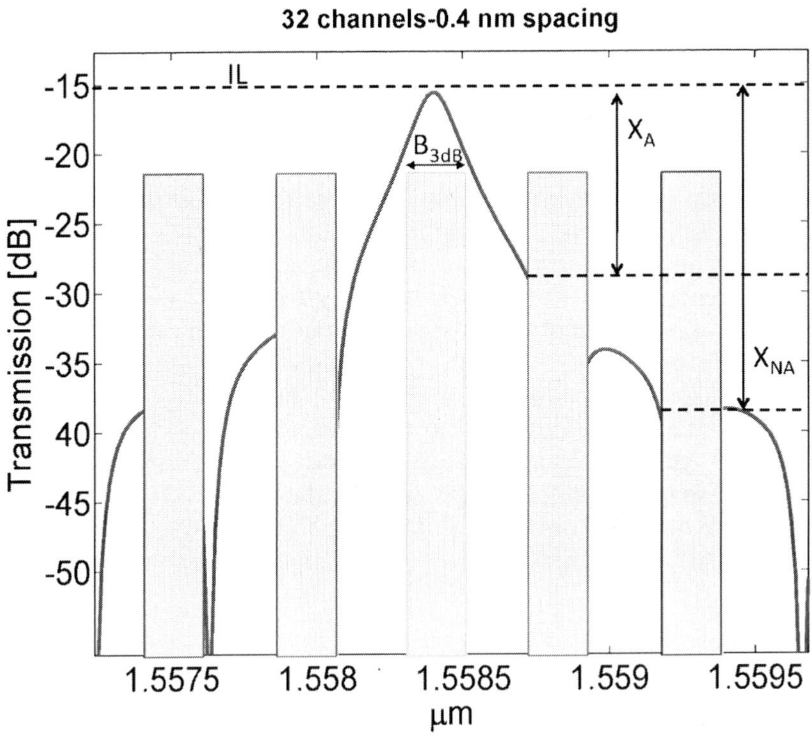


**Figure 4.** Backplane transfer function of a certain TX-RX pair.

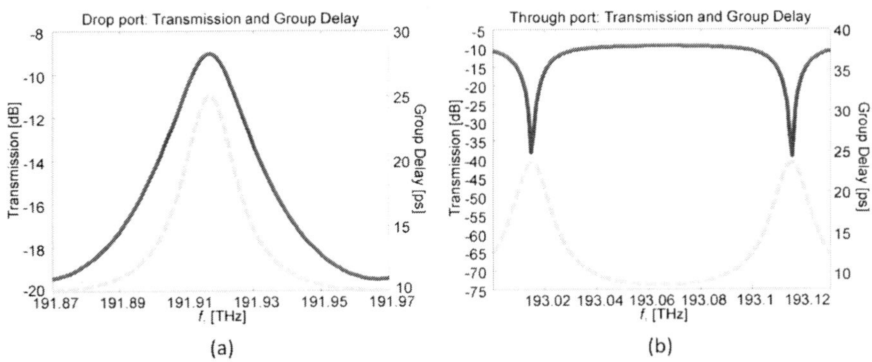


**Figure 5.** Transfer function "-" and group delay "- -" at the (**a**) drop port and (**b**) through port of an RR.

In Figure 6 we report the transfer functions just before the RSC of the architecture in Figure 2 with $S = 4$ planes and $N/S = 4$ TX/RX per plane. The first picture (architecture receiver plane 1) is given by two drop port pass-through. In fact, it has the highest peak of the group delay and the smallest bandwidth. It is worth noting that compared to Figure 5a, the group delay is almost doubled while retaining its shape and the insertion loss is increased of about 1 dB. The same considerations apply for the reception plane 3, as signals traverse two through ports. The reception plane 2 is reached after traversing a combination of one drop and one through port. In fact, it is half way from a totally flat response (reception plane 3) and a Gaussian shaped transmission and group delay (reception plane 1). Finally, the response at plane 4 has a shape, which is driven by the last drop port signals have to pass through. Note that signals are received after traversing another drop port at the specific RSC, thus, the final transmission and group delay are once more "drop port-shaped" before the RX. This is shown in Figure 7 for all the RXs at plane in the architecture reported in Figure with $S = 4$ planes and $N/S = 4$ TX/RX per plane.

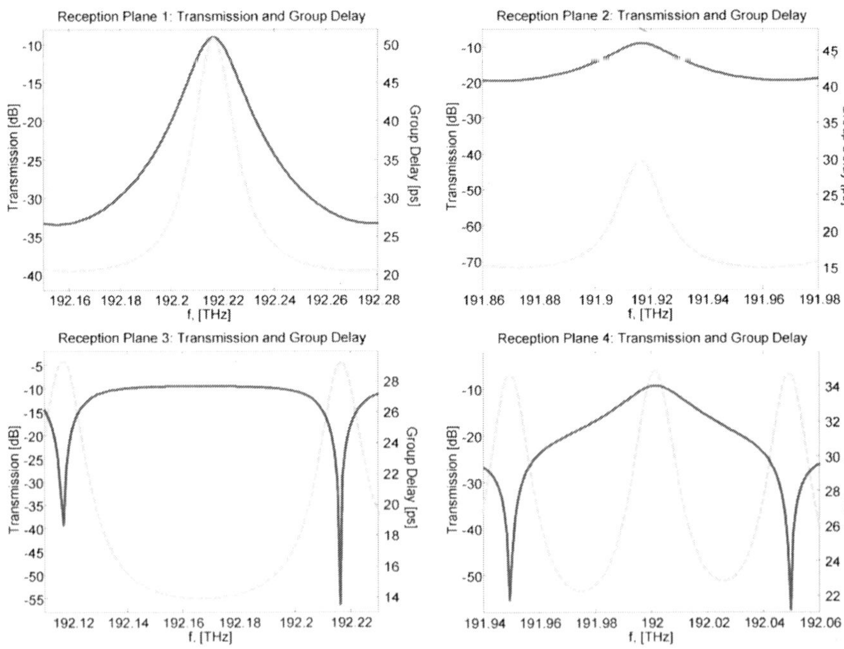


**Figure 6.** Transfer function "-" and group delay "- -" after the PSC at different planes.

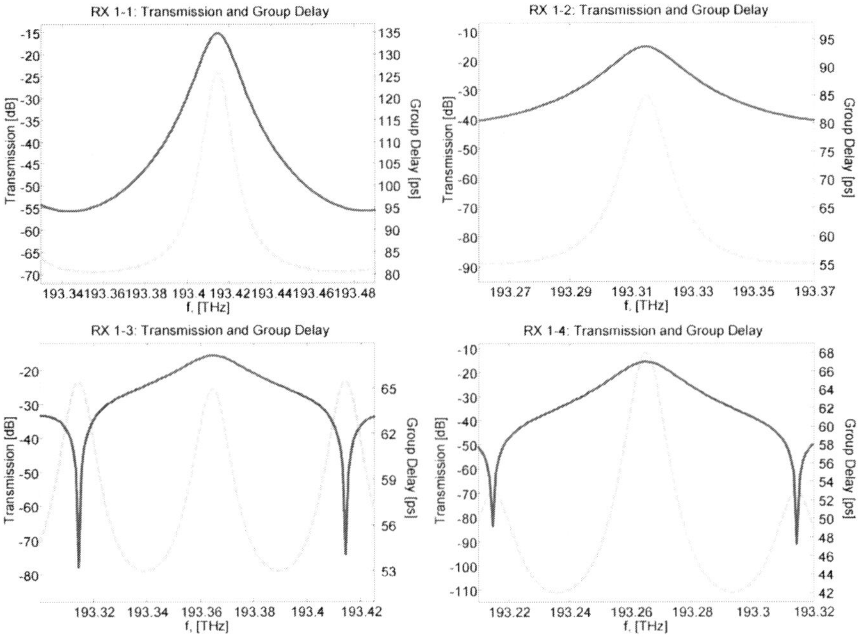


**Figure 7.** Transfer function "-" and group "- -" delay at RXs after the RSC at plane 1.

The group delay gives a hint of the pulse spread at the RX. In fact, it is the reciprocal of the *group velocity* (*i.e.*, the propagation velocity of a group of waves whose frequencies are distributed over an infinitesimally small bandwidth centered on frequency $f_0$) multiplied by the propagation distance. If we suppose $f_1$ and $f_2$ being the leading edge and the tail frequency of a spectrum centered around $f_0$, the pulse spread can be determined by taking the difference between the group delay at frequency $f_0$ and that at either $f_1$ or $f_2$ [11].

# 3. ILLUSTRATIVE NUMERICAL RESULTS

In this section, we report numerical results of our performance assessement procedure applied to the proposed architecture. We have investigated the feasibility of our backplane design considering typical implementation issues: number of line cards, switching element-round trip losses, frequency drifting due to thermal variations and waveguide-crossing effects.

## 3.1. Number of Optical Channels (λ)

We have evaluated the performance parameters for the following optical-channel configurations: 16 × 10 Gb/s (100 GHz spacing), 32 × 10 Gb/s (50 GHz spacing), and 64 × 10 Gb/s (25 GHz spacing), *i.e.*, $N$ = 16, 32, 64, $S$ = 4 and $N/S$ = 4, 8, 16). We have designed each RR with a −3 dB bandwidth set to 20 GHz.

Insertion loss increases when the number of channels scales up, due to star coupler/splitter losses that increase with $N/S$(see Table 1). The variability among TX-RX transfer-function peaks is due to the increasing number of cascaded RRs in the RSC (*i.e.*, $N/S$ = 4, 8, 16). The $B_m$ usually corresponds to a RX on the first output plane as signals traverse two drop functions in the PSC, whereas $B_M$ corresponds to a RX on the third plane as signals traverse two through functions. The analysis shows that performance parameters are critically low at 25 GHz spacing, while they are fair at 50 GHz and even better at 100 GHz. In the following we show results for 32 channels, since a typical core router hosts 32 LCs at 10 Gb/s [12].

**Table 1.** Increasing channels at fixed spectrum.

| λ Chan@GHz | <$IL$> [dB] | $IL_m$ [dB] | $IL_M$ [dB] | $B_m$ [GHz] | $B_M$ [GHz] | $X_A$ [dB] | $X_{NA}$ [dB] | $X_{FL}$ [dB] |
|---|---|---|---|---|---|---|---|---|
| 16@100 | −12.5 | −12 | −14 | 10.2 | 20.9 | −21.2 | −27.2 | −32 |
| 32@50 | −15.8 | −15 | −17.6 | 10.4 | 20.1 | −14 | −20.8 | −29 |
| 64@25 | −19.8 | −18 | −22.5 | 10.5 | 17.6 | −7.2 | −15.2 | −18 |

## 3.2. Roundtrip Loss of Ring Resonators (α)

The effect of roundtrip loss on the RR transfer function is to increase the drop insertion loss and, when various rings are cascaded, also to widen the bandwidth (*i.e.*, the filter becomes less selective). Table 2 reports the values of the investigated parameters. The impact of the paths through the PSC is dramatic. $B_M$ and $B_m$ remarkably increase, causing an increase of XT. It is worth noting that loss also makes the variability of *IL* across different channels to increase due to the low selectivity of filters and high *IL* at the drop port.

**Table 2.** Increasing loss/turn, 32@50 GHz.

| $\alpha$ [dB/turn] | <*IL*> [dB] | $IL_m$ [dB] | $IL_M$ [dB] | $B_m$ [GHz] | $B_M$ [GHz] | $X_A$ [dB] | $X_{NA}$ [dB] | $X_{FL}$ [dB] |
|---|---|---|---|---|---|---|---|---|
| 0 | −15.8 | −15 | −17.6 | 10.4 | 20.1 | −14 | −20.8 | −29 |
| 0.2 | −20.2 | −19.5 | −21.5 | 12.2 | 31.4 | −9.7 | −17.3 | −24 |
| 0.4 | −23 | −22.6 | −24.5 | 13.5 | 41.2 | −9.4 | −14.8 | −22 |
| 0.6 | −25.9 | −24.7 | −27.1 | 14.9 | 51.1 | −7.9 | −13 | −20 |
| 0.8 | −28 | −26.4 | −29.4 | 14.9 | 57 | −6.7 | −12.2 | −17 |
| 1.0 | −29.6 | −27.8 | −31.4 | 15.9 | 62 | −5.9 | −11.2 | −14 |

## 3.3. Thermal Variations (ΔT)

The change in the refractive index of a material as a function of the variation of temperature is known as the thermo-optic effect. An unintentional waveguide temperature variation turns into a resonance-frequency drifting: e.g., $\Delta T = 1$ °C, corresponds to 10 GHz frequency shift (in silicon) [13]. We have investigated the impact of such thermal perturbation on the overall backplane transfer function when a value within the range ±0.5, ±1, and ±1.5 °C of thermal variation is applied to any RR with uniform probability distribution. Table 3 shows that a thermal drifting of ±1.5 °C has a detrimental impact. In fact, the transfer functions are highly distorted and the variability on *IL* (measured as gap between $IL_m$ and $IL_M$) increases from almost 2 to 6 dB. Our architecture is clearly sensitive to thermal variations higher than 1.5 °C.

**Table 3.** Resonance frequency drifting, 32@50 GHz.

| $\Delta T$ [°C] | <*IL*> [dB] | $IL_m$ [dB] | $IL_M$ [dB] | $B_m$ [GHz] | $B_M$ [GHz] | $X_A$ [dB] | $X_{NA}$ [dB] | $X_{FL}$ [dB] |
|---|---|---|---|---|---|---|---|---|
| 0 | −15.8 | −15 | −17.6 | 10.4 | 20.1 | −14 | −20.8 | −29 |
| ±0.5 | −15.9 | −15.3 | −17.8 | 11.9 | 19.8 | −12.7 | −19.7 | −29 |
| ±1.0 | −16.2 | −16 | −18.4 | 11.6 | 20.1 | −12.2 | −20.4 | −27 |
| ±1.5 | −16.5 | −15.3 | −21.6 | 10.7 | 20.3 | −11.6 | −15.7 | −23 |

## 3.4. Waveguide Crossing

There are three major impairment contributions related to waveguide crossing: Crossing Insertion Loss (CIL), Crossing Reflected Power (CRP) and Crossing XT (CXT) power, the latter due to adjacent waveguide power coupling [14]. Such impairments depend on the crossing angle. We consider values that are related to the orthogonal waveguide crossing, which represents the best-case, since such a condition can be achieved by a careful geometric design of the waveguides.

CIL increases the variability of *IL* among channel resonance peaks since signals directed to different Output planes may traverse a different number of waveguide crossings. The gap between $IL_m$ and $IL_M$ increases from almost 3 to 4 dB when setting CIL up to 1 dB; nevertheless, XT level values scale down along with peak loss.

CRP above 2% is highly detrimental. It causes high variability of channel *IL*, resonance frequency shifting and severe distorsion. Moreover, high values of XT have been found, the worst case is $X_A$ = −10 dB and $X_{NA}$ = −15 dB. The most detrimental impairment is the waveguide CXT power. In addition, in this case, values higher than 2% may impair feasibility. $<IL>$ varies from −15 dB of the 0% case to −21 dB of the 15% and the worst-case gap between $IL_m$ and $IL_M$ is 11 dB.

## 3.5. Technology Remarks

In the proposed architecture, the *FSR* of RRs in RSC drives the choice of the technology to be used (see Design Section). As for the PSC of our reference architecture (32 chann./50 GHz, 2 stages), the *FSR* of last-stage RR is 200 GHz while for the RSC, the *FSR* of each RR is 1.6 THz, corresponding to rings with radius = 7 μm, thus, the choice of silicon becomes mandatory [7]. Smaller rings, that is, wider *FSR*, are difficult to achieve in practice.

# 4. BER PERFORMANCE ANALYSIS

Direct detection for transmission at 10 Gb/s of 32 OOK-NRZ signals spaced at 50 GHz is considered. For BER performance analysis the impulse response used for pulse shaping was chosen with a rise and fall time of 35% the bit duration. As is known, the effect of filtering operations performed by a certain TX-RX pair on the input signal and the resulting distortions can be estimated by means of the eye diagram [15]. In order to evaluate the distortions introduced by optical filtering, we consider the case where the detected signal is dominated by the photo-detector thermal noise. Under this assumption the $Q$factor can be calculated from the eye diagram as:

$$Q = \frac{\mu_1 - \mu_0}{\sigma_1 + \sigma_0} \tag{1}$$

where $\mu_1$ and $\mu_0$ are the average voltages associated to the levels 1 and 0 at the eye center, respectively, while $\sigma_1$ and $\sigma_0$ are the effective standard deviations associated to the levels 1 and 0, respectively. The $Q$ factor allows for estimating the BER associated to the received signal after its conversion into the electrical domain by using the formula:

$$\text{BER} = \frac{1}{2}\text{erfc}\left(\frac{Q}{\sqrt{2}}\right) \tag{2}$$

Each signal transmitted at a specific wavelength has been filtered by the PSC transfer function and then selected by the corresponding RSC transfer function. The eye diagram has been evaluated for each of the transfer functions resulting from all the TX-RX pair combinations. Figure 8 shows the best, the worse and the average BER performance versus the electrical Signal-to-Noise Ratio (SNR) obtained considering the case of roundtrip loss $\alpha = 0$. The electrical SNR is given by the ratio of the signal power and noise power after electrical filtering with the same bandwidth as that of the RSC. As a reference, in the Figure, the BER performance obtained in the ideal case, *i.e.*, without signal distortions, is also reported. From the curves we observe that the SNR loss of the average case with respect to that without (w/o) distortions varies between 6 dB, at BER = $10^{-2}$, and 8 dB, at BER = $10^{-10}$. In the same range of BERs, the SNR loss of the best channel with respect to that w/o distortions is comprised between 3.5 and 4 dB. The SNR gap between the best and the worse performance associated to channel #10 and channel #7, respectively, varies between 4 dB, at BER = $10^{-2}$, and 5 dB, at BER = $10^{-10}$.

Figure 9 reports the best and the worst performance associated to channel #10 and channel #7, respectivily, for increasing values of the roundtrip loss $\alpha$ = 0, 0.2, 0.4, 0.6, 0.8, 1. From the figure we observe that for a given value of $\alpha$ the SNR gap between the best and the worse channel is approximately always the same. Moreover, an SNR loss of approximately 5 dB can be observed between $\alpha = 0$ and $a = 1$.

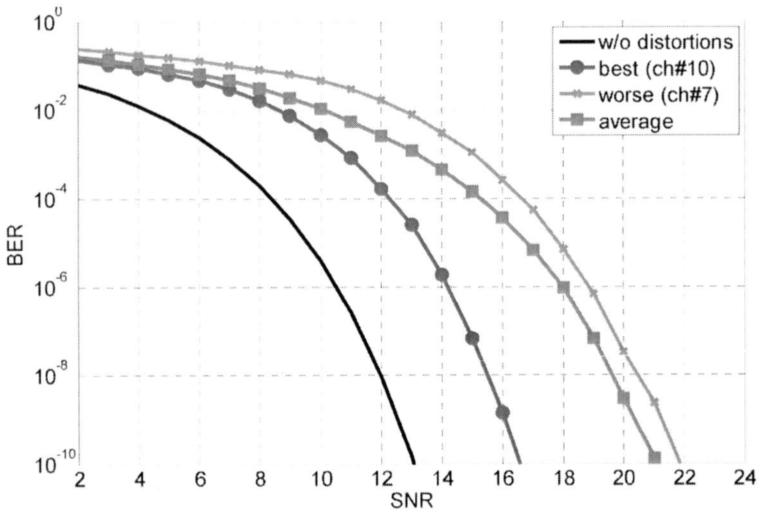


**Figure 8.** BER *vs.* SNR for transmission at 10 Gb/s of NRZ-OOK channels with roundtrip loss $\alpha$ = 0. The best performance was obtained for channel #10, the worse for channel #7 and the average considering the mean over the 32 channels.

The effect of XT on the BER is shown in Figure 10. We have verified that in order to evaluate the effect of XT on the channel of interest it is enough to consider only the effect of the two more adjacent signals. From Figure 10 we observe that the SNR loss of the average case compared to that w/o distortions increases of only 0.2 dB at BER = $10^{-2}$ with respect to the situation without XT considered in Figure 8 while it increases up to 1.3 dB at BER = $10^{-10}$. The loss in SNR of the best case with respect to the ideal case w/o distortions is comprised now between 4.5 dB at $10^{-2}$ and 6 dB at $10^{-10}$. It is worth noting that the SNR gap between the best and the worst case remains the same as that without XT.

In Figure 11 the effect of XT on the best and the worse performance is reported for increasing values of the roundtrip losses $\alpha$ = 0, 0.2, 0.4, 0.6, 0.8, 1. In contrast to the analogous situation without XT considered in Figure 9, in this case we observe that for the values of $\alpha$ equal to 0.6, 0.8 and 1 the BER presents an error floor. This effect arises because the presence of some configurations of interference that cause the closure of the eye even in the situation where noise is absent. By comparing the curves reported in Figure 9 and Figure 11 at a given BER it is possible to evaluate the degradation introduced by XT for a fixed value of $\alpha$.

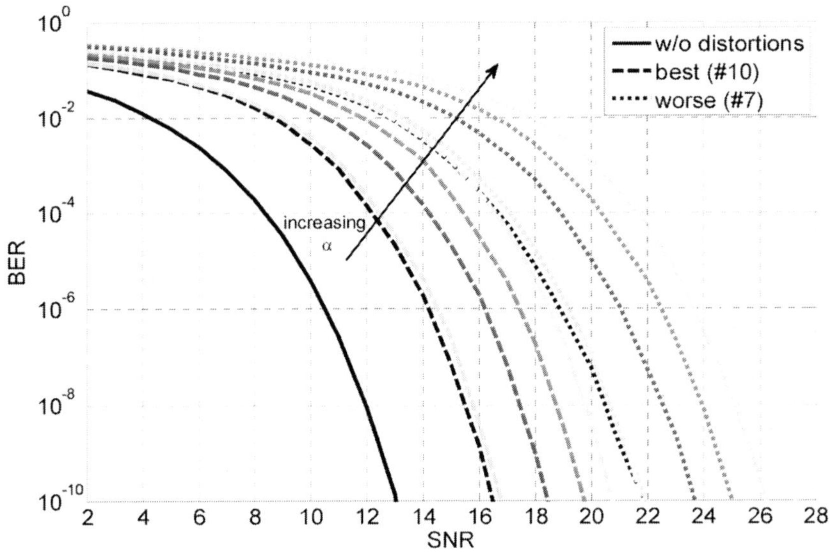


**Figure 9.** BER *vs.* SNR for transmission at 10 Gb/s of 32 NRZ-OOK channels with increasing roundtrip losses $\alpha$ = 0, 0.2, 0.4, 0.6, 0.8, 1. Dashes curves: best channel (#10); dotted curves: worse channel (#7).

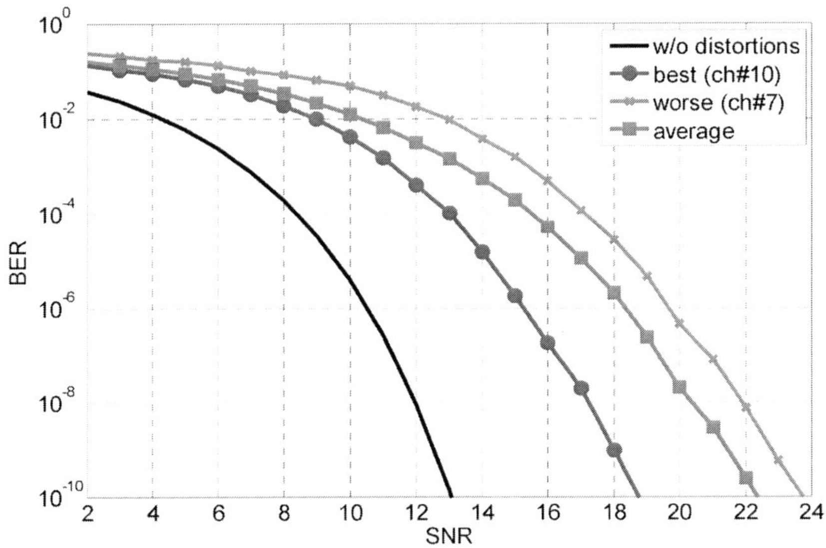


**Figure 10.** BER *vs.* SNR for transmission at 10 Gb/s of 32 NRZ-OOK channels with roundtrip loss $\alpha = 0$ in presence of XT.

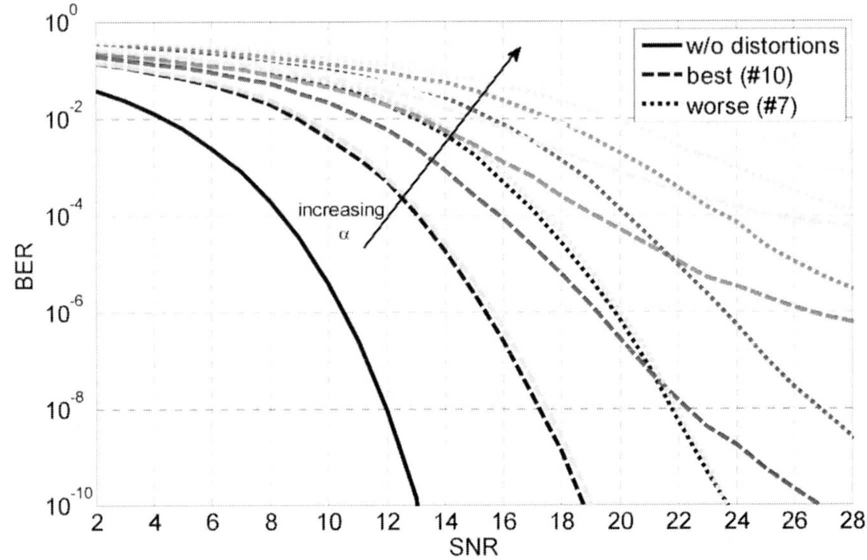


**Figure 11.** BER *vs.* SNR for transmission at 10 Gb/s of 32 NRZ-OOK channels with increasing roundtrip losses $\alpha$ = 0, 0.2, 0.4, 0.6, 0.8, 1 in presence of XT. Dashes curves: best channel (#10); dotted curves: worse channel (#7).

It is worth observing that for input signals with rates higher that 10 Gb/s, *i.e.*, 40, 100, and 160 Gb/s an increase of modulation order is required in order to satisfy the bandwidth limitation imposed by the optical filtering. As is well known, high order modulation formats require coherent demodulation and are more sensitive to signal distortions introduced by optical filters and XT. Therefore, in order to achieve a performance that is comparable with that of 10 Gb/s transmission an appropriate signal processing is required.

# 5. CONCLUSIONS

We have presented a new RR-based switching fabric and a novel approach to assess architecture performance. Our transfer function analysis allows us to accurately capture the impact of each design issue with respect to the ideal scenario. This capability is not achievable in power budget and XT power penalty evaluations. In this work, we have analyzed the OI architecture in terms of optical performance parameters and signal distorsions introduced by the filtering operations performed by PSC and corresponding RSC for a given value of the wavelength of the input signal. From the observation of the eye diagrams for all the TX-RX pairs in case of OOK-NRZ transmisison at 10 Gb/s, performance analysis has been worked out by estimating the BER both without and with XT *vs.* SNR of the proposed architecture.

## ACKNOWLEDGMENTS

The research leading to these results has received funding from the European Union Seventh Framework Programme (FP7/2007-2013) under grant agreement n. 257740 (Network of Excellence "TREND"). This manuscript is an invited paper as part of the Special Issue on "All Optical Networks for Communications" published in *Photonics*.

## AUTHOR CONTRIBUTIONS

Giuseppe Rizzelli, Domenico Siracusa and Guido Maier jointly designed the switching architecture. Maurizio Magarini defined the analysis procedure used for BER computation, while Mehmood Alam performed the BER simulations; Giuseppe Rizzelli also carried out the transfer-function evaluation of the architecture, exploiting integrated-optics analysis software and supervision provided by Andrea Melloni.

## CONFLICTS OF INTEREST

The authors declare no conflict of interest.

## REFERENCES

1. Korotky, S.K. Traffic trends: Drivers and measures of cost-effective and energy-efficient technologies and architectures for backbone optical networks. In Proceedings of the Optical Fiber

Communication Conference and Exposition and the National Fiber Optic Engineers Conference, Los Angeles, CA, USA, 4–8 March 2012; p. OM2G.1.

2. Bianco, A.; Carta, E.; Cuda, D.; Finochietto, J.M.; Neri, F. An optical interconnection architecture for large packet switches. In Proceedings of the International Conference on Transparent Optical Networks, Rome, Italy, 1–5 July 2007; pp. 226–229.

3. Siracusa, D.; Maier, G.; Linzalata, V.; Pattavina, A. Scalability of optical interconnections based on the arrayed waveguide grating in high capacity routers. In Proceedings of the IEEE International Conference on Optical Network Design and Modeling, Bologna, Italy, 8–10 February 2011; pp. 1–6.

4. Klein, E.J. Densely Integrated Microring-Resonator Based Components for Fiber-to-the-Home Applications. Ph.D. Thesis, University of Twente, Enschede, The Netherlands, 2007.

5. Melloni, A.; Martinelli, M. The ring-based resonant router. In Proceedings of the ICTON 2003; Warsaw, Poland: 29 June–3 July 2003.

6. Cuda, D.; Gaudino, R.; Gavilanes, G.A.; Neri, F.; Maier, G.; Raffaelli, C.; Savi, M. Capacity/cost tradeoffs in optical switching fabrics for terabit packet switches. In Proceedings of the IEEE International Conference on Optical Network Design and Modeling, Braunschweig, Germany, 18–20 February 2009; 2009; pp. 1–6.

7. Rizzelli, G.; Siracusa, D.; Maier, G.; Magarini, M.; Melloni, A. Performance of ring-resonator based optical backplane in high capacity routers. In Proceedings of the 15th International Conference on Transparent Optical Networks, Cartagena, Spain, 23–27 June 2013; pp. 1–4.

8. Melloni, A.; Costa, R.; Cusmai, G.; Morichetti, F. The role of index contrast in dielectric optical waveguides. *Int. J. Mater. Prod. Technol.* **2009**, *34*, 421–437.

9. Filarete srl, Italy. Milano. Available online: http://www.aspicdesign.com (accessed on 11 April 2014).

10. Melati, D.; Morichetti, F.; Canciamilla, A.; Roncelli, D.; Soares, F.; Bakker, A.; Melloni, A. Validation of the building-block-based approach for the design of photonic integrated circuits. *J. Lightwave Technol.* **2012**, *30*, 3610–3616.

11. "Dispersion". In *Fundamentals of Optical Fibers: Wiley Series in Pure and Applied Optics*; Buck, J.A., Ed.; Wiley-Interscience, John Wiley & Sons: Hoboken, NJ, USA, 2004.

12. T Series Core Routers. Available online: http://www.juniper.net/ (accessed on 11 April 2014).

13. Knights, A.P.; Jessop, P.E. Silicon waveguides for integrated optics. In *Optical Waveguides: From Theory to Applied Technologies*; Calvo, M.L., Lakshminarayanan, V., Eds.; Taylor and Francis: London, UK, 2007; pp. 231–270.

14. Bukkems, H.G. Minimization of the loss of intersecting waveguides in inP-based photonic integrated circuits.*Photonics Technol. Lett.* **1999**, *11*, 1420–1422.

**15.** Downie, J.D. Relationship of Q penalty to eye-closure penalty for NRZ and RZ signals with signal-dependent noise. *J. Lightwave Technol.* **2005**, *23*, 2031–2038.

7

# TRANSPORT SCHEMES FOR FIBER-WIRELESS TECHNOLOGY: TRANSMISSION PERFORMANCE AND ENERGY EFFICIENCY

**Christina Lim *, Yizhuo Yang and Ampalavanapillai Nirmalathas**

Department of Electrical and Electronic Engineering,
University of Melbourne, VIC 3010, Australia

## ABSTRACT

Fiber-wireless technology has been actively researched as a otential candidate for next generation broadband wireless signal distribution. Despite the popularity, this hybrid scheme has many technical challenges that impede the uptake and commercial deployment. One of the inherent issues is the transport of the wireless signals over a predominantly digital optical network in today's telecommunication infrastructure. Many different approaches have been introduced and demonstrated with digitized RF transport of the wireless signals being the most compatible with the existing optical fiber networks. In this paper, we review our work in the area of digitized RF transport to address the inherent issues related to analog transport in the fiber-wireless links and compare the transmission performance and energy efficiency with the other transport strategies.

## KEYWORDS

fiber-wireless; radio-over-fiber; digitized RF transport

# 1. INTRODUCTION

Fiber-wireless technology has gained significant momentum as the "last mile connectivity" in the last two decades in order to address the increasing demand for larger capacity and faster transmission speeds, fuelled by the proliferation of affordable smart phones and tablets packed data-intensive multimedia applications [1,2,3]. In addition, this hybrid technology is able to leverage on existing optical fiber infrastructure and, hence, is able to unify the telecommunication backhaul infrastructure by using a common backhaul network to support both wired and wireless services. Figure 1illustrates a typical infrastructure of a fiber-wireless system with a central office (CO) serving multiple base stations (BSs) interconnected by high speed optical fiber based backhaul while the BSs serve end users wirelessly. It is well-established that centralized control in fiber-wireless system enables simple, low-cost design antenna base stations to be widely deployed where most hardware intelligence are located in a central location.

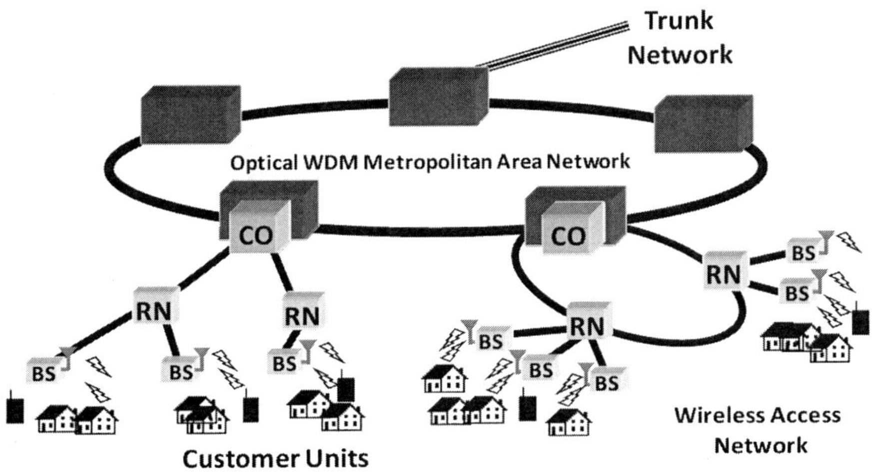


**Figure 1.** Fiber-wireless network.

One strategy to establish centralized control is to optically distributing the wireless signals at the wireless carrier frequencies in an analog fashion. Despite the elegance of the hybrid fiber-wireless infrastructure, the performance is limited by the inherent properties of analog optical link [3,4]. Although many techniques have been proposed to improve the performance of analog optical links for the transport of wireless signals including linearization [5,6,7,8,9,10], predistortion [11], and modulation depth optimization [12,13,14,15], they demand for complex implementation with precise control which negates the cost effectiveness of fiber-wireless scheme.

We have previously demonstrated an alternative transport scheme based on digitization of the wireless signals using bandpass sampling technique [16,17] to improve the fiber-wireless link dynamic range and to overcome the inherent issues of analog optical link. In this paper, we review the work we have carried out in this area in particular focusing on strategies to improve the performance, capacity and energy efficiency of fiber-wireless links with respect to other transport schemes. The paper is organized as follows. Section 2 describes and reviews the different transport schemes with respect to their roles and the base station configurations in fiber-wireless link, and Section 3 presents the experimental investigations comparing analog optical transport *versus* digitized transport based on bandpass sampling technique for the transmission of broadband wireless OFDM signals. Section 4 governs the study and quantification of energy consumption for different transport schemes with respect to base station site power consumption and wireless cell size.

## 2. WIRELESS SIGNALS TRANSPORT STRATEGIES FOR FIBER-WIRELESS LINK

To understand how wireless signals are distributed in a fiber-wireless link, Figure 2 shows the different link configuration for a simple point-to-point transmission connecting a CO and an antenna BS. Regardless of the wireless carrier frequencies, the wireless signals can be distributed based on five transport schemes namely RF-over-fiber (RFoF), IF-over-fiber (IFoF), Baseband-over-fiber (BBoF), Digitized RF-over-fiber (DRFoF) and Digitized IF-over-fiber (DIFoF) as shown in Figure 2.

The simplest scheme for transporting wireless signals via an optical fiber backhaul network is to directly transport the wireless signals at the wireless carrier frequency without any further frequency processing at the base station via RF-over-fiber transport scheme (Figure 2a). In this configuration, the wireless signals are modulated onto an optical carrier and the optically modulated wireless signals are transported in an analog format over the optical link. RF-over-fiber has the advantage of simple base station design with added benefits of centralized control, wireless signal transparency and multi-band wireless support. However one of the major drawbacks is the needs for optical devices with speeds matching that of the wireless carrier frequency which becomes more stringent for wireless signals beyond the microwave bands. It is also well-established that optically modulated wireless signals suffers from fiber chromatic dispersion which degrades the received RF power and limits the fiber transmission distance [18,19].

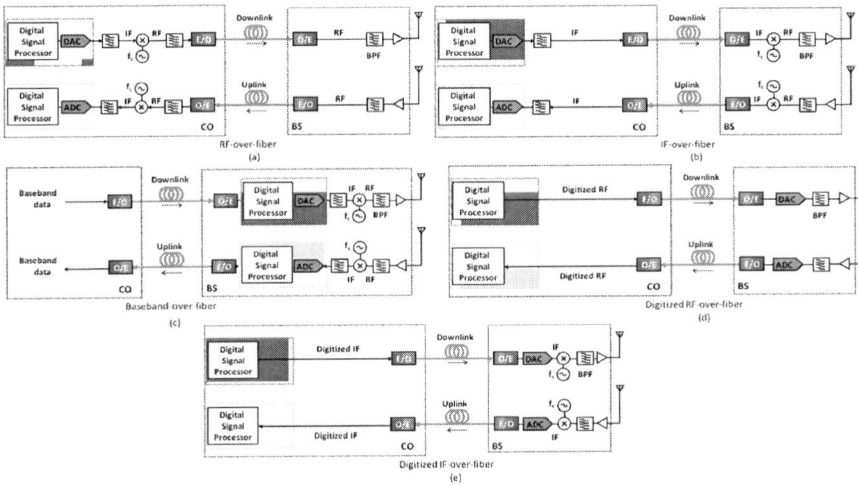


**Figure 2.** Schematic showing the transport scheme for (**a**) RF-over-Fiber (RFoF) (**b**) IF-over-Fiber (IFoF) (**c**) Baseband-over-Fiber (BBoF) (**d**) Digitized RF-over-Fiber (DRFoF) (**e**) Digitized IF-over-Fiber (DIFoF).

In contrast, IF-over-fiber transport scheme transports the wireless signals at a lower intermediate frequency (IF) over the optical platform (Figure 2b) which in turn relaxes the requirements for high-speed optoelectronic hardware. The optical distribution of the IF signals also has a much reduced fiber chromatic dispersion impact. However this is at the expense of a more complex antenna BS design where it necessitates a stable local oscillator (LO) and mixers for frequency translation purposes.

Figure 2c shows the architecture for baseband-over-fiber transport where the wireless signal is transported as baseband data optically and the wireless signal is processed entirely at the antenna BS. This transport scheme is compatible to existing optical networks where data are transmitted in a predominantly digital environment. Despite the simplicity in the optical domain, the antenna BS for baseband-over-transport has to house additional hardware to electronically process the wireless signals which increases the BS complexity while decreases the link transparency. In the recent years, a new transport scheme based on digitization of the wireless signals have been proposed and demonstrated [16] for fiber-wireless application that exploits the superior digital optical links while maintaining a relatively simple BS design. As opposed to baseband-over-fiber transport where the wireless signals are transported at the information level and all signal processing are carried out in the antenna BS, digitized wireless transport essentially digitized the wireless signals before the optical transport. Digitized wireless transport can be further classified into Digitized RF-over-fiber (DRFoF) transport and Digitized IF-over-fiber (DIFoF) transport.

Digitized RF-over-fiber scheme as illustrated in Figure 2d digitizes the wireless signal to produce a sampled digital data stream in a serial format that can directly modulate the optical source. This enables the use of digital photonic link to transport the wireless signals overcoming the inherent issues related to analog photonic links. Since the wireless signals are transported in a digital format, the dynamic range of the optical link is independent of the fiber transmission distance until the received signal goes beyond the link sensitivity. In this scheme, the complex signal processing functionality is carried out in the CO while only a minimal set of frontend components consisting of the analog-to-digital (ADC) and digital-to-analog (DAC) converters are located in the antenna BS. The link is able to maintain its transparency as the BS does not require full processing of the wireless signals. The key enabler for this scheme is the ADC/DAC technology. To digitize the wireless signals, the analog bandwidth of the ADC must be larger than the wireless carrier frequency and with sufficient sampling rate and resolution. This requirement becomes more stringent for wireless signals beyond the microwave bands. To overcome this limitation, digitized IF-over-fiber is introduced whereby the wireless signal is first downconverted to an IF frequency to cater for the limitations of current ADC technology. This scheme is depicted in Figure 2e.

Open base station architecture initiative (OBSAI) [20] and Common Public Radio Interface (CPRI) [21] are two initiatives to create an open market for cellular networks by providing specifications for interface of wireless base stations. OBSAI and CPRI are based on digitized IF-over-fiber scheme and the base station architecture is shown in Figure 3a. In the downlink direction, the optical digital datastream is recovered using a photodetector (PD) and digitally processed before the analog signal at an IF frequency is reconstructed using a DAC. The wireless signal is then generated after two stages of frequency upconversion process. Likewise in the uplink direction, the wireless signal is downconverted using two frequency translation stages before the IF signal is digitized, processed and modulated onto an optical carrier to be transported back to the CO.

We can further relax the requirements on the ADC technology by using bandpass sampling technique [22] rather than direct sampling, especially given that for most wireless applications the required information bandwidth is only a small fraction of the wireless carrier frequency. Bandpass sampling undersamples the wireless signal using a much lower sampling frequency that is comparable to the wireless information bandwidth instead of the wireless carrier frequency [22]. Therefore upon sampling, it will generate an infinite number of image replicas of the wireless signal. The sampling frequency has to be carefully chosen to minimize noise aliasing from adjacent images. Since an infinite number of image replicas are generated, the wireless signal can be recovered using appropriate bandpass filter. Hence we are able to perform frequency translation without using LO and mixers. Shown in Figure 3b is the BS architecture of a fiber-wireless link incorporating digitized IF-over-fiber transport based on bandpass sampling. The benefit of using bandpass sampling reflects in the uplink path where no frequency translation stage is required which can further simplify the base station design.

# 3. TRANSMISSION EXPERIMENT COMPARING ANALOG AND DIGITIZED RF-OVER-FIBER TRANSPORT

Numerous transmission demonstrations have been carried out for different transport schemes to quantify the performance of the fiber-wireless link with the main focus on increasing the transmission capacity, improving the link dynamic range and transmission distance [23,24,25,26,27]. One strategy to increase the transmission capacity is to efficiently utilize the wireless bandwidth by incorporating advanced modulation format in conjunction with broadband orthogonal frequency-division-multiplexing (OFDM). The demand for larger transmission capacity is driven by the emergence and proliferation of bandwidth-hungry applications such as real-time high-definition video streaming and graphic intensive online gaming.

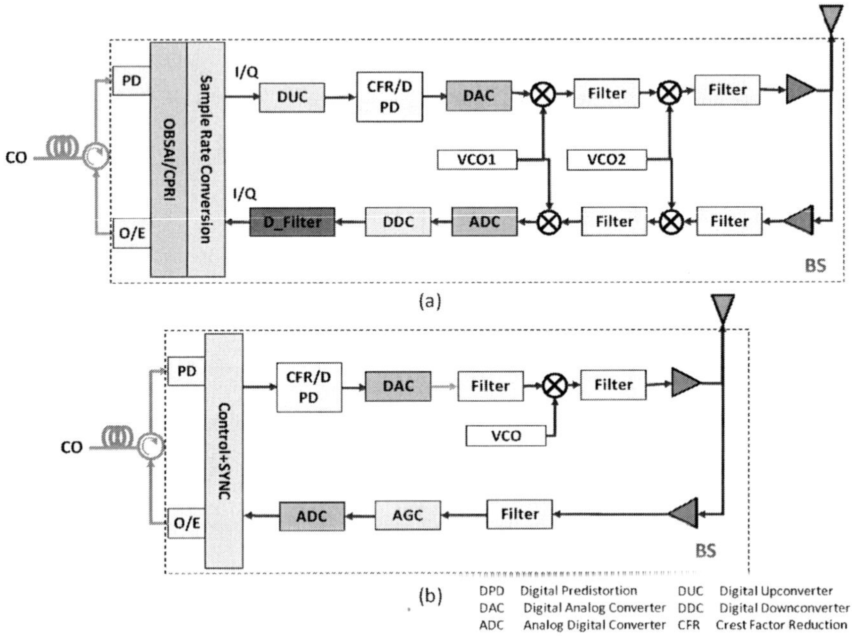


**Figure 3.** Possible base station configuration for (**a**) digitized IF-over-Fiber transport and (**b**) digitized IF-over-Fiber transport incorporating bandpass sampling technique.

OFDM with its inherent advantages of spectral efficiency, immunity to multipath fading and dispersion tolerance, is currently the leading modulation for current wireless standards ranging from 3GPP Long Term Evolution (LTE) to personal area network ultra-wideband (UWB). The multi-carrier nature of OFDM demands for highly linear transmission link with the ability to handle large peak-to-average power (PAPR). This requirement becomes more stringent for wideband (>1 GHz) signals. Although many different demonstrations have been carried out for the distribution of OFDM signals over fiber-wireless links using analog optical transport [28,29,30,31,32,33], the linearity requirement has necessitates the use of low modulation depth which leads to large received DC power at the photodetector and hence, power wastage. On the other hand, digitized transport may emerge as an ideal candidate for the distribution of broadband OFDM signals in fiber-wireless links which overcomes the nonlinearity issue associated with analog transport. However, the needs for highspeed ADC and DAC coupled with excessive optical bit rate after digitization may render the digitized transport scheme unattractive. Given the pros and cons of the analog and digitized transport, we have previously carried out an experimental comparison for the distribution of a 1 GHz OFDM signal using both analog and digitized RF-over-fiber transport schemes [34].

## 3.1. Experimental Demonstration

Shown in Figure 4a is the experimental setup for the uplink broadband wireless OFDM signal transmission over a fiber-wireless link incorporating digitized RF-over-fiber transport [34]. In this demonstration, the baseband (BB) OFDM was programmed in Matlab and physically generated using an arbitrary waveform generated (AWG). To avoid using high-speed I/Q mixers, we make corresponding frequency components conjugate complex to ensure Q equal to zero all the time. The signal bandwidth was limited to 1 GHz and there were 128 carriers with 80 being modulated with 4-QAM data, resulting in a data rate of 1.25 Gb/s. The OFDM signal was upconverted to 10 GHz using a local oscillator (LO) to generate the 10 GHz broadband OFDM wireless signal. In the BS, the OFDM signal at 10 GHz was bandpass sampled using an 8-bit and 1 GSample/s ADC in a digital sampling oscilloscope (DSO). After bandpass sampling, the samples were quantized, serialized and processed offline before the digital data was generated using an AWG. Here the 10 GHz OFDM signal was converted to an 8 Gbps digital data stream. For this demonstration, a DFB laser in conjunction with a Mach-Zehnder (MZM) was used to modulate the digital data stream onto an optical carrier before it was transported over 20 km of standard fiber (SMF). In the CO, the optical signal was detected with a PIN photodetector and the digital data stream was processed offline.

For comparison, the same 10 GHz OFDM signal was distributed using analog RF-over-fiber transport scheme and the experimental setup is shown in Figure 4b. The upconverted OFDM signal was externally modulated onto the optical carrier using a MZM before transported over 20 km of SMF. Upon reception at the CO, the received 10 GHz OFDM was downconverted to baseband and processed offline. In order to maintain good link linearity, the peak-to-peak driving voltage of the RF signal into the MZM was fixed at one tenth of the MZM switching voltage (Vpi). This corresponds to a modulation depth of 0.16.

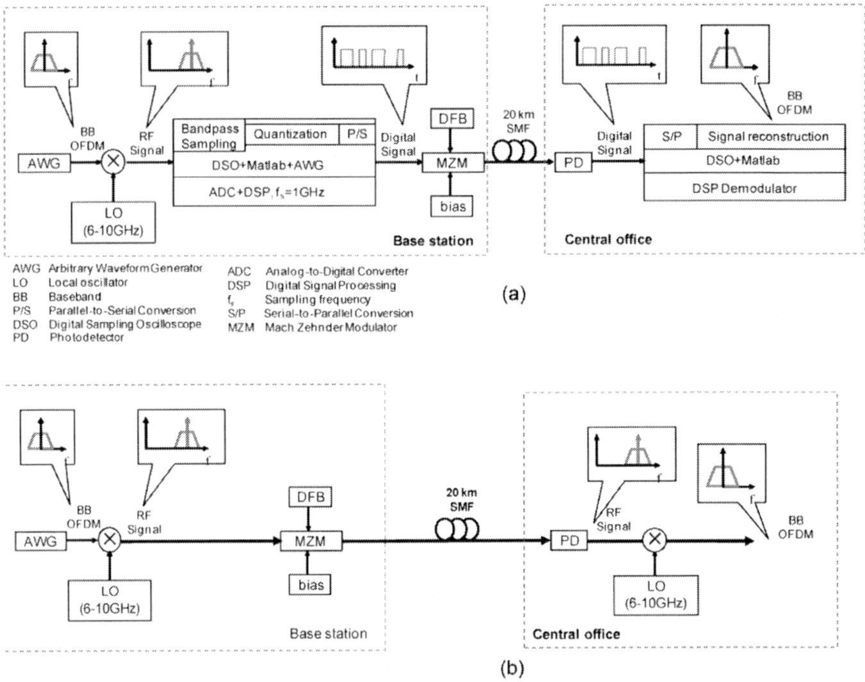


**Figure 4.** Experimental setup for (**a**) digitized RF-over-fiber transport and (**b**) analog RF-over-fiber transport of wideband OFDM signal. © [2008] IEEE. Reprinted, with permission, from *IEEE OSA Lightwave Technology, Journal.*

## 3.2. Experimental Results and Discussions

Shown in Figure 5a is the measured error vector magnitude (EVM) plotted as a function of received optical power (dBm) for digitized RF-over-transport after transmission over 20 km of SMF for a bit resolution of 8 bits. Also plotted are results for an ADC bit resolution of 4 and 6 bits [34]. The dotted line indicates error-free reception for 4-QAM signal (corresponding to a bit-error-rate of $10^{-12}$). The results indicate that the performance of the link is not sensitive to the received optical power until the received optical power goes beyond the link sensitivity whereby the performance degrades drastically with decreasing optical power. For a 4-bit resolution, a minimum received optical power of −13 dBm is needed to achieve error-free transmission. The same trend was seen for 6-bit and 8-bit resolution but with better EVM performance and lower received optical power requirement (−13.5 dBm and −15 dBm).

Figure 5b shows the measured EVM for the analog RF-over-fiber transport of the OFDM signal for transmission over 0, 10 and 20 km of SMF [34]. The performance of the analog link degrades steadily with increasing transmission distance and with decreasing received optical power. Error-free reception is only possible for a received optical power of >−11 dBm for a transmission distance of 10 km and is not achievable for 20 km of SMF for received optical power <−10 dBm. Comparing the two results from Figure 5a,b, we can see that for an error-free transmission over 20 km of SMF analog RF-over-fiber requires higher received optical power compared (>−10 dBm) to digitized RF-over-fiber transport (−15 dBm for 8-bit resolution). Even for a 4-bit ADC resolution, optical bit rate of 4 Gb/s and received optical power of −13 dBm, the digitized RF-over-fiber link outperforms the analog link distributing the same OFDM signal at 10 GHz with an received optical power of −10 dBm.

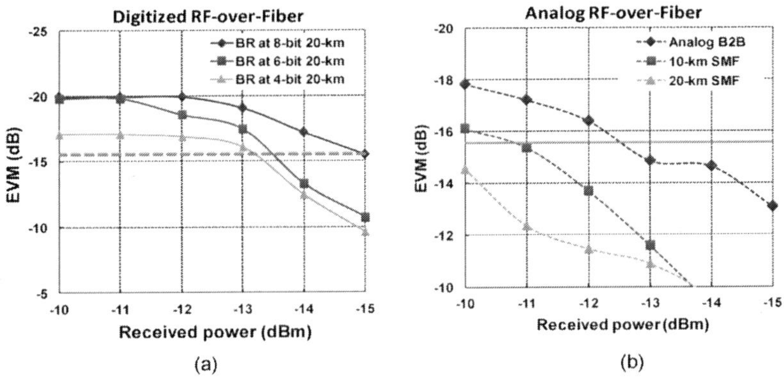


**Figure 5.** Error vector magnitude measured for (**a**) digitized RF-over-fiber transport and (**b**) analog RF-over-fiber transport over 20 km of fiber. © [2008] IEEE. Reprinted, with permission, from *IEEE OSA Lightwave Technology, Journal.*

# 4. ENERGY CONSUMPTION OF DIFFERENT TRANSPORT SCHEMES

Despite the fact that digitized transport is more superior in terms of dynamic range performance compared to analog transport, other factors such as energy consumption have to be taken into consideration. Telecommunication networks have been identified as one of the major contributors to the overall ICT energy consumption with wireless access network being the largest contributor to $CO_2$ emissions [35,36,37]. With the wireless base station contributing approximately two-thirds of the $CO_2$ emission, it is imperative to first reduce and optimize the energy consumption of wireless base station in order to minimize the carbon footprint [35,36,37,38,39]. On the other hand, the proliferation of smart devices in recent years has put a massive burden on our current mobile networks. The most straightforward upgrade is to reduce the cell size and increase the number of mobile base stations. However this will further increase the proportion of base station energy consumption, further compounding the energy crisis. To address this issue we introduce fiber-wireless technology with optical fiber backhauling the small cells with majority of the processing power located in a central location. This architecture results in a simple, compact and energy-efficient base station.

## 4.1. Single Base Station Energy Consumption for Different Transport Schemes

To quantify the energy consumption of the fiber-wireless network, we investigate the energy consumption per square meter for the transport schemes discussed in Section 2. We have previously presented a generalized power consumption model for fiber-wireless base stations [40] that supports a variety of wireless transport schemes. Shown in Figure 6 is the conventional base station architecture based on baseband-over-fiber transmission scheme. It consists of Baseband Unit (BBU), Data Converter Unit, RF unit, Power Amplifier, Feeder, Clock Management, Power Supply and Battery Backup, and Cooling System. The power consumption of this base station can be calculated using Equation (1) and Table 1 shows the power consumption of each unit approximated using commercially available devices [40].

$$P_{BBoF} = N_{sector} \times (P_{Tx} / \mu_{PA} / L_{feeder} + P_{BBU} + P_{DDC/DUC} + P_{ADC/DAC} + P_{DPD} + P_{RFU} + P_{M}) \times (1 + L_{PS}) \times (1 + \mu_{C}) \tag{1}$$

Based on the assumption that the power for each unit remains the same regardless of the transport scheme, the model can be further applied to analog RF-over-fiber (RFoF), analog IF-over-fiber (IFoF), digitized RF (DRFoF) and IF-over-fiber (DIFoF) schemes where the detailed base station architectures are shown in Figure 7a–d. It is also noted that in this study, the same ADC/DAC specifications were used for both DRFoF and DIFoF. The power consumption of the ADC/DAC is predominately determined by the bit resolution used which is the same for both DRFoF and DIFoF in this study. Plotted inFigure 8 is the power consumption per base station as a function of base station transmitting power for different fiber-wireless transport schemes [40]. From the results, it can be seen that for single base station, baseband-over-fiber transport consumes the most power while digitized RF/IF-over-fiber transport exhibit an advantage in energy-saving especially at high transmitting power.

**Table 1.** Parameters used in base station model.

| Parameter | Estimated Value | Parameter | | Estimated Value |
|---|---|---|---|---|
| Power consumption of baseband unit ($P_{BBU}$) | 58 W | PA efficiency ($\mu_{PA}$) | BBoF, DIFoF, DRFoF | 25% |
| | | | IFoF, RFoF | 15% |
| Power consumption of DUC/DDC ($P_{DUC/DDC}$) | 3 W | Transmitting power ($P_{TX}$) | | 40 W |
| Power consumption of ADC/DAC ($P_{ADC/DAC}$) | 2 W | Feeder loss ($L_{feeder}$) | BBoF | 0.5 |
| | | | IFoF, RFoF, DIFoF, DRFoF | 1 |
| Power consumption of digital pre-distortion ($P_{DPD}$) | 5 W | Power supply loss ($L_{PS}$) | | 0.15 |
| Power consumption of RF unit ($P_{RFU}$) | 2 W | Cooling efficiency ($\mu_C$) | | 0.2 |
| Power consumption of clock management ($P_C$) | 1 W | Number of sectors ($N_{sector}$) | | 3 |

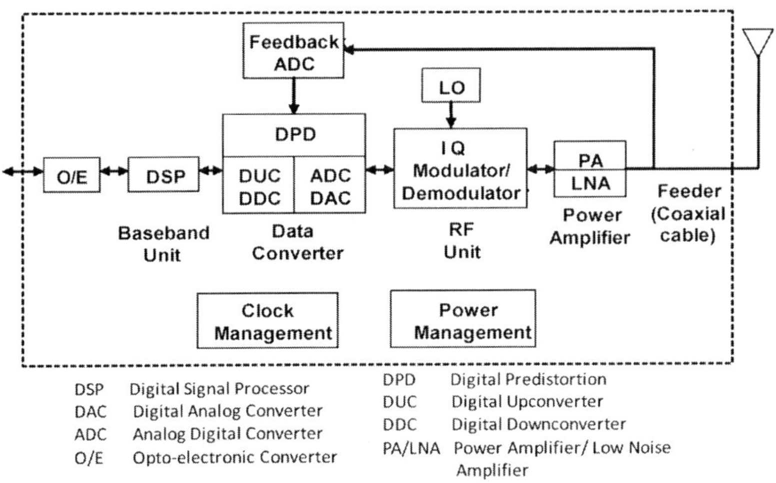


**Figure 6.** Conventional base station architecture incorporating BBoF transport scheme.

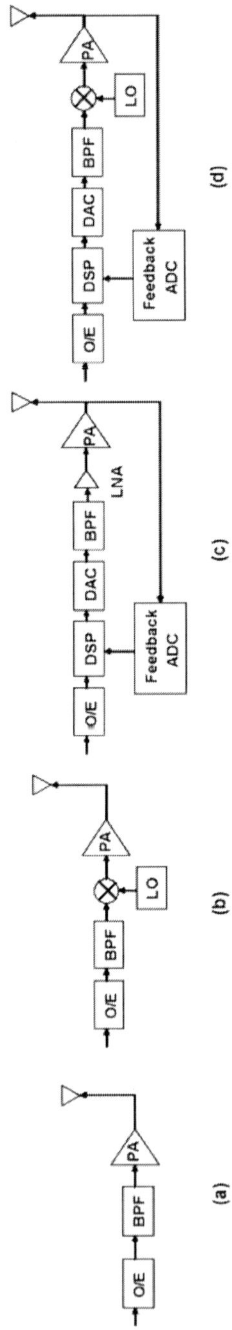


**Figure 7.** Base station architectures for (**a**) Analog RF-over-fiber (RFoF) (**b**) Analog IF-over-fiber (IFoF) (**c**) Digitized RF-over-fiber (DRFoF) and (**d**) Digitized IF-over-fiber (DIFoF).

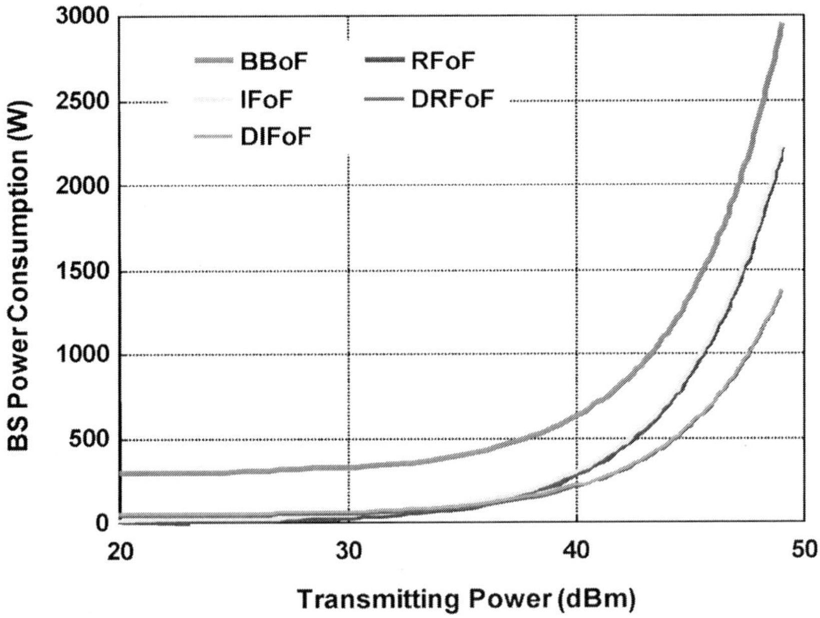


**Figure 8.** Base station power consumption as a function of transmitting power for the five different transport schemes investigated.

## 4.2. Energy Consumption per Unit Coverage

To establish a more realistic and practical evaluation on power consumption for fiber-wireless network, we have also studied the energy consumption per base station coverage [41]. Here we model the mobile wireless network as an infinite regular grid of sites characterized by the size distance, generating equally sized hexagonal cell structures of side length *d* as shown in Figure 9a. Each hexagonal cell represents the coverage of a base station which is located in the center of the cell at a distance of $3\sqrt{3}/2d^2$. To ensure that the entire cell is covered, the transmitting power ($P_{TX}$) has to be greater than the customer's receiver sensitivity ($P_{RX}$) after taking into account the propagation loss ($PL$) over a distance of *d*. The $PL$quantifies the wireless signal deterioration due to path loss, shadowing and multipath fading which can be formulated into Equation (2) [42]. The value is dependent on man-made structures [42,43] and hence, is different for different areas. Using the estimated parameters in [44] and Equation (2), we are able to estimate the propagation losses for rural and urban areas which are quantified by Equations (3) and (4). Figure 9b shows the maximum cell side lengths ($d$) at different transmitting power in both rural and urban areas for a typical receiver sensitivity of −120 dBm. It is obvious that the cell size is much larger for rural area compared to urban area for the same base station transmitting power. Since the propagation models are different for urban and rural areas, we will investigate the power consumption per square meter as a function of cell size for both rural and urban areas separately.

$$PL = PL(d_0, f) + 10n\log(d/d_0) \quad (2)$$

$$PL_{rural}(dB) = 49 + 43.5\log(d) \quad (3)$$

$$PL_{urban}(dB) = 77 + 48\log(d) \quad (4)$$

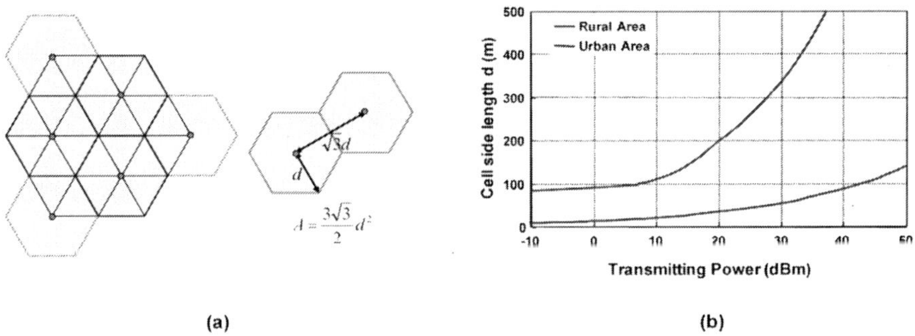


**Figure 9.** (**a**) Hexagonal cell coverage and the corresponding length measurement (**b**) Calculated cell side length *versus* transmitting power for rural and urban areas (With kind permission from Springer Science and Business Media).

## 4.2.1. Rural Area

Figure 10a,b show the power consumption per square km plotted as a function of base station transmission power (Figure 10a) and cell size (Figure 10b) for the five fiber-wireless transport schemes investigated [41]. The power consumption per square km for the five transport schemes investigated follows the same trend, where it first reduces and then increases with transmitting power (increasing cell size). There is an optimal transmitting power (cell size) associate with each transport scheme to achieve the lowest energy consumption. The optimal transmitting power for baseband-over-fiber transport scheme is ~40 dBm; for digitized RF/IF-over-fiber is ~33 dBm; and for analog RF/IF-over-fiber is ~22 dBm. Both results show that baseband-over-fiber transport scheme consumes the most power with regards to every square km. The power consumption per square km for the digitized and analog transport crosses at a transmitting power of ~34 dBm, which indicates that analog RF/IF-over-fiber transport scheme is more energy-efficient when the base station transmitting power is <34 dBm and digitized RF/IF-over-fiber is more efficient for transmitting power >34 dBm. This can be attributed to the additional power required for digital signal processing for digitized transport which is regardless of the base station transmitting power. For current macro base station used in rural areas, the transmitting power is at least 40 dBm [45] and this is indicated by the red line in Figure 10a,b. Hence for the cell size used in the rural area at present, the digitized RF/IF-over-fiber transport schemes are the most energy-efficient to service the rural area.

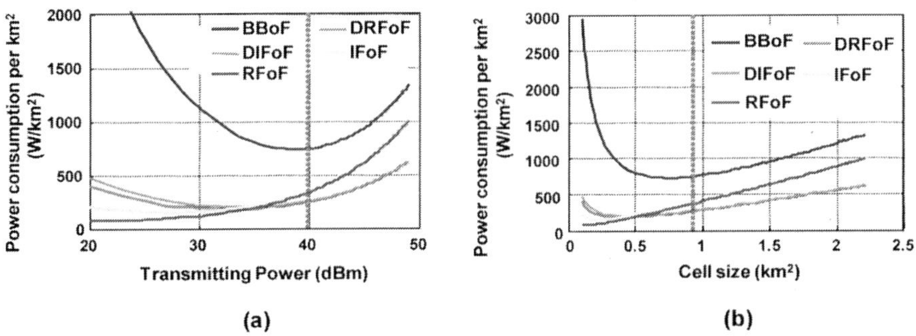


**Figure 10.** Power consumption per square km coverage for rural area for the five different transport schemes plotted as a function of (**a**) base station transmitting power and (**b**) cell size (With kind permission from Springer Science and Business Media).

### 4.2.2. Urban Area

Figure 11a,b show the power consumption per square km for the urban areas. Compared to Figure 10a,b, the results are significantly different even though they follow a similar trend. In the urban areas, the baseband-over-fiber scheme is still the most energy inefficient. Due to the high density of mobile users in the urban areas, micro base stations are commonly used with smaller cell size and lower transmitting power (<25 dBm) [45]. This falls on the left side of the red dashed line (transmitting power = 25 dBm) in Figure 11a,b. Therefore for the more dense urban areas where microcells are used, analog transport schemes are more energy-efficient compared to digitized transport; with analog RF-over-fiber being the most energy-efficient among the five schemes investigated.

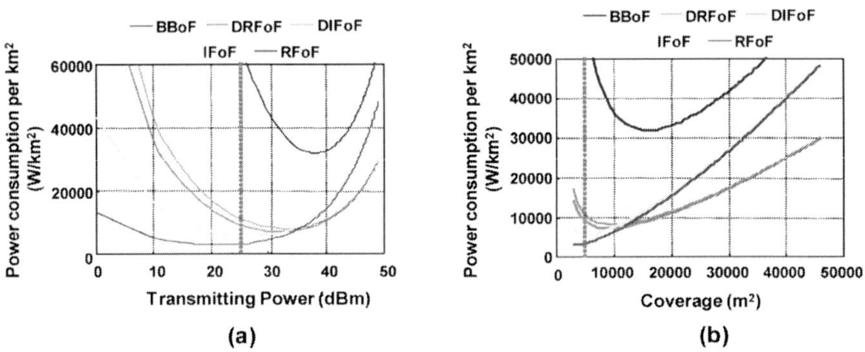


**Figure 11.** Power consumption per square km coverage for urban area for the five different transport schemes plotted as a function of (**a**) base station transmitting power and (**b**) cell size (With kind permission from Springer Science and Business Media).

# 5. CONCLUSIONS

This paper provides a comprehensive summary and review of some of the research that has been carried out by our research group in the area of fiber-wireless systems; in particular focusing on wireless signal transport schemes and power consumption. The main focus of these research activities has been directed towards the realization of high-performance and energy-efficient fiber-wireless links. The study shows that digitized transport scheme based on bandpass sampling has the potential to simplify the base station architecture while maintaining high link performance. We have demonstrated the transport of 1 GHz bandwidth OFDM signal at 6–10 GHz RF frequency via digitized transport. Error-free reception was achieved even at 4-bit ADC resolution and it outperformed the analog transport counterpart. We have also investigated the power consumption of the base station under different transport schemes for both rural and urban coverage. The study showed that although a digitized transport scheme resulted in better link performance, analog transport is more energy-efficient for urban areas using micro base stations with a transmitting power of <25 dBm. Digitized transport is more energy-efficient for rural environments with a transmitting power of >40 dBm.

## ACKNOWLEDGEMENTS

This research work was supported by the Australian Research Council Future Fellowship (FT0991110) and Discovery Project Grants (DP0987767).

## AUTHORS CONTRIBUTIONS

This review article was jointly written and proof-read by Christina Lim, Yizhuo Yang and Ampalavapillai Nirmalathas. The experimental results and modelling work were carried out by Yizhuo Yang who is the postdoctoral research staff employed with full supervision from Christina Lim and Ampalavanapillai Nirmalathas who are the chief investigators for this project. Christina Lim and Ampalavanapillai Nirmalathas are responsible for the progress of this project.

## CONFLICTS OF INTEREST

The authors declare no conflict of interest.

## REFERENCES

1. Capmany, J.; Novak, D. Microwave photonics combines two worlds. *Nat. Photonics* **2007**, *1*, 319–330.

2. Nirmalathas, A.; Lim, C.; Novak, D.; Waterhouse, R.B. Millimeter Waves in Communication Systems, Innovative Technology Series Information Systems and Networks. In *Millimeter Waves in*

*Communication Systems*, 1st ed.; Michel, N., Ed.; Hermes Penton Science Ltd.: London, UK, 2002; pp. 43–67.

3. Lim, C.; Nirmalathas, A.; Bakaul, M.; Gamage, P.; Lee, K.L.; Yang, Y.; Novak, D.; Waterhouse, R. Fiber-wireless networks and subsystem technologies. *IEEE J. Lightw. Technol.* **2010**, *28*, 390–405.

4. Kurniawan, T.; Nirmalathas, A.; Lim, C.; Novak, D.; Waterhouse, R. Performance analysis of optimized millimeter-wave fiber radio links. *IEEE Trans. Microw. Theory Tech.* **2006**, *54*, 921–928.

5. Ismail, T.; Liu, C.-P.; Mitchell, J.E.; Seeds, A.J. High-dynamic-range wireless-over-fiber link using feedforward linearization. *IEEE J. Lightw. Technol.* **2007**, *25*, 3274–3282.

6. Lee, S.H.; Kang, J.M.; Won, Y.Y.; Kwon, H.C.; Han, S.K. Linearization of RoF optical source by using light-injected gain modulation. *Proc. Microw. Photonics* **2005**.

7. Novak, D.; Clark, T.; O'Connor, S.; Oursler, D.; Waterhouse, R. High performance, compact RF photonic transmitter with feedforward linearization. In Proceedings of Military Communication Conference 2010 (Milcom2010), San Jose, CA, USA, 31 October–3 November 2010; pp. 880–884.

8. Shah, A.R.; Jalali, B. Adaptive equalisation for broadband predistortion linearisation of optical transmitters. *IEEE Proc. Optoelectron.* **2005**, *152*, 16–32.

9. Djupsjobacka, A. A linearization concept for integrated-optic modulators. *IEEE Photonics Technol. Lett.* **1992**, *4*, 869–872.

10. Skeie, H.; Johnson, R. Linearization of electro-optic modulators by a cascade coupling of phase modulating electrodes.*Proc. SPIE Int. Soc. Opt. Eng.* **1991**, *1583*, 153–164.

11. Roselli, L.; Borgioni, V.; Zepparelli, F.; Ambrosi, F.; Comez, M.; Faccin, P.; Cassini, A. Analog laser predistortion for multiservice radio-over-fiber systems. *IEEE J. Lightw. Technol.* **2003**, *21*, 1211–1223.

12. LaGasse, M.J.; Charczenko, W.; Hamilton, M.C.; Thaniyavarn, S. Optical carrier filtering for high dynamic range fibre optic links. *Electron. Lett.* **1994**, *30*, 2157–2158.

13. Esman, R.D.; Williams, K.J. Wideband efficiency improvement of fiber optic systems by carrier subtraction. *IEEE Photonics Technol. Lett.* **1995**, *7*, 218–220.

14. Attygalle, M.; Lim, C.; Pendock, G.J.; Nirmalathas, A.; Edvell, G. Transmission improvement in fiber wireless links using fiber Bragg gratings. *IEEE Photonics Technol. Lett.* **2005**, *17*, 190–192.

15. Lim, C.; Attygalle, M.; Nirmalathas, A.; Novak, D.; Waterhouse, R. Optical modulation depth analysis for improving transmission performance in fiber-radio links. *IEEE Trans. Microw. Theory Tech.* **2006**, *54*, 2181–2187.

16. Gamage, P.A.; Nirmalathas, A.; Lim, C.; Wong, E.; Novak, D.; Waterhouse, R. Experimental demonstration of digitized RF transport over optical fiber links. In Proceedings of the 2008 IEEE International Topical Meeting on Microwave Photonics jointly held with the 2008 Asia-Pacific Microwave Photonics Conference (MWP2008/APMP2008), Gold Coast, Queensland, Australia, 9 September–3 October 2008; pp. 15–18.

17. Nirmalathas, A.; Gamage, P.A.; Lim, C.; Novak, D.; Waterhouse, R.; Yang, Y. Digitized RF transmission over fiber.*IEEE Microw. Mag.* **2009**, *10*, 75–81.

18. Meslener, G.J. Chromatic dispersion induced distortion of modulated monochromatic light employing direct detection. *IEEE J. Quantum Electron.* **1984**, *20*, 1208–1216.

19. Schmuck, H. Comparison of optical millimeter-wave system concepts with regard to chromatic dispersion. *Electron. Lett.* **1995**, *31*, 1848–1849.

20. OBSAI System Specification v2.0. Available online: http://www.obsai.com (accessed on 17 December 2013).

21. CPRI Specification v6.0. Available online: http://www.cpri.info (accessed on 17 December 2013).

22. Vaughan, R.G.; Scott, N.L.; White, D.R. The theory of bandpass sampling. *IEEE Trans. Acoust. Speech Signal Process.***1991**, *39*, 1973–1984.

23. Lim, C.; Nirmalathas, A.; Novak, D.; Waterhouse, R.; Yoffe, G. Millimeter-wave broadband fiber-wireless system incorporating

baseband data transmission over fiber and remote LO delivery. *IEEE J. Lightw. Technol.* **2000**, *18*, 1355–1363.

24. Jiang, W.-J.; Yang, H.; Yang, Y.-M.; Lin, C.-T.; Ng'oma, A. 40 Gb/s RoF signal transmission with 10 m wireless distance at 60 GHz. In Proceedings of the Optical Fiber Communication Conference, Los Angeles, CA, USA, 4–8 March 2012.

25. Kanno, A.; Inagaki, K.; Morohashi, I.; Sakamoto, T.; Kuri, T.; Hosako, I.; Kawanishi, T.; Yoshida, Y.; Kitayama, K. 40 Gb/s W-band (75–110 GHz) 16-QAM radio-over-fiber signal generation and its wireless transmission. *Opt. Exp.* **2011**,*19*, B56–B63.

26. Zibar, D.; Sambaraju, R.; Caballero, A.; Herrera, J.; Westergren, U.; Walber, A.; Jensen, J.B.; Marti, J.; Monroy, I.T. High-capacity wireless signal generation and demodulation in 75- to 110-GHz band employing all-optical OFDM.*IEEE Photonics Technol. Lett.* **2011**, *23*, 810–812.

27. Zhang, J.; Yu, J.; Chi, N.; Dong, Z.; Li, X.; Chang, G.K. Multichannel 120 Gb/s data transmission over 2x2 MIMO fiber-wireless link at W-band. *IEEE Photonics Technol. Lett.* **2013**, *25*, 780–783.

28. Ng'oma, A.; Shih, P.-T.; George, J.; Annunziata, F.; Sauer, M.; Lin, C.-T.; Jiang, W.-J.; Chen, J.; Chi, S. 21 Gbps OFDM wireless signal transmission at 60 GHz using a simple IMDD radio-over-fiber system. In Proceedings of the Optical Fiber Communication Conference, San Diego, CA, USA, 21–25 March 2010.

29. Liu, C.; Jian, W.; Chien, H.-C.; Chowdhury, A.; Chang, G.-K. Experimental analyses and optimization of equalization techniques for 60-GHz OFDM radio-over-fiber system. In Proceedings of the Optical Fiber Communication Conference, San Diego, CA, USA, 21–25 March 2010.

30. Lombard, P.; Le Guennec, Y.; Maury, G.; Novakov, E.; Cabon, B. Optical distribution and upconversion of MB-OFDM in ultrawide-band-over-fiber systems. *IEEE OSA J. Lightw. Technol.* **2009**, *27*, 1072–1078.

31. Hraimel, B.; Zhang, X.; Liu, T.; Xu, T.; Nie, Q.; Shen, D. Performance enhancement of an OFDM ultra-wideband transmission-over-fiber link using a linearized mixed-polarization single-drive x-cut mach-zehnder modulator. *IEEE Trans. Microw. Theory Tech.* **2012**, *60*, 3328–3338.

32. Lebedev, A.; Pang, X.; Vegas Olmos, J.J.; Beltran, M.; Llorente, R.; Forchhammer, S.; Tafur Monroy, I. Feasibility study and experimental verification of simplified fiber-supported 60-GHz picocell mobile backhaul links. *IEEE Photonics J.***2013**.

33. Jazayerifar, M.; Cabon, B.; Salehi, J.A. Transmission of multi-band OFDM and impulse radio ultra-wideband signals over single mode fiber. *IEEE OSA J. Lightw. Technol.* **2008**, *26*, 2594–2603.

34. Yang, Y.; Lim, C.; Nirmalathas, A. Digitized RF-over-fiber technique as an efficient solution for wideband wireless OFDM delivery. In Proceedings of the Optical Fiber Communication Conference, Los Angeles, CA, USA, 4–8 March 2012.

35. Ericsson (2007) Sustainable Energy Use in Mobile Communications, White Paper. Available online: http:// www.telecomtv.com (accessed on 17 December 2013).

36. Pickavet, M.; Vereecken, W.; Demeyer, S.; Audenaert, P.; Vermeulen, B.; Develder, C.; Colle, D.; Dhoedt, B.; Demeester, P. Worldwide energy needs for ICT: The rise of power-aware networking. In Proceedings of the IEEE 2nd International Symposium on ANTS, Mumbai, Maharashtra, India, 15–17 December 2008.

37. Etoh, M.; Ohya, T.; Nakayama, Y. Energy consumption issues on mobile network systems. In Proceedings of the International Symposium Issues on Mobile Network Systems, Turku, Finland, 28 July–1 August 2008; pp. 365–368.

38. Deruyck, M.; Vereecken, W.; Tanghe, E.; Joseph, W.; Pickavet, M.; Martens, L.; Demeester, P. Power consumption in wireless access networks. In Proceedings of the 2010 European Wireless Conference, Lucca, Italy, 12–15 April 2010; pp. 924–931.

39. Chowdhury, P.; Tornatore, M.; Sarkar, S.; Mukherjee, B. Building a green wireless-optical broadband access network (WOBAN). *IEEE J. Lightw. Technol.* **2010**, *28*, 2219–2229.

40. Yang, Y.; Lim, C.; Nirmalathas, A. Comparison of energy consumption of integrated optical-wireless access networks. In Proceedings of the Optical Fiber Communication Conference, Los Angeles, CA, USA, 6–10 March 2011.

41. Yang, Y.; Lim, C.; Nirmalathas, A. Radio-over-fiber as the energy efficient backhaul option for mobile base stations. In Proceedings of the International Topical Meeting on Microwave Photonics and

Asia-Pacific Microwave Photonics (MWP/APMP2011), Singapore, 18–21 October 2011; pp. 242–245.

42. Goldsmith, A. *Wireless Communications*; Cambridge University Press: New York, NY, USA, 2005.

43. Richter, F.; Fehske, A.J.; Fettweis, G.P. Energy efficiency aspects of base station deployment strategies for cellular networks. In Proceedings of the Vehicular Technology Conference Fall (VTC 2009-Fall), Anchorage, AK, USA, 20–23 September 2009.

44. Lee, W.C.Y. *Mobile Cellular Telecommunications Systems*; McGraw-Hill Book Company: New York, NY, USA, 1990.

45. Fehske, A.J.; Richter, F.; Fettweis, G.P. Energy efficiency improvements through micro sites in cellular mobile radio networks. In Proceedings of the GlobeCom Workshops, Honolulu, HI, USA, 30 November–4 December 2009.

8

# PASSIVE OPTICAL ACCESS NETWORKS: STATE OF THE ART AND FUTURE EVOLUTION

**Tommaso Muciaccia, Fabio Gargano and Vittorio M. N. Passaro ***

Photonics Research Group, Department of Electrical and Information Engineering, Politecnico di Bari, via Edoardo Orabona n. 4, 70125 Bari, Italy

## ABSTRACT

In the very last years, optical access networks are growing very rapidly, from both the network operators and the research interests points of view. Fiber To The Home (FTTH) is already a reality in plenty of real contexts and there has been a further stimulus to the proposal of new solutions and the investigation of new possibilities, in order to optimize network performance and reduce capital and operational expenditure. A complete and systematic overview of passive optical access networks is presented in this paper, concerning both the hot research topics and the main operative issues about the design guidelines and the deployment of Passive Optical Networks (PON) architectures, nowadays the most commonly implemented approach to realize optical fiber links in the access networks. A comparison of advantages and disadvantages of different multiplexing techniques is discussed, with specific reference to WDM-based networks, almost universally considered as the enabling technology for future proof bandwidth requirements. An exhaustive summary is also given about the-state-of-the-art of modulation and encoding techniques recently proposed by the scientific community, as well as the open challenges (such as colorless and coolerless ONUs) for telecom companies and international standardization compliance.

## KEYWORDS

access networks; PON; FTTH; NG-PON2; NG-PON3; WDM-PON; colorless ONU


## 1. STANDARDIZATION EVOLUTION AND APPLICATION SCENARIOS OF PASSIVE OPTICAL ACCESS NETWORKS

Nowadays, the deployment of optical access networks (OAN) represents one of the most important technological challenges for telecommunications operators. Huge investments have already been planned for partial or total replacement of traditional copper-based access networks, which remains a precious asset for incumbent operators but, at the same time, has become a bottleneck for the provisioning of ultra-broadband services to residential and business customers.

As a consequence, national and international strategies to deploy OANs have been designed, too. In 2009, the United States have activated the ambitious program "Connecting America: The National Broadband Plan" to provide 100 Mb/s downstream and 50 Mb/s upstream connections to one hundred millions American households and 1 Gb/s connections to local communities by 2020. In 2010, the European Union has set as one of the main targets in the "Digital Agenda" that, by 2020, 100% of Europeans should have broadband access of above 30 Mb/s and 50% or more of European households should have access speeds above 100 Mb/s [1].

In the last few years, several approaches have been defined to deploy next generation access (NGA) networks. Various Fiber To The X (FTTX) solutions differ for the different fiber termination point in the optical distribution network (ODN), e.g., the local exchange in Fiber To The Cabinet (FTTC) solutions, or the boundary of living spaces in Fiber To The Home (FTTH) solutions [2].

Naturally, FTTH networks are by far the best performance ones, even if they are as well the most expensive ones to be rolled out. Although trenching constitutes the largest cost (~70%) to realize underground infrastructure [3], the remainder of costs (~30%) depends on chosen technological solutions: from this it follows the great importance of accurate techno-economic analysis to evaluate advantages and disadvantages of each implementation proposal, taking into account capital and operational expenditures. Detailed models have been proposed to analyze framework and scenarios, in order to describe the financial perspective and technical viability of a telecommunication investment project [4].

Moreover, besides technical options, other issues like regulatory and competitive aspects should be taken into account. Open access-based competition is generally required in modern scenarios: local loop unbundling is wholesaled by the network owner, typically through a regulated approach [5]. Infrastructure sharing can be developed at different levels, as conduit and collocation facilities, physical layer unbundling, through dark fiber leasing or optical layer unbundling, data link layer unbundling (bit stream services), e.g., through the implementation of virtual local area networks (VLAN), network layer unbundling, through policy-based routing to multiple internet service providers (ISP) [6].

Passive optical networks (PON) are actually considered the most cost-effective way to deploy FTTH networks. In fact, PONs are point-to-multipoint (P2MP) networks without any active equipment in the splitting remote node (RN). On the other hand, active optical networks (AON) are P2MP networks as well, but they require optical-electrical-optical conversion and Medium Access Control (MAC) switching in the RN. In literature, AONs are much less represented than PONs because of their limited use in real context and because they are less complex from the optical domain point of view, since they are based on single wavelength point-to-point links [7]. Another possibility to deploy FTTH OANs is the point-to-point P2P physical link from the central office (CO) to each user, but it is a more costly and less common solution, because no parts of physical infrastructure are shared. On the contrary, PONs and AONs reduce both roll-out per-user time and costs, especially in greenfield environments. Compared to AONs, PONs ensure higher reliability, simpler maintenance and reduced power consumption because of the absence of active elements in the ODN, so implying lower operational expenditure.

It is worth noting that reach-extended PONs (called long reach PON, LR-PON) have also been proposed recently. According to the preceding definition, these networks can still be considered "passive" since they use optical amplifiers and not reshaping, retiming and retransmitting (3R) regeneration, and therefore any conversion of the optical signal to the electrical domain is not needed. The presence of amplifiers in the ODN will require an additional power consumption and consequent operative cost, but it will enable higher power budget in order to extend the maximum available reach or the splitting ratio (and, therefore, the number of end users) of a single PON, allowing to consolidate the number of COs.

Since the greatest advantage of PONs is the infrastructure sharing, this means that signals for and from each user must be combined at the physical level through multiplexing techniques in downstream and multiple access techniques in upstream. Time division multiplexing (TDM) and time division multiple access (TDMA) are by far the most commonly adopted solutions assigning a time slot to each downstream and upstream channel. Up to now, TDM-PONs are the only PON model completely standardized by ITU-T, even if other possibilities have already been proposed, studied, optimized, tested and will be standardized in the very next years.

In the last fifteen years, the Full Services Access Network (FSAN) group has developed technical specifications that have been the basis for ITU-T standards. Broadband PON (BPON) has been the first complete standard defined by ITU-T in 2001 (reorganized in ITU-T G.983.1 ÷ G.983.5 in 2005) which takes origin on ATM PON (APON) and provides 155.52 ÷ 1244.16 Mb/s downstream bitrate and 155.52 ÷ 622.08 Mb/s upstream bitrate, with 20 km available reach and up to 64 end-users [8].

In 2004, IEEE has introduced Ethernet-PON (EPON), an alternative standard (IEEE 802.3ah later included in the overall standard IEEE 802.3) using Ethernet as transport protocol instead of less performing ATM protocol and providing for symmetric 1.25 Gb/s [9]. EPON has rapidly become the most common standard in Asia and, in 2009, has evolved in 10G-EPON (IEEE 802.3av) providing for 10 Gb/s downstream and 1 Gb/s upstream bitrate [10].

ITU-T response to EPON is the Gigabit-Capable PON (GPON). This standard (ITU-T G.984) has been introduced in 2004 and has encountered great success in Europe and America. It supports various bitrate options but it typically provides for 2488 Mb/s downstream bitrate and 1244 Mb/s upstream bitrate, with 30 km available reach and up to 128 end-users [11]. In 2010, XG-PON has been standardized (ITU-T G.987) as the GPON natural evolution, providing for 10 Gb/s downstream bitrate and 2.5–10 Gb/s upstream bitrate, with 20–60 km available reach and 64–128 end-users (longest available reach and highest number of end-users are conflicting requirements) [12].

PON international standards are summarized in Table 1.

**Table 1.** PON International Standards.

| Technology | Standard | Year | Downstream Bitrate [Gb/s] | Upstream Bitrate [Gb/s] |
|---|---|---|---|---|
| BPON | ITU-T G.983.1 ÷ G.983.5 | 2001 | ⩽1.25 | ⩽0.625 |
| EPON | IEEE 802.3ah | 2004 | 1.25 | 1.25 |
| 10G-EPON | IEEE 802.3av | 2009 | 10 | 1 |
| GPON | ITU-T G.984 | 2004 | 2.5 | 1.25 |
| XGPON | ITU-T G.987 | 2010 | ⩽10 | 10 |
| NG-PON2 | ITU-T G.989.1 (specifications) | 2013 | 40 | 10 |

In the roadmap defined by the FSAN group, describing the evolution of OANs in the next years, XG-PON has already been selected as the best candidate for NG-PON1 solutions, *i.e.*, the next generation PONs. The following step will be represented by NG-PON2 solutions whose system requirements have already been standardized by ITU-T G.989.1 in 2013. NG-PON2 must support at least 40 Gb/s aggregate downstream capacity and more than 10 Gb/s in the upstream direction, with 40 km maximum passive fiber reach and up to 60 km by eventually using reach extenders, with more than 64 end-users connected by these networks. Moreover, NG-PON2 solutions should ensure compatibility with legacy systems GPON and XG-PON and with existing ODNs, saving long-lived infrastructure investments [13].

According to FSAN evolution path, which can be seen in Figure 1, coexistence with legacy solutions is absolutely advisable in brownfield contests where cost-effective deployment could be actuated through gradual migration to new standards [14]. On the other hand, in greenfield contests (e.g., rural areas) compatibility with legacy ODNs is not mandatory and therefore completely innovative solutions, overcoming traditional time division multiplexing, could be adopted. These new approaches could be eventually defined in future NG-PON3 standards.

## 2. PON ARCHITECTURE AND DESIGN GUIDELINES

In Figure 2 an extremely simplified PON architecture is depicted, which could be referred to any PON solution. According to terminology adopted since ITU-T G.983 standard, a generic PON tree is constituted by an optical line termination (OLT) placed in the CO and providing network-side interface of the OAN which, connected through ODN to *N* optical network units (ONUs), provides the user-side interface. In the case of FTTH networks, each ONU is coincident with an optical network termination (ONT), because it includes the user port function. The ODN can include a certain number of splitting stages (typically one or two) placed in RNs (remote node). The components located in a RN have generally multiplexing and de-multiplexing functions, so they could be power splitters or more complex devices, but still rigorously "passive", with the meaning illustrated in the previous paragraph. RN could be the ideal place where an optical amplifier in LR-PONs can be installed. The RNs split the ODN in two or more fiber spans: feeder fiber is the one next to the CO; distribution fiber is the one exiting each port of the first splitting stage; drops are the optical link reaching each single users and is generally included in vertical cabling inside buildings.

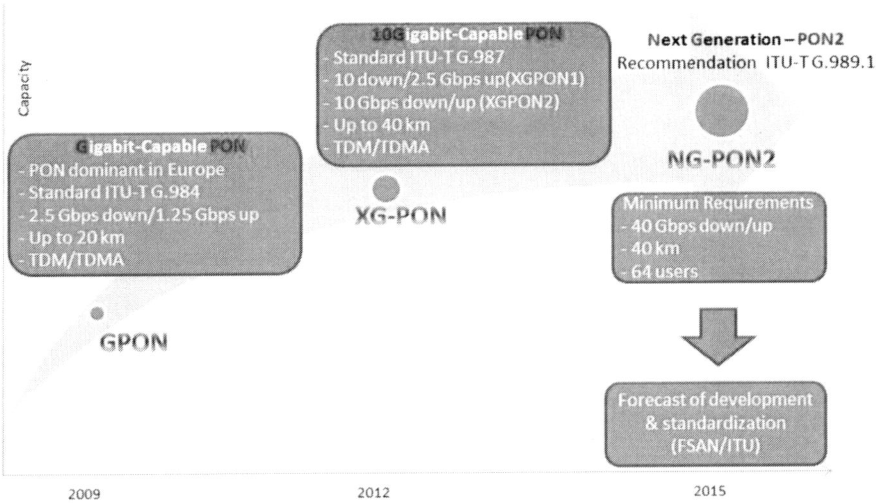


**Figure 1.** FSAN evolution roadmap.

Both the ONUs and the OLT are equipped with optical transceivers providing for generation, modulation, detection and demodulation of optical signals.

Laser diodes are commonly used as optical sources; they are generally preferred to LEDs because they allow the emission of high powers (typically up to 100 mW) and the possibility to be directly modulated at high frequencies (typically up to 25 GHz). In TDM-PONs, conventional Fabry-Pérot lasers are actually employed because they represent the cheapest solution. They are not suitable for WDM-PONs because of their multimodality due to great gain bandwidth. In these networks, distributed feedback (DFB) lasers could be preferred because of their good high-speed modulation properties due to their narrow bandwidth (less than a few megahertz). Unfortunately, they are still quite expensive and should require a thermoelectric cooler (TEC) module necessary to compensate the wavelength shift of about 0.1 nm/°C. VCSELs (vertical cavity surface emitting lasers) should be potentially more suitable for low-cost mass production, but this kind of sources emitting at 1550 nm are not still commercially available because their optical and thermal properties at this wavelength are not yet improved enough.

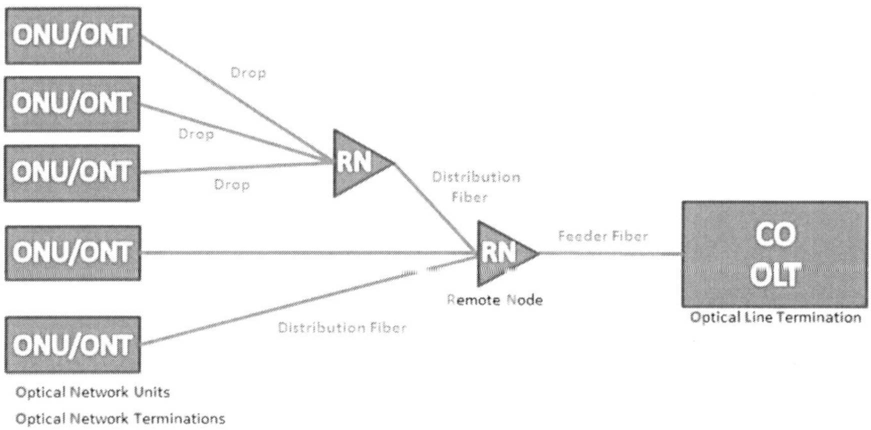


**Figure 2.** Schematic representation of a PON.

Although some semiconductor lasers could be modulated up to 40 Gb/s, systems operating at 10 Gb/s or higher bitrates should use external modulation. In fact, chirping effect would become no more negligible and would cause spectral broadening. This aspect, combined with fiber group velocity dispersion (GVD), should limit the available reach. Both$LiNbO_3$ Mach-Zehnder interferometers and electro-absorption modulators can be used in order to increase both bandwidth and extinction ratio and reduce the insertion losses. Their typical performances and basic properties are summarized inTable 2 [15].

Optical receivers in PONs can employ PIN photodiodes or APDs (avalanche photodiodes). The latter can achieve a sensitivity of two orders magnitude higher than the former. In order to improve the overall receiver sensitivity, the bandwidth of the trans-impedance amplifier is commonly designed to be 0.7 times the bitrate, which is a reasonable trade-off between noise suppression by filtering and inter-symbol interference due to limited bandwidth. Some of the most meaningful parameters of PIN and APD are compared in the following Table 3 [15,16].

**Table 2.** Modulators overview.

| | Mach-Zehnder Interferometer | Electro-Absorption Modulator |
|---|---|---|
| Extinction Ratio | 20 dB | 20 dB |
| Modulation bandwidth | 100 GHz | 50 GHz |
| Modulation voltage | <10 V | <2 V |
| Insertion loss | Low | Moderate |
| Size | Large | Small |
| Monolithic integration | No | Yes |

**Table 3.** Photodetectors overview.

| | PIN | APD |
|---|---|---|
| Responsivity | 1 A/W | 100 A/W |
| Bandwidth | 10 GHz | 40 GHz |
| Overall sensitivity | −22 dBm | −30 dBm |

In the following Table 4, the most frequently employed optical fiber are compared about core diameter, cut-off wavelength (the lower limit for single mode operation), attenuation ($\alpha_{fiber}$) and dispersion ($D$) coefficients.

In TDM-PONs power splitters are placed in the RN, where a 1:2 splitter can be made with Y-branch or MMI (Multimode Interferometer) planar lightwave circuits or it could be realized by fusing two pieces of fiber with tapered core diameter to achieve weakly field confinement. A generic 1:N splitter can be obtained by cascading multiple 1:2 splitters. Power splitter/combiners strongly limits the power budget because of the high insertion losses $IL_{splitter} = \log_2 N$ [dB].

In WDM-based PONs the RN is equipped with array waveguide gratings (AWG) providing for wavelength routing in the optical domain. Unlike power splitters, AWGs have fixed insertion loss (typically, from 2.5 to 6 dB) regardless of the number of channels. Moreover, their spectrum periodicity allows their use in different optical wavelength ranges, e.g., the upstream and the downstream band, making it possible to be simultaneously used both as multiplexer and as demultiplexer. On the other hand, if full duplexing is realized at the same wavelength, the same AWG port cannot be assigned for both upstream and downstream, so a 2:N AWG must be used at the RN. An important issue for the use of AWG in PONs is a-thermal operation to ensure real-time wavelength consistency. It could be obtained by the usage of compensating negative thermo-optic coefficient or by mechanically movable plates.

**Table 4.** Optical fibers overview.

| Denomination | ITU-T Standard | Core Diameter [μm] | Cut-Off Wavelength | $\alpha_{fiber}$ | $D$ |
|---|---|---|---|---|---|
| Graded-index multimode | G.651 | 50 ± 3 | - | 2 ÷ 4 dB/km | - |
| Standard single mode | G.652 | 8.6 ÷ 9.5@1300 nm | 1270 nm | 0.5 ÷ 1 dB/km | 3.5 ÷ 20 ps/(nm·km) |
| Dispersion-shifted single mode | G.653 | 7 ÷ 8.3@1550 nm | 1270 nm | <0.5 dB/km | <3.5 ps/(nm·km) |
| Non-Zero-Dispersion single mode | G.655 | 9 ÷ 11@1550 nm | 1480 nm | ≤0.35 dB/km | ≤6 ps/(nm·km) |

The RN is also the ideal place to locate an eventual optical amplifier. In LR-PONs, in-line amplifier configuration is preferable to booster solution basically for two reasons: a booster amplifier would require very high saturation power not easily achievable and the optical signal reaches the amplifier input port when it has already been attenuated through the feeder fiber, avoiding the non-linear effects induced by high powers.

Optical amplifiers suitable for LR-PONs are erbium-doped fiber amplifiers (EDFAs), semiconductor optical amplifiers (SOAs) and Raman amplifiers (RAs). EDFAs can only operate near 1.55 μm. In order to amplify signals in the fiber second window, particular fluoride fibers doped with praseodymium (ZBLAN fibers) could be theoretically used, but their cost is still a strong limitation. RAs seem to represent a better solution because of the great flexibility in choosing amplification wavelength range, even if large pump power is required to obtain large gain. Also travelling-wave SOAs offer high flexibility and can provide large gain bandwidth, although their technology has not yet reached sufficient maturity. Hybrid configurations have also been proposed, e.g.,

SOA-Raman hybrid amplifier (SRHA) has been presented to obtain larger bandwidth by exploiting complementary spectra of both amplifiers.

The use of optical amplifiers could allow relaxed conditions about power budget, which is a very important design parameter, since most PONs are power-limited systems. Given a specified amplifier gain $G_A$, overall channel losses $C_L$, and photo-receiver sensitivity $S_{Rx}$ needed to ensure a predefined BER, the required power budget is defined as follows:

$$P_{budget} = P_{Tx} - S_{Rx} > C_L - G_A + M_S \quad (1)$$

where $M_S$ is the system margin allocated to maintain performance during the entire system lifetime. $C_L$ must take into account fiber, splitter, slice and connector losses and insertion losses of all the devices placed in the ODN.

Group velocity dispersion (GVD) can be evaluated through the dispersion length $L_D$, *i.e.* the distance over which a Gaussian pulse broadens by a factor of $\sqrt{2}$ due to GVD and in absence of any other effect:

$$L_D = \frac{2\pi c}{4\lambda^2 {f_{sym}}^2 |D|} \quad 2)$$

where a net channel bandwidth $B = 2f_{sym}$ is considered, being $f_{sym}$ the symbol rate. Equation (2) clearly shows how high symbol rates increase sensitivity to dispersion. Links operating at higher line rates need dispersion compensation devices to compensate the chromatic dispersion of the optical path. Several types of dispersion accommodation technologies are used, including passive dispersion compensation, self-phase modulation, prechirp and dispersion supported transmission [17]. Adaptive chromatic dispersion compensators are described in the recommendation ITU-T G.667 [18].

Polarization mode dispersion (PMD) can be evaluated through the PMD length, *i.e.*, the distance over which the polarization mode dispersion effects are no more negligible:

$$L_{PMD} = \frac{1}{4PMD^2 \cdot {f_{sym}}^2} \quad (3)$$

Long links operating at high symbol rate can require the use of polarization mode dispersion compensators (PMDCs) whose features are described in the recommendation ITU-T G.666 [19]. Both line PMDCs, in the middle of the transmission link, and PMDC receivers, with compensating functionalities integrated in the terminal, can be used [17].

Chirping-inducing self-phase modulation (SPM) can be neglected if the following condition is satisfied:

$$P_{Tx} < \frac{\alpha_{fiber}}{\gamma_{NL}} \quad (4)$$

where $\alpha_{fiber}$ is the fiber attenuation coefficient and the non-linearity parameter is defined as:

$$\gamma_{NL} = \frac{2\pi}{\lambda} \frac{n_2}{A_{eff}} \quad (5)$$

being $A_{eff}$ the optical fiber modal area, $n_2$ the second order nonlinear core fiber refractive index and $\lambda$ the operative wavelength. Cross-phase modulation (XPM) can generally be considered negligible when the channel spacing is quite large (such as in CWDM wavelength grid), while four-wave mixing (FWM) can be ignored if the dispersion parameter is nonzero. Non-linear effects can be mitigated by using large effective area fibers (LEAFs), for example holey fibers. A micro-structured fiber, reported in 2004, shows an effective area of 130 $\mu m^2$, despite a quite large dispersion, 27 ps/(nm·km) and a loss of 0.48 dB/km [20].

# 3. OPERATIVE ISSUES

In this paragraph some operative issues will be briefly described: power consumption, security, monitoring and resiliency. Operative problems related to power consumption are strictly dependent on the exponential growth of services and user bandwidth requirements. Nowadays, about 10% of global $CO_2$ emission is due to ICT systems and activities, but this value is expected to increase in the next few years. The contribution of access networks to overall consumption of wireline telecommunication systems is about 70%, so it is much higher than in backbone and metropolitan networks, due mostly to the high number of user units [21]. For this reason, stand-by modality and dynamic techniques have been proposed and implemented to optimize the use of energy. Of course, OLT in the CO also gives an important contribution to power consumptions. In order to give a benchmark, the maximum power consumption of an OLT supplying 256 users in a commercial TDM-PON has been evaluated to be 200 W, implying that the power consumption per user is about 0.8 W [22].

Nowadays, costs associated to energetic provisioning represent a high component of operative expenditure and the respect of recent strict standards is another contribution to overall outflow. In 2007, European Council has set a new target by 2020 to reduce gas carbon emission by 20% compared to emission levels in 1990 [23]. Moreover, gas carbon emission associated with OANs is related not only to network systems operations but also to their fabrication, transport, distribution and removal. Although the contribution of systems operation is strongly predominant, deployment-related energy consumption should not be ignored especially in FTTH PONs where a single fiber drop per user is needed. In a 15-years lifetime, neglecting all ONU contributions, 80% of overall energy consumption is related to production, transport and distribution of network assets, whose 83% is represented by passive ones [24].

Another diriment operative issue is related to the security in OANs. In most PON architectures, broadcast nature of downstream traffic makes information potentially accessible to all users on the PON tree. Actually, two typologies of security methods are adopted, the former is based on an authentication request sent from the OLT to the ONU through a password inside the data stream, the latter consists of encryption of downstream traffic.

TDM-PONs employ Advanced Encryption Standard (AES), therefore only the frame payload is encrypted and keys distribution is repeated several times during the connection. In spite of security countermeasures, networks are still vulnerable to some attack typologies, such as the presence of a false OLT, man-in-the-middle and denial-of-service attack [25]. Researchers will have to work on these challenges over the next few years.

Network monitoring and troubleshooting is a further important issue during the lifetime of the systems, from the installation testing to maintenance operations. Optical time domain reflectometry (OTDR) is a pulse-echo technique, based on Rayleigh back-scattering, used for decades to measure optical length and attenuation coefficient, localize and characterize splices, connectors, splitters, bend losses, faults, breakdowns and any kind of discontinuity along the optical link.

The presence of splitters in PONs makes critical the use of conventional OTDR techniques, because it determines a strong localized attenuation of back-scattering signal to be detected and monitored. In addition to this, discontinuities existing on single PON drops can produce reflective and non-reflective events overlapped with the ones associated to each other PON drop, and they can even overlap to ghost events, due to multiple reflections.

In order to overcome the P2MP problem, bidirectional measures can be performed or partial PON analysis can be carried on [26], making individual measure for each single branch [27].

Anyway, these open issues can be partially solved, in a more efficient way, by several expedients and innovative techniques such as the use of reference reflectors, innovative elaboration algorithms [28], multi-wavelength approach, polarization modulator [27], Brillouin frequency shift assignment (BFSA) to each drop, optical coding (OC) assignment to each ONU/ONT through passive optical out-of-band encoder. BFSA and OC techniques are actually important research topics [29]. Moreover, tunable OTDR [27] and similar techniques based on comparison between a reference amplified spontaneous emission signal and its reflection from an optical reflector have been proposed for WDM-PONs [30].

Network monitoring is strictly related to reliability and resiliency issues. Typical failures in PONs could be due to fiber link interruptions (e.g., cable cuts), performance degradation of splitter and AWGs or malfunctions of OLT and ONUs transceivers associated with aging or environmental factors.

The intensive deployment of FTTH PONs will lead to an increased concentration of customers, which will have an impact on network reliability and service availability. Moreover, if LR-PONs will increase the service area and allow the consolidation of the number of COs, on the other hand it will increase the number of customer potentially affected by a network failure [31].

Network downtime can be evaluated by multiplying mean time to repair (MTTR) and failures in time (FIT). Typically a network downtime is of the order of several tens of minutes per year [15].

Although the occurrence of failures is generally out of control of the physical network owner, their effects on the delivered services are dependent on the efficacy and efficiency of monitoring techniques and on the network topologies adopted by the provider. Protection switching is the solution to provide high reliability. It is based on the use of redundancies in network topologies to allow traffic rerouting around the deteriorated element. Automatic protection switching (APS) techniques and protocols can be introduced, in place of forced protection switching, in order to enable automatic restoration and reduce MTTR abruptly. Self-healing OANs (SHOANs) include automatic detection of failures (such as loss of signal, loss of frame, signal degrade) and APS triggered by the fault detection.

Optical switches could be adopted to enable commutation from the working path to the protection path in case of failure: either unidirectional or bidirectional schemes can be adopted. Although PONs are P2MP networks, their single spans (e.g., feeder fiber or single drops) can be considered P2P links where essentially two basic redundancy principles can be adopted: 1 + 1 and 1:1 protection schemes. In a 1 + 1 scheme, when the fault is detected, the traffic is switched from the working fiber to the protection fiber and working fiber becomes active again when the fault is cleared. In a 1:1 scheme, on the other hand, even when the fault is cleared, traffic is still directed through the protection fiber, which actually becomes the working link.

More complex N:M protection techniques have been recently proposed for PONs to enable flexible design of cost and reliability and the elimination of substantially unused backup equipment. These new solutions employ 2D and 3D MEMS switch [32]. In 2011, a SHOAN has been proposed in which two or more optical access architectures are partners of each other, and they are interconnected by elementary optical crossbar switches [33].

If fibers have to be duplicated, the ring topology have to be preferred to the most common tree topology because it allows a diversification of optical and/or physical path.

Current standards relate to unprotected tree topology, but sometimes they suggest several protection switching solutions. For example, BPON and GPON standards specify four reference models for resiliency and redundancy: type A, where only the optical fibers are duplicated; type B, where the OLT and the feeder fiber are duplicated (splitter at the RN has two ports on the OLT side); type C, where also the drops are duplicated in order to enable recovery from failures also on the ONU side; type D, where a combination of type B and C schemes, allowing the mixing of duplicated and non-duplicated ONUs. GPON should support 50 ms protection times [11].

IEEE standards does not suggest any protection model, but several protection schemes have been proposed in literature in the last years, also avoiding redundancy of equipment [34].

# 4. MULTIPLEXING TECHNIQUES

TDM-PONs and WDM-PONs have already been mentioned in this paper. In particular, TDM-PONs standards have been overviewed in the first paragraph. TDM and WDM are the most discussed examples of multiplexing techniques used in PONs. Then, the object of this paragraph is just the comparison of these and other multiplexing techniques which have a strong impact on the definition of the general architecture presented in Figure 2 and, thereby, on its development costs. Main advantages and disadvantages of each technique are summarized in Table 5.

In TDM-PONs a time slot is assigned to each user to transmit and receive data on a shared optical carrier, typically allocated in the third optical fiber window for downstream and in the second optical fiber window for upstream. Complex algorithms of ranging and dynamic bandwidth assignment are generally used to regulate TDMA and avoid packet collision [35]. TDM-PON solutions can count on mature technology and cheap components. Therefore, they are the simplest and fastest way to deploy PONs according to a short-term technical and economic analysis.

The main drawback for TDM-PONs is the hard-limited scalability because of the compelling trade-off between number of users and per-user bandwidth. An increase in the user bit rate is quite difficult to be achieved because each ONU receiver works at the system bit rate, which is much higher than the effective user bit rate. Even by using high speed digital signal processors (DSP) or field-programmable gate arrays (FPGA), a further increase in the bit rate to more than 10 Gbit/s (as provided by XG-PON) will inevitably collide with electronics limitations.

Up to now, bandwidth sharing is not perceived by users as a bottleneck but, in the next few years, new services will challenge the statistical multiplexing. Conventional burst-experience services, such as browsing, mailing and file-sharing, will step down in favor of video and multimedia services, generating a deeply different traffic typology based on continuous stream. Another scaling possibility is reducing the number of users on a single PON tree, thereby lowering the splitting ratio, but it would require more fiber to be rolled-out and so it would not be properly a cost-effective solution. Another drawback for TDM-PONs is represented by the power dividing splitters that decrease the optical power in ODN, so constituting energy inefficiency.

**Table 5.** Multiplexing techniques for PONs.

| Technique | Main Advantages | Main Drawbacks |
|---|---|---|
| TDM | Mature technology<br>Cheap components | Limited scalability<br>Energy inefficiency |
| WDM | Minimum bandwidth guarantee<br>Network security<br>Transparency to modulation formats and bitrate<br>Simplified unbundling<br>Scalability | Power consumption in CO<br>Floor space availability in CO<br>Photonics technology not yet consolidated |
| OCDM | Soft-limited technique<br>Time synchronization not required<br>Intrinsic data confidentiality | Multi-access interference<br>Multiplexing techniques not yet consolidated |
| OFDM | High spectral efficiency<br>Scalability | Complex receivers<br>Multiplexing techniques not yet consolidated |
| PDM | Information capacity increasing | Complex receivers |
| SDM | Information capacity increasing | Expensive deployment<br>Standardization needed |
| TWDM | Gradual migration from commercial solutions | Limited scalability<br>Energy inefficiency |

On the other hand, the introduction of wavelength division multiplexing (WDM) in PONs is not only driven by increasing the bandwidth requirements but it would lead to considerable improvements for future generation PONs. A WDM-PON provides P2P logical link to each user by assigning one or two operative wavelengths to each user. This network typology allows symmetrical services with reserved bandwidth, minimum bandwidth guarantee, simplified techniques to provide network security and transparency to modulation formats and bitrate [36].

In modern competitive scenarios, where passive infrastructures can be shared among several operators, WDM would provide simplified unbundling for regulatory compliance to ensure equivalence of input (EOI). Since optical interfaces can be kept physically separated, a channel set can be assigned to each operator to realize its own optical links.

WDM technology has been deployed for decades by telecom operators for backbone and transport networks: coarse WDM (CWDM) and dense

WDM (DWDM) and their wavelength plan are respectively defined by ITU-T G.694.2 [37] and ITU-T G.694.1 [38]. Further channel bandwidth reduction can be performed to increase the number of users on the PON tree. Ultra-dense WDM (UDWDM) solutions have already been proposed and ensure network scalability for long-term evolution. WDM-PONs could favor the all-optical merger of OANs with metro networks by using ring architectures and reconfigurable optical add-drop multiplexers (ROADMs) for wavelength routing, such an architecture allowing high scalability for the entire system and active traffic adaptation [39].

In addition to this, WDM-PONs can provide flexible and various services. A new architecture has been recently proposed to support wireless access traffic, optical virtual passive network (VPN) communications and conventional wired services [40].

Of course, WDM-PONs will require more complex devices compared to conventional TDM-PONs, therefore technological maturity of high-performance photonic devices will be required. To this aim, the evolution of silicon photonics will be crucial to obtain high reliability and low cost for large-scale application. The essential components required will be a-thermal AWGs, WDM transmitter arrays at the OLT, narrow-band optical filters and colorless and coolerless ONUs.

One of the main drawbacks for WDM-PONs is the need for a higher number of transceivers at the OLT compared to TDM-PONs. This implies larger floor space availability in the CO and higher power consumption in the CO. Anyway, in order to evaluate a fair comparison between WDM- and TDM-PONs, it is worth noting that WDM-PONs would be by far more energetically efficient if equal user bitrate is considered, because ODN losses are reduced by the replacement of power splitters with AWGs in the RNs.

PONs based on optical code division multiplexing (OCDM) represent a possible alternative to WDM-PONs for future-proof band consuming services. OCDM is a soft-limited multiplexing technique. Once given the overall optical bandwidth and user bitrate, the number of users is not strictly imposed but it depends on the orthogonality of codes assigned to each channel. M-sequence codes, Kasami codes, Gold codes and Hadamard codes have been proposed in literature, because they exhibit good auto-correlation and cross-correlation properties. Since different channels can only be quasi-orthogonal, multi-access interference (MAI) must be taken into account, through signal-to-noise ratio (SNR) degradation:

$$SNR = \frac{S}{(K-1)S + \sigma^2} = \frac{1}{(K-1) + \frac{\sigma^2}{S}} \tag{6}$$

where $K$ is the number of contemporary users transmitting on the same OCDM-PON tree and $S$ and $\sigma^2$ are respectively their signal and noise average power [41].

Different implementations of OCDM-PONs have been proposed, mostly gatherable into two main categories: spectrum-sliced and time-sliced techniques. Spectrum-sliced solutions are based on the assignment of a spectral signature to each data-flow, where the signature is obtained through a combination of spectral slices originated by filtering a broadband light source (BLS) through narrow-band optical filters (e.g., fiber Bragg gratings). For this reason, these solutions are also denominated in literature as spectral amplitude coding OCDM (SAC-OCDM). A SAC-OCDM PON has been proposed in 2012, employing dynamic cyclic shift (DCS) codes and AND subtraction technique to suppress the effect of MAI [42]. However, spectrum-sliced solutions have some drawbacks, as the use of BLSs limits the modulation speed and, since optical bandwidth is quite large, dispersion issues become relevant. In time-sliced techniques a signature sequence is impressed on signals in the time domain. This could happen in the electrical or in the optical domain. In the first case, the multiplexing technique is often referred as CDM over fiber (COF), but in this paper it has been classified under OCDM since it is based on the same working principle, although by applying a different implementation. As the second possibility, super-structured fiber Bragg gratings (SS-FBG) have been proposed. A SS-FBG (see Figure 3) is a standard FBG with rapidly varying spatial refractive index of uniform amplitude and pitch on which a slow spatial refractive index modulation is superimposed [41]. By connecting a pulsed optical source, an optical circulator and a SS-FBG time-sliced phase encoder/decoder can be implemented, and a convolution between the input impulse and the SS-FBG impulse response (*i.e.*, the signature) can be obtained.

The greatest advantage obtained by such a configuration is that neither time synchronization nor precise optical timing is required. Moreover, intrinsic data confidentiality is achieved, since each user multiplies and de-multiplies the data-stream with a different code.

Coexistence of TDM and OCDM in the electrical domain on the same ODN has been demonstrated in 2009 [48]. This solution is particularly interesting because it allows coexistence with GPON legacy systems and combination of best-effort and guaranteed-bandwidth services [43].

PONs based on orthogonal frequency division multiplexing (OFDM) are up to now theoretically the most performing networks, since they can provide up to 100 Gb/s per user and easy scalability. In OFDM-PONs orthogonal sub-bands are modulated through advanced modulation techniques. High channel spectral efficiency can be obtained through sub-band overlapping. On the other hand, OFDM-PONs would require very complex receivers based on high-speed DSPs and FPGAs.

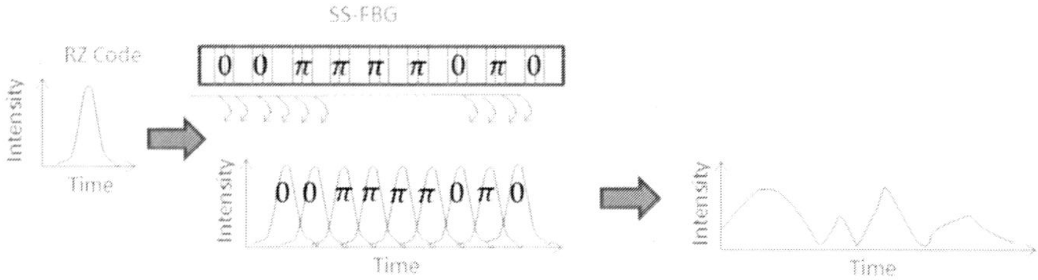


**Figure 3.** Working principle of an SS-FBG.

PONs based on polarization division multiplexing (PDM) and space division multiplexing (SDM) represent more unconventional solutions, actually far from commercial development. PDM-PONs are based on the exploitation of both orthogonal polarizations at the same wavelength. Although it would increase the information capacity of the fiber, since polarization state rotates along fiber propagation, it would be quite difficult to distinguish the two contributions at the receivers.

SDM-PONs can be implemented in several ways, *i.e.*, through multiple bonded fibers (but it would be trivial and expensive), through multi-core fibers (MCF) or mode division multiplexing (MDM) in multimodal fibers [44]. PONs using MCFs and MDMs have not yet been studied in deep by university and industry researchers, probably because they are not seen as an attractive opportunity by operators. Moreover, they would need standardization to be implemented in a competitive context.

Beyond the aforementioned multiplexing techniques, it is worth noting that hybrid configurations exploiting the advantages of different technologies have not only been proposed in literature, but they probably represent the best way to upgrade the actual network capacity.

TWDM-PON architecture is a hybrid TDM-WDM solution, which has received great interest by the scientific community and international organizations for standardization. In ITU-T G.989.1, TWDM-PON has already been recommended as one of the possible technological solutions especially, for gradual migration scenarios to allow the reuse of existing ODNs [11]. In a TWDM-PON several TDM systems (e.g., GPON or XGPON) are stacked through CWDM or DWDM, then scheduling algorithms will be needed to reconfigure wavelengths and time slots dynamically, which will be the evolution of actual DBAs [44].

A hybrid DWDM-TDM LR-PON has been presented in 2011 with record performance in terms of reach (135.1 km), number of users (8192 ONUs) and aggregate capacity (symmetric 320 Gb/s), using in the architecture an EDFA to provide the gain to overcome the large fiber and splitting losses [45].

TWDM-PON certainly represents the more cost-effective architecture for very next years, but they cannot give a long-term solution for the same reason explained about TDM-PON disadvantages. They will be probably replaced by pure WDM-PONs, when WDM technology will become more mature and cheap.

WDM-OCDM architectures have been also proposed, but they still are in an embryonic stage, with a transmission capacity of 1.24 THz reached in 2010, and spectral efficiency of 0.41 bit/s/Hz [46].

In 2012, 1.2 Tb/s symmetric WDM-OFDM-PON has been demonstrated. Over 90 km have been reached with 1:32 passive split and 10/1.25 Gb/s guaranteed peak rate in downstream and upstream direction, respectively [47]. Also energy-efficient WDM-OFDM-PONs have been experimentally demonstrated in 2012. The architecture proposed uses an opto-mechanic switch at the OLT to share the OFDM module among several users as a solution for a context with low traffic demand (e.g., non-peak hours of the day) [48].

Other configurations will probably be proposed in the next years for new generation PONs. Certainly new proposals will be evaluated on the base of power consumption, co-existence with legacy systems, spectral flexibility and resilience of the network. In Figure 4 the probable evolutionary scenario for PONs is sketched. In the next years several TDM systems, such as GPON and XG-PON, will be stacked through CWDM multiplexing, therefore TWDM PONs will replace conventional TDM-PONs especially in denser metropolitan areas. OCDM- and OFDM-PONs can be expected to be developed in specific contexts where standard compliance is not requested, e.g., for trial deployments or scientific laboratories.

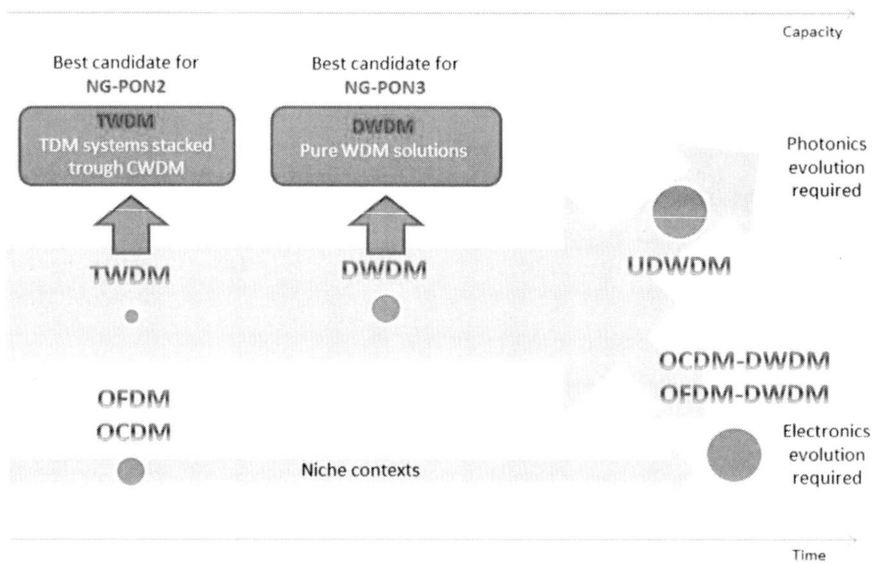


**Figure 4.** PON evolutionary scenario.

DWDM can be expected to be the main long-term solution for multiplexing in future PONs and, therefore, DWDM-PONs can be considered the strongest candidates for future NG-PON3 standards.

In order to further increase the number of users on a single PON tree, two future trends can be predicted: a progressive channel bandwidth shrinking in favor of UDWDM solutions or the adoption of innovative multiplexing techniques (e.g., OCDM and OFDM) over DWDM wavelength grid. The first possibility will be mainly bound by the technological improvements of photonic devices (e.g., sources, filters, AWG, *etc.*), the second possibility will essentially depend on electronics evolution in order to develop low-cost and high-speed DSPs and FPGAs.

# 5. MODULATION AND ENCODING TECHNIQUES

Once defined the multiplexing techniques to be employed in the PON, appropriate modulation and encoding solutions must be chosen for the various multiplied channels. In the last few years, increasing research efforts have been involved in finding modulation with higher performances while trying to keep simple the required components and overall architecture.

The technique conventionally adopted is the intensity modulation with direct detection (IM-DD), which is based on on-off keying (OOK). Up to now, it is used in every commercial TDM-PON standard. In fact, this solution is by far the simpler and more cost effective but, on the other hand, it has low spectral efficiency and scalability. Then, transceivers become expensive if high bitrate is required and physical damages appear. If direct modulation schemes are employed, chirping effect, due to temporary carrier density change in the laser active region, produces undesired phase changes and consequent frequency modulation and GVD, if Equation (4) is no longer satisfied.

The emerging of DWDM- and UDWDM-PONs will require an increasing spectral efficiency due to channel-limited bandwidth. Therefore, advanced multilevel modulations have been proposed to increase the bitrate while keeping low the signal bandwidth. If an alphabet having M symbols is used, the relationship between symbol rate $f_{sym}$ and bit rate $f_b$ is ruled by the well-known equation:

$$f_{sym} = \frac{f_b\ [bit/s]}{log_2 M}\ [sym/s] \quad (7)$$

Innovative modulations will require an improvement in optical equipment to be employed in future PONs. Especially high linearity and dynamic range will be needed for optical receivers.

M-ary amplitude shift keying (M-ASK), M-ary phase shift keying (M-PSK) and M-ary amplitude phase shift keying (M-APSK) have been proposed. Some constellation diagrams are depicted in Figure 5. In particular M-APSK can obtain an optimal distribution of symbols in the complex plane in terms of transmitted power minimizing. Rectangular constellation diagrams (*i.e.*, quadrature amplitude modulation, QAM) or concentric ring constellation diagrams can be implemented. The k-th symbol in M-APSK constellation is described as follows:

$$s_k = A_k e^{j\vartheta_k} \quad (8)$$

If $A_k$ is constant an M-PSK constellation diagram is obtained with constant envelop and theoretical insensitivity to attenuation, if $\vartheta_k$ is invariant an M-ASK (also called pulse amplitude modulation, PAM) constellation diagram is obtained which would need a simple envelope detector. M-PSK and M-APSK can require two typologies of optical receivers, *i.e.*, coherent receivers (homodyne or heterodyne) which need local oscillator synchronized with the transmitted optical carrier, 90° degrees optical hybrid, photodiodes, trans-impedance amplifiers (TIA) and DSP, and incoherent receivers, which need only 90° degrees optical hybrid for phase detection, an additional photodiode for amplitude level detection, TIAs and DSP.

Coherent receivers can obtain better performance (about 3 dB SNR) and are suitable for long haul transmissions, even if they have been proposed for WDM-PONs, too [49]. In general, incoherent receivers are naturally less expensive and more suitable for PONs and OANs. On the other hand, incoherent solutions need the pilot carrier transmission on the orthogonal mode compared to signal propagation mode or the use of differential encoding (in this last case, self-coherent detection is implemented).

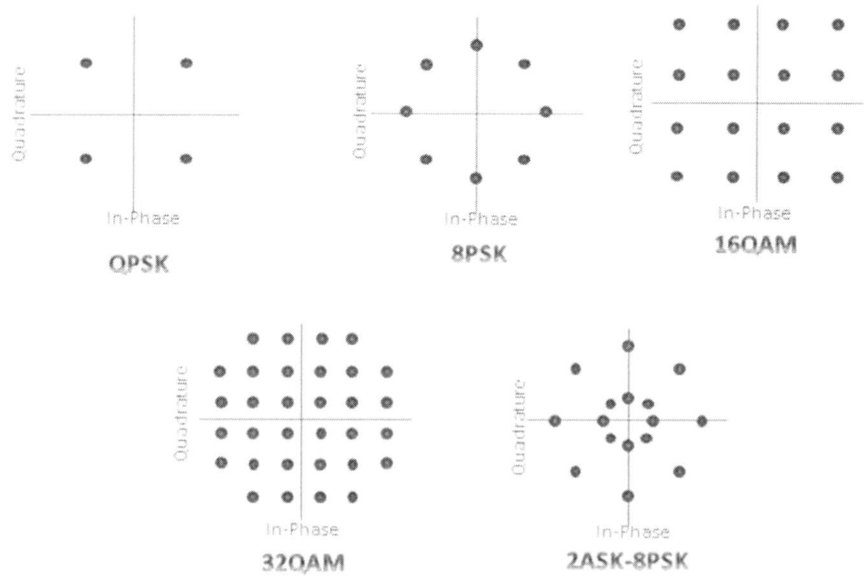


**Figure 5.** Advanced modulation constellation diagrams.

The use of differential encoding could be useful also for the use of the same wavelength for both downstream and upstream transmission, which is an option to enhance the bandwidth usage. Differential-PSK (DPSK) constant envelop, for example, makes it easy to re-modulate at the ONU the downstream signal into the OOK upstream signal. Even if some problems related to Rayleigh backscattering arise, symmetrical operation at 10 Gb/s over 25 km reach has been experimentally demonstrated [50]. A symmetric 10 Gb/s WDM-PON has been recently proved to reach 80 km distances without using any remote amplification, by 4-PSK (QPSK) in the upstream direction [51].

An optimized differential-QPSK (DQPSK) receiver in silicon with germanium photodiodes and a 2 × 4 MMI 90° hybrid has been presented and tested for 10 Gb/s and 20 Gb/s [52].

Receiver complexity is enhanced by the necessity of a clock recovery system for signal demodulation. Methods proposed up to now for this purpose can be classified into two categories, electro-optical methods, which exploit FWM, EAMs or non-linear optical mirror (NOLM), and all-optical methods, which are the best performing ones and allow demodulation speed of the order of several Gb/s.

In order to optimize the bandwidth allocation, single sideband modulation (SSB) is used since both upper and lower sideband share the same informative content. In this way, spectral efficiency is doubled, but complex filters with theoretical null transition band are required.

Frequency shift keying (FSK) and polarization shift keying (PolSK) have been also recently proposed. In particular, a WDM-OFDM-PON architecture has been demonstrated in 2010, which uses a multicast 2.5 Gb/s PolSK signal in overlay on unicast signals [53].

Modulation techniques are strictly related to encoding techniques, which deal with the way data are formatted to make the systems more power efficient, spectrally efficient, noise or jitter immune, resilient to other undesired effects.

Non-return to zero (NRZ) and return to zero (RZ) encoding are by far the most conventional solutions for IM-DD or M-ASK modulations, where RZ would be preferable for dispersion resilience. Since optical pulses have the same length independently from the transmitted bit, dispersion produces the same effects on each pulse so they can be analogically or digitally compensated. RZ drawback is a larger bandwidth required if compared to NRZ encoding.

Inverse RZ (IRZ) has been proposed for downstream data encoding because the power is transmitted both for mark and space symbol, so signal power is higher and better distributed in time domain to allow re-modulation on the same wavelength from the ONU with higher extinction ratio. At the receiver, an inverting TIA after the photodiode is sufficient to rebuild the RZ sequence. Modified duo-binary RZ (MDRZ) and carrier-suppressed RZ (CSRZ) have been also proposed to improve the receiver quality factor in high bit rate WDM-PONs [54].

Duo-binary IRZ has been demonstrated to be particularly dispersion-tolerant. In fact, this encoding technique has been proposed for 80 km reach WDM-PON with 10 Gb/s downstream and 2.5 Gb/s upstream [55]. Dicode and modified duo-binary encoding have been proposed in 2012 for mitigating both Rayleigh backscattering and chromatic dispersion [56].

# 6. MAIN TECHNOLOGICAL OPEN CHALLENGES

One of the main obstacles to be removed for large-scale application of new generation PONs is the cost of ONUs, which will require complex optical and electronic systems. From the optical point of view, the main interesting challenge is related to the use of colorless ONUs for cost-effective solutions in WDM-PONs.

Colorless ONUs would provide for wavelength independent operation. Colored wavelength specific operation would be difficult to be managed for inventory issues, and wavelength selective operation would be expensive, considering the state-of-the-art of available technology.

In fact, tunable sources have been proposed. For example, a tuning mechanism for VCSELs uses a micro-electro-mechanical system (MEMS) structure that changes the cavity length through the electrostatic control with a few microsecond tuning speed and 20 nm tuning range. VCSELs could be a strong candidate for OANs but only if their fabrication techniques will mature.

Colorless ONUs should also be coolerless, that is they would not require thermo-electric cooler (TEC) modules in order to reduce the power consumption. In fact, necessity of TEC controllers to allow thermal stability is another drawback of tunable sources, whose emission wavelength is intrinsically sensitive to temperature variations.

Several approaches have been proposed to realize colorless and coolerless ONUs. The wavelength of these optical sources is not determined by the gain media but by external factors, such as optical filters or injected signals, in order to reduce the consequences of temperature changes or aging effects at the ONU.

The use of tunable filters to obtain colorless operation actually is not a hot research topic, because it would provide a simple solution but being bulky and expensive [31]. Nevertheless, some interesting proposals have been advanced in the last years in this field. A thermally tunable filter, fabricated with integrated micro-ring resonators with specific polymers (PMATRIFE/PVCi), has been proposed in 2012, having a tested range of 120 μm to 180 μm, free spectral range (FSR) from 2.0 μm to 1.37 μm; extinction ratio around 10 dB to 20 dB for each of the two ports (drop and out), thermal coefficient measured as −0.11 nm/°C, maximum time estimated to reach equilibrium temperature 40 ms. Considering the modeled response time, the electrical consumption is assumed to be very low [57].

Spectrum-sliced BLS (SS-BLS) solution is obtained by slicing the spectrum of a BLS with a narrow-band optical filter in such a way that a unique wavelength for each channel is achieved. The optical filtering could be implemented by an AWG. This technique has several disadvantages since a narrow part of the spectrum is used and the signal power is degraded, thereby an optical booster should be used to restore the power budget. BLSs are quite inexpensive (a super-luminescent diode would be sufficient), but their spectrum limits the number of users and crosstalk limitation could also occur.

Solutions based on the use of injection-locked Fabry-Pérot (IL-FP) laser diodes have also been proposed. Since Fabry-Pérot laser cavities can have the property to be multimodal, if an external optical signal is injected into the cavity, only one mode can be excited. Therefore an external reference seed is sent from the CO to the ONU to synchronize the user transceiver on a well-defined wavelength. Also this technique is quite inexpensive but backscattering problems could require additional fibers to send the seed and the number of users would be limited by locking range.

Most of colorless ONUs solutions are based on the use of SOAs and reflective SOAs (RSOAs), whose use is considered strategic because they can modulate and amplify the signal simultaneously instead of using separate devices, which would lead to extra manufacturing cost. RSOA are preferable because the signal travels twice through the device allowing double gain. Conventional RSOA bandwidth in direct modulation is about 2 GHz and it depends on the carrier lifetime. This feature makes RSOA employable in 1.25 Gb/s and 2.5 Gb/s systems keeping low the inter-symbolic interference, even if amplified stimulated emission (ASE) noise should be taken into account. Recent works have shown RSOA operation at 10 Gb/s without electronic equalization. In fact, a RSOA with fiber-to-fiber 22.8 dB gain, 3 dB ASE and 30 nm bandwidth has been demonstrated [58]. RSOA-based systems performance is actually limited by RSOA limited saturation power and chirping effect which limits the power budget [59].

Two possibilities have been proposed to use RSOAs in ONUs. The former is to modulate through RSOA carrier transmitted from the CO (backscattering problems can require dual feeder fibers), the latter is to re-modulate through RSOA the downstream data from the CO (reduced extinction ratio can require dual feeder fibers, too). In the former the upstream wavelength grid is generated at the central office (this is called wavelength seeding), *e.g.*, through a SS-BLS. In the latter, full-duplex communication is achieved [60]. Various techniques can be adopted, such as aforementioned encoding and modulation schemes [51], different modulations techniques at OLT and at ONU (e.g., ASK/FSK), or the same ASK modulation technique at OLT and at ONU but with different power offset levels.

Currently RSOAs are still expensive for large-scale applications in PONs and they require TEC modules, so they actually do not represent a coolerless solution. Anyway, some RSOA based on new materials have been proposed to avoid TECs, making them less expensive and power consuming. Other options, which have been considered for similar purposes, are based on the use of EAMs and reflective EAMs (REAM) [61].

# 7. CONCLUSIONS

PON deployment issues and future solutions have been widely overviewed, existing standards have been compared and FSAN roadmap has been illustrated. Terminology and architectural aspects have been presented. Several simple design guidelines have been given to take into account undesired effects, such as GVD, PMD and SPM. Power consumption, security and network monitoring and troubleshooting issues have been briefly described as the main operative issues in PONs. Actually adopted countermeasures and open challenges have been also presented.

Although WDM appears as the most future proof technology, a comparison of all PON multiplexing techniques, such as OCDM, OFDM, PDM, SDM, has been proposed, summarizing their main advantages and drawbacks. Some possible trends for future evolution of PONs have been forecasted on the base of the complexity required by innovative multiplexing techniques.

Modulation and encoding techniques for future PONs have been presented as well. Multilevel modulation schemes, such as M-APSK, have been found as necessary evolution to make WDM-PONs scalable by reducing the channel bandwidth. Moreover, IRZ and other advanced codes have been briefly overviewed to stress their importance to make networks more resilient to undesired effects. Finally, several solutions for colorless and coolerless ONUs have been reported, such as SS-BLS, IL-FP and RSOA-based techniques.

Some intriguing references to recent scientific proposals and results have been suggested to give some examples of the huge research efforts involved in the very last years to propose innovative PON solutions in order to fulfill the requirement of future bandwidth-hungry telecommunication services.

## CONFLICT OF INTEREST

The authors declare no conflict of interest.

## REFERENCES

1. Ciccarella, G.; Ghibaudo, C.L.; Guarino, F. NGAN nel mondo: Piani e sostenibilità economica. *Notiziario Tecnico—Telecom Italia* **2012**, *11*, 10–23.

2. Katsianis, D.; Rokkas, T.; Neokosmidis, I.; Tselekounis, M.; Varoutas, D.; Zacharopoulos, I.; Bartzoudi, A. Risks Associated with Next Generation Access Networks Investment Scenarios. *IEEE Netw.* **2012**, *26*, 11–17.

3. Casier, K.; Vanhaverbeke, J.; Lannoo, B.; van Ooteghem, J.; Verbrugge, S.; Meersman, R.; Colle, D.; Pickavet, M.; Demeester, P.

Economics of FTTH: A comparative study between active and passive optical networks. In Proceedings of the 48th FITCE Congress, Prague, Czech Republic, 3–5 September 2009.

4. Kantor, M.; Wajda, K.; Lannoo, B.; Casier, K.; Verbrugge, S.; Pickavet, M.; Wosinska, L.; Chen, J.; Mitcsenkov, A. General Framework for Techno-Economic Analysis of Next Generation Access Networks. In Proceedings of the 12th International Conference on Transparent Optical Networks (ICTON), Munich, Germany, 27 June–1 July 2010.

5. Banerjee, A.; Sirbu, M. Towards technologically and competitively neutral Fiber To The Home (FTTH) infrastructure. In Proceedings of the 31st Research Conference on Communication, Information and Internet Policy (TPRC), Arlington, Virginia, 19–21 September 2003.

6. Banerjee, A.; Sirbu, M. Fiber To The Premise (FTTP) industry structure: implications of a wholesale-retail split. In Proceedings of 2nd International Conference on the Broadband Networks, Boston, MA, USA, 7 October 2005.

7. Larsen, C.P.; Gavler, A.; Wang, K. Comparison of Active and Passive Optical Access Networks. In Proceedings of 9th Conference on Telecommunications Internet and Media Techno Economics (CTTE), Ghent, Belgium, 7–9 June 2010.

8. International Telecommunication Union. Broadband optical access systems based on Passive Optical Networks (PON), Recommendation ITU-T G.983.1. 2005.

9. IEEE. Local and metropolitan area networks—Part 3: CSMA/CD Access Method and Physical Layer Specifications Amendment: Media Access Control Parameters, Physical Layers, and Management Parameters for Subscriber Access Networks, Standard for Information Technology IEEE 802.3ah. 2004.

10. IEEE. Local and metropolitan area networks—Part 3: CSMA/CD Access Method and Physical Layer Specifications Amendment 1: Physical Layer Specifications and Management Parameters for 10 Gb/s Passive Optical Networks, Standard for Information Technology IEEE 802.3av. 2009.

11. International Telecommunication Union. Gigabit-capable passive optical networks (GPON): General characteristics, Recommendation ITU-T G.984.1. 2008.

12. International Telecommunication Union. 10-Gigabit-capable passive optical network (XG-PON) systems: Definitions, abbreviations and acronyms, Recommendation ITU-T G.987. 2012.

13. International Telecommunication Union. 40-Gigabit-capable passive optical networks (NG-PON2): General requirements, Recommendation ITU-T G.989.1. 2013.

14. Chanclou, P.; Cui, A.; Geilhardt, F.; Nakamura, H.; Nesset, D. Network Operator Requirements for the Next Generation of Optical Access Networks. *IEEE Netw.* **2012**, *26*, 8–14.

15. Kazovsky, L.G.; Cheng, N.; Shaw, W.; Gutierrez, D.; Wong, S. *Broadband Optical Access Networks*; John Wiley & Sons: Hoboken, NJ, USA, 2011.

16. Kharraz, O.; Forsyth, D. Performance comparisons between PIN and APD photodetectors for use in optical communication systems. *Opt. Int. J. Light Electron Opt.* **2013**, *124*, 1493–1498.

17. International Telecommunication Union. Optical fibres, cables and systems. In *ITU-T Manual 2009*; 2010.

18. International Telecommunication Union. Characteristics of adaptive chromatic dispersion compensators, Recommendation ITU-T G.667. 2006.

19. International Telecommunication Union. Characteristics of polarization mode dispersion compensators and of receivers that compensate for polarization mode dispersion, Recommendation ITU-T G.666. 2011.

20. Nielsen, M.D.; Jacobsen, C.; Mortensen, N.A.; Folkenberg, J.R.; Simonsen, H.R. Low-loss photonic crystal fibers for transmission systems and their dispersion properties. *Opt. Express* **2004**, *12*, 1372–1376.

21. Chowdhury, P.; Tornatore, M.; Sarkar, S.; Mukherjee, B. Building a Green Wireless-Optical Broadband Access Network (WOBAN). *J. Lightwave Technol.* **2010**, *28*, 2219–2229.

22. Ennser, K.; Devlin, B.; Mangeni, S. Towards Greener Optical Access Networks. In Proceedings of the 12th International Conference on Transparent Optical Networks (ICTON), Munich, Germany, 27 June–1 July 2010.

23. European Commission. Communication Europe 2020: A strategy for smart, sustainable and inclusive growth. In the Communication from the Commission Europe 2020, Brussels, Belgium, 3 March 2010.

24. Ecobilian, S.A. *Developing a Generic Approach for FTTH Solutions Using LCA Methodology*; FTTH Council Europe: Paris, France; February; 2007.

25. Jacobs, S. *Security management of next generation telecommunication networks and services*; Wiley-IEEE Press: Hoboken, NJ, USA, 2013.

26. Gaede, S.; Schaeffer, C.G. Overview of Optical Monitoring in Next Generation Optical Access Networks. In Proceedings of the Photonic Networks ITG Symposium, Leipzig, Germany, 2–3 May 2011.

27. Park, J.; Baik, J.; Lee, C. Fault-detection technique in a WDM-PON. *Opt. Express* **2007**, *15*, 1461–1466.

28. Mendes Lima, G.F.; Lamounier, E.A.; Cardoso, A.; Santos Peretta, I.; Barcelos, S.; Sadaiti Muramoto, W.; Rigon, E. A TEO-Based Algorithm to Detect Events Over OTDR Measurements in FTTH PON Networks. *Lat. Am. IEEE Trans.* **2013**, *11*, 886–891.

29. Rad, M.M.; Fouli, K.; Fathallah, H.A.; Rusch, L.A.; Maier, M. Passive Optical Network Monitoring: Challenges and Requirements. *IEEE Commun. Mag.* **2011**, *42*, s45–s52.

30. Lee, K.; Kang, S.; Lim, D.S.; Lee, H.K.; Sorin, W.V. Fiber Link Loss Monitoring Scheme in Bidirectional WDM Transmission Using ASE-Injected FP-LD. *IEEE Photonics Technol. Lett.* **2006**, *18*, 523–525.

31. Hülsermann, R.; Breuer, D.; Lange, C. Impact of network reliability on network costs in next generation access networks. In Proceedings of the 12th International Conference on Transparent Optical Networks (ICTON), Munich, Germany, 27 June–1 July 2010.

32. Mitsui, T.; Sakamoto, T.; Hara, K.; Yoshimoto, N. Flexible and scalable PON protection architecture using N:M redundancy toward Next Generation Access Network. In Proceedings of the 17th Asia-Pacific Conference on Communications (APCC), Sabah, Malaysian, 2–5 October 2011.

33. Huang, A.; Liu, S.; Xie, L.; Chen, Z.; Mukherjee, B. Self-healing optical access networks (SHOAN) operated by optical switching technologies. *IEEE Trans. Netw. Serv. Manag.* **2011**, *8*, 234–244.

34. P'ng, W.T.; Abdullah, M.K.; Khatun, S.; Ahmad-Anas, S.B.; Shaari, S. Protection networking of Ethernet PON FTTH access network. In Proceedings of the 7th International Conference on Laser and Fiber-Optical Networks Modeling, Yalta, Crimean, 15–17 September 2005.

35. De Bortoli, M.; Mercinelli, R.; Solina, P.; Tofanelli, A. Tecnologie ottiche per l'accesso: Le soluzioni Passive Optical Network. *Notiziario Tecnico—Telecom Italia* **2004**, *1*, 104–119.

36. Grobe, K.; Elbers, J. PON in Adolescence: From TDMA to WDM-PON. *IEEE Commun. Mag.* **2008**, *46*, 26–34.

37. International Telecommunication Union. Spectral grids for WDM applications: CWDM wavelength grid, Recommendation ITU-T G.694.2. 2003.

38. International Telecommunication Union. Spectral grids for WDM applications: DWDM frequency grid, Recommendation ITU-T G.694.1. 2002.

39. Segarra, J.; Sales, V.; Prat, J. Agile Reconfigurable and Traffic Adapted All-Optical Access-Metro Networks. In Proceeding of the 11th International Conference on Transparent Optical Networks ICTON, Azores, Portugal, 28 June–2 July 2009.

40. Wu, Y.; Ye, T.; Zhang, L.; Hu, X.; Li, X.; Su, Y. A cost-effective WDM-PON architecture simultaneously supporting wired, wireless and optical VPN services. *Opt. Commun.* **2011**, *284*, 1139–1145.

41. Kitayama, K.; Wang, X.; Wada, N. OCDMA over WDM PON-Solution Path to Gigabit-Symmetric FTTH. *J. Lightwave Technol.* **2006**, *24*, 1654–1662.

42. Abda, T.H.; Aljunida, S.A.; Fadhila, H.A.; Ahmad, R.; Arief, A.R. Design and Evaluation Fiber-to-the-Home Network Based on DCS Code. In Proceedings of the IEEE 3rd International Conference on Photonics (ICP), Penang, Malaysia, 1–3 October 2012.

43. Tamai, H.; Sarashina, M.; Iwamura, H.; Kashima, M.; Gupta, G.C.; Ushikubo, T.; Kamijoh, T.; Chanclou, P.; Genay, N.; Landousies, B.; *et al.* First Demonstration of Coexistence of Standard Gigabit TDM-PON and Code Division Multiplexed PON Architectures toward Next

Generation Access Network. *J. Lightwave Technol.* **2009**, *27*, 292–298.

44. Hernández, M.; Arcia, A.; Alvizu, R.; Huerta, M.; Bolivar, S. A Review of XDMA-WDM-PON for Next Generation Optical Access Networks. In Proceedings of the Global Information Infrastructure and Networking Symposium (GIIS), Choroni, Venezuela, 17–19 December 2012.

45. Ossieur, P.; Antony, C.; Clarke, A.M.; Naughton, A.; Krimmel, H.G.; Chang, Y.; Ford, C.; Borghesani, A.; Moodie, D.; Poustie, A.; *et al.* A 135-km 8192-Split Carrier Distributed DWDM-TDMA PON with 2 × 32 × 10 Gb/s Capacity. *J. Lightwave Technol.* **2011**, *29*, 463–474.

46. Cincotti, G.; Kataoka, N.; Wada, N.; Kitayama, K. Future passive optical networks: Can CDM beat WDM? In Proceedings of the Future Network & Mobile Summit Conference, Florence, Italy, 16–18 June 2010.

47. Cvijetic, N.; Cvijetic, M.; Huang, M.; Ip, E.; Huang, Y.; Wang, T. Terabit Optical Access Networks Based on WDM-OFDMA-PON. *J. Lightwave Technol.* **2012**, *30*, 493–503.

48. Hu, X.; Zhang, L.; Cao, P.; Wang, K.; Su, Y. Energy-efficient WDM-OFDM-PON employing shared OFDM modulation modules in optical line terminal. *Opt. Express* **2012**, *20*, 8072–8077.

49. Chung, Y.C. High-Speed Coherent WDM PON for Next-Generation Access Network. In Proceedings of the 15th International Conference on Transparent Optical Networks (ICTON), Cartagena, Spain, 23–27 June 2013.

50. Chiuchiarelli, A.; Proietti, R.; Presi, M.; Choudhury, P.; Contestabile, G.; Ciaramella, E. Symmetric 10 Gb/s WDM-PON based on a Cross Wavelength-Reusing Scheme to Avoid Rayleigh Backscattering and Maximize Band Usage. In Proceedings of the LEOS Annual Meeting Conference, Belek-Antalya, Turky, 4–8 October 2009.

51. Cho, K.Y.; Hong, U.H.; Jung, S.P.; Takushima, Y.; Agata, A.; Sano, T.; Horiuchi, Y.; Suzuki, M.; Chung, Y.C. Long-reach 10-Gb/s RSOA-based WDM PON employing QPSK signal and coherent receive. *Opt. Express* **2012**, *20*, 15353–15358.

52. Aamer, M.; Sotiropoulos, N.; Brimont, A.; Fedeli, J.; Marris-Morini, D.; Cassan, E.; Vivien, L.; Ribaud, K.; Grosse, P.; Hartmann, J.M.; *et al.* A

Silicon Differential Receiver with Zero-Biased Balanced Detection for Access Networks. *J. Lightwave Technol.* **2013**, *25*, 1207–1210.

53. Liu, B.; Xin, X.; Zhang, L.; Yu, J.; Zhang, Q.; Yu, V. A WDM-OFDM-PON architecture with centralized lightwave and PolSK-modulated multicast overlay. *Opt. Express* **2010**, *18*, 2137–2143.

54. Malti Sharma, M.; Sheetal, A. Comparison of CSRZ, DRZ and MDRZ Modulation Formats for High Bit Rate WDM-PON System using AWG. *Int. J. Emerg. Technol. Adv. Eng.* **2012**, *2*, 83–87.

55. Liu, Z.; Xu, J.; Qiu, Y.; Chan, C. An 80-km-Reach Centralized-light-source WDM PON utilizing Inverse-RZ Duobinary Downstream Signals. In Proceedings of the 36th European Conference and Exhibition on Optical Communication (ECOC), Torino, Italy, 19–23 September 2010.

56. Guo, Q.; Tran, A.V. Mitigation of Rayleigh noise and dispersion in REAM-based WDM-PON using spectrum shaping codes. *Opt. Express* **2012**, *20*, B452–B461.

57. Maalouf, A.; Bastianelli, C.; Mahé, F.; Belmiloudi, A.; Gadonna, M.; Bosc, D. Improved integrated resonators in polymer technology for tunable filter. *Proc. SPIE* **2012**.

58. Kim, H.; Choi, B.; Kim, K.; Kim, D.; Kwon, O.; Oh, D. Improvement of modulation bandwidth in multisection RSOA for colorless WDM-PON. *Opt. Express* **2009**, *17*, 16372–16378.

59. Borghesani, A. Reflective Based Active Semiconductor Components for Next Generation Optical Access Networks. In Proceedings of the 36th European Conference and Exhibition on Optical Communication (ECOC), Torino, Italy, 19–23 September 2010.

60. Presi, M.; Proietti, R.; Prince, K.; Contestabile, G.; Ciaramella, E. A 80 km reach fully passive WDM-PON based on reflective ONUs. *Opt. Express* **2008**, *16*, 19043–19048.

61. Cavaliere, F.; Ponzini, F.; Presi, M.; Ciaramella, E. Migration towards High Speed Optical Access Enabled by WDM Techniques. In Proceedings of the Communications and Photonics Conference and Exhibition (ACP), Shanghai, China, 2–6 November 2009.

9

# SELF-HOMODYNE DETECTION IN OPTICAL COMMUNICATION SYSTEMS

**Benjamin J. Puttnam [1,*], Ruben S. Luís [1], José Manuel Delgado Mendinueta [1], Jun Sakaguchi [1], Werner Klaus [1], Yukiyoshi Kamio [1], Moriya Nakamura [2], Naoya Wada [1], Yoshinari Awaji [1], Atsushi Kanno [1], Tetsuya Kawanishi [1] and Tetsuya Miyazaki [1]**

[1] National Institute of Information and Communications Technology (NICT), 4-2-1 Nukui-kita, Koganei, Tokyo 184-8795, Japan

[2] Department of Electronics and Bioinformatics, School of Science and Technology, Meiji University, 1-1-1 Higashimita, Tama-ku, Kawasaki 214-8571, Japan

## ABSTRACT

We review work on self-homodyne detection (SHD) for optical communication systems. SHD uses a transmitted pilot-tone (PT), originating from the transmitter laser, to exploit phase noise cancellation at a coherent receiver and to enable transmitter linewidth tolerance and potential energy savings. We give an overview of SHD performance, outlining the key contributors to the optical signal-to-noise ratio penalty compared to equivalent intradyne systems, and summarize the advantages, differences and similarities between schemes using polarization-division multiplexed PTs (PDM-SHD) and those using space-division multiplexed PTs (SDM-SHD). For PDM-SHD, we review the extensive work on the transmission of advanced modulation formats and techniques to minimize the trade-off with spectral efficiency, as well as recent work on digital SHD, where the SHD receiver is combined with an polarization-diversity ID front-end receiver to provide both polarization and modulation format alignment. We then focus on SDM-SHD systems, describing experimental results using multi-core fibers (MCFs) with up to 19 cores, including high capacity transmission with broad-linewidth lasers and experiments incorporating SDM-SHD in networking. Additionally, we discuss the requirement for polarization tracking of the PTs at the receiver and path length alignment and review some variants of SHD before outlining the future challenges of self-homodyne optical transmission and gaps in current knowledge.

## KEYWORDS

self-homodyne coherent detection; coherent optical systems; multi-core fiber

## 1. INTRODUCTION

Self-homodyne detection (SHD) by using a transmitted pilot-tone (PT) as the local oscillator (LO) for coherent reception has long been proposed and investigated as a useful technique to minimize the impact of laser phase noise in optical communication systems [1,2,3]. The laser phase noise cancellation property of SHD systems arises from having the signal and LO originating from the same laser and preserving their coherence throughout the optical transmission path. This can relax requirements for narrow linewidth lasers [4,5,6,7], particularly when using phase-noise-sensitive, spectrally efficient, high-order modulation formats [8,9,10] or orthogonal frequency division multiplexing (OFDM) [11]. SHD may also reduce the complexity of the receiver and consequently provide energy savings by reducing the required digital signal processing (DSP) resources [12,13].

SHD using a PT polarization-division multiplexed to the data (PDM-SHD) has been demonstrated for a range of advanced modulation formats, including quaternary-amplitude modulation (QAM) with a constellation of 64 symbols [8,9] and OFDM [11]. However, one drawback of PDM-SHD is the trade-off of the spectral efficiency (SE) compared to polarization-division multiplexed modulation formats. Since it is only possible to modulate a single polarization with data signals, the SE of PDM-SHD systems can be reduced by up to 50% compared to PDM transmission using intradyne detection (ID), although techniques, such as spectral interleaving of the PT and signal, may be used to improve the SE [14,15]. For this reason, SHD has recently been applied to space-division multiplexing (SDM) systems and specifically to multi-core fiber (MCF) technology [16,17,18]. SDM-SHD systems allow transmission of the PT through one dedicated SDM channel with the remainder used for signal-bearing channels. In this case, the SE reduction compared to an equivalent ID system becomes inversely proportional to the number of SDM channels and becomes comparable to the SE of ID systems for a high number of spatial channels. In SHD schemes, the path length alignment between signal and PT can impact phase noise cancellation, particularly in the presence of fiber non-linearities [3]. For SDM systems using MCF, it is envisaged that different cores of the same fiber experience the same environmental disturbances during transmission, and thus, path length variations between channels are minimized compared to a system comprising parallel single core or ribbon fibers [19].

In this paper, we review existing work on SHD and draw conclusions about the feasibility and gaps in current knowledge. In Section 2, we give an overview of SHD performance, outline the key contributors to the optical signal-to-noise ratio (OSNR) penalty and explain the advantages, differences and similarities between PDM-SHD and SDM-SHD. We also discuss some variants of SHD and the requirement for polarization tracking of the PT. In Section 3, we focus on PDM-SHD, reviewing the extensive work on using SHD to enable the transmission of advanced modulation formats, techniques to minimize the trade-off with spectral efficiency and recent work on digital SHD, where the SHD receiver is combined with an ID front-end receiver to provide both polarization and modulation format alignment. In Section 4, we focus on SDM systems and the differences with PDM-SHD. We review experimental results using MCF with up to 19 cores, including high capacity transmission using wide-linewidth lasers and experiments incorporating SHD in networking and discuss the requirement for path length alignment. Finally, in Section 5, we outline the future challenges of self-homodyne detection and research work required to make such systems attractive for commercial optical communication networking applications.

# 2. OVERVIEW OF SELF-HOMODYNE DETECTION

## 2.1. Concept and Early Work

The first experimental demonstration recognizable as SHD in optical fiber transmission was reported by Miyazaki and Kubota in 2005 in a development of homodyne coherent transmission systems, which, in the absence of sufficiently fast DSP, required optical phase locked loops [20] to lock the frequency and phase of the locally generated LO to the incoming signal. The original concept used a polarization multiplexed PT to demodulate 20 Gb/s QPSK. As described in Section 3.1, the same approach was used for transmission demonstrations of high-order modulation formats with minimal modification before the advent of SDM and MCF technology led to the investigation of SHD with the PT transmitted on one core of a MCF. Schematics of both kinds of SHD schemes are shown in Figure 1.

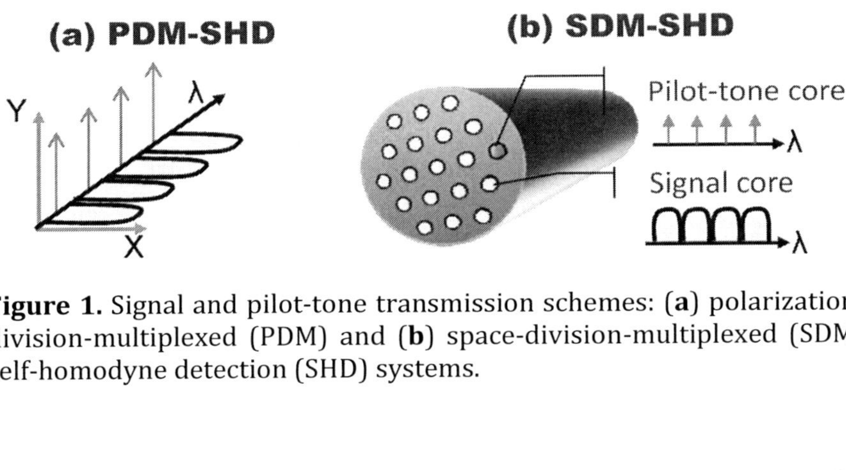


**Figure 1.** Signal and pilot-tone transmission schemes: (**a**) polarization-division-multiplexed (PDM) and (**b**) space-division-multiplexed (SDM) self-homodyne detection (SHD) systems.

Although the implementation details of both schemes vary, the motivation, benefits and drawbacks of SHD are common. SHD offers the potential of removing the dependence of the laser linewidth on transmission performance by taking advantage of phase noise cancellation between the signal and LO that originate from the same laser. In addition, SHD reduces the receiver complexity both in terms of hardware and DSP resources by eliminating the LO from the coherent receiver, removing the need for frequency offset compensation and reducing the requirement for carrier phase recovery by several orders of magnitude, as discussed in Section 2.3. These benefits are traded off with the additional noise the PT accumulates during signal transmission and results in a receiver penalty, discussed in Section 2.2. Additionally, implementation difficulties arise, since it becomes necessary to maintain some correlation in the optical path lengths traversed by the interacting signal and PT (see Section 4.3), and effort must also be made to track or control the PT polarization at the receiver input, as discussed in Section 2.4.

## 2.2. Penalty Compared to Intradyne Detection

In an SHD system, both the PT and signal accumulate amplified spontaneous emission (ASE) noise along the transmission path, resulting in a reduced post-detection, signal-to-noise ratio (SNR), compared to an equivalent ID system with similar signal optical SNR (OSNR) and a nearly noiseless, receiver-side generated LO [21]. The characteristics of noise after coherent detection in a SHD system differ from those of an ID system, due to additional contributions generated by the interactions of the PT and the signal noise, as well as the PT noise and signal noise. These yield beat noise contributions similarly to those observed in direct-detection systems [22]. In this subsection, we describe the performance of SDM-SHD systems to provide a general performance model, which is later adapted to the case of PDM-SHD systems.

To derive terms for the OSNR, consider the detection of a single or polarization-multiplexed signal and its corresponding PT at the input of a polarization-diverse coherent receiver. For simplicity, it is assumed that both the signal and PT polarization states are aligned with the optical axis of the receiver polarization beam splitter (PBS), which may in practice require an active control loop, as discussed in Section 2.4. The electrical current associated with the coherent detection of one of the signal polarization components may then be approximated by:

$$i_r = 2\mathrm{R}\left(e_s\, e_p^* + n_s\, e_p^* + e_s\, n_p^* + n_s\, n_p^*\right) * h \tag{1}$$

where $e_s$ and $e_p$ are the projections of the complex envelopes of the signal and PT fields over the analyzed polarization axis and the terms $n_s$ and $n_p$ are the corresponding ASE noise contributions of the signal and PT, respectively. R and $h$ are the responsivity of the photodetectors and the impulse response of the receiver front-end, respectively. The terms within parenthesis on the right-hand side of Equation (1) correspond to the detected signal, the signal ASE-PT beat noise, the signal-PT ASE beat noise and the signal ASE-PT ASE beat noise, respectively. Equation (1) assumes that ASE noise dominates the system performance, disregarding other noise sources, like thermal or shot noise. As an approximation, it will be assumed that the terms in Equation (1) are statistically independent, in which case, the variance of the electrical current conditioned to the detection of a given symbol with nominal power, $P_i$, may be approximated by:

$$\sigma_i^2 = \sigma_{p-ns}^2 + \sigma_{s-np,i}^2 + \sigma_{ns-np}^2 \tag{2}$$

The terms on the right-hand side of Equation (2) are the variances of the signal ASE-PT beat noise, the signal-PT ASE beat noise and the signal ASE-PT ASE beat noise, respectively. Note that in ID systems, only the term, $\sigma^2_{p\text{-}ns}$, is present and may be approximated by [23]:

$$\sigma_{p-ns}^2 = 4\mathrm{R}^2\, R_s\, N_s\, P_p \tag{3}$$

where $R_s$, $N_s$ and $P_p$ are the signal symbol rate, the power spectral density (PSD) of the signal ASE and the average PT power, respectively. The signal dependent term, $\sigma^2_{s\text{-}np,I}$, may be estimated assuming negligible inter-symbol interference, as:

$$\sigma_{s-np,i}^2 = 2\mathrm{R}^2\, W_p\, N_p\, P_i/p \tag{4}$$

where $W_p$, $N_p$ are the full-width at half maximum of the PT filter and the PSD of the PT ASE, respectively. The term, $p$, takes the values of one or two for single-polarization or polarization-multiplexed signals, respectively. Finally, $\sigma^2_{ns\text{-}np}$ may be approximated using the rectangular filter approach [24] from directly-detected systems as:

$$\sigma_{ns-np}^2 = 4\mathrm{R}^2\, N_s\, N_p\, W_p\, R_s \tag{5}$$

With Equations (2–5), one may estimate the SNR for SDM-SHD as [21]:

$$\mathrm{SNR}_{SH} = \frac{2B_{\mathrm{ref}}}{pR_s}\mathrm{OSNR}_s\left(1 + \frac{W_p}{2pR_s}\frac{\mathrm{OSNR}_s}{\mathrm{OSNR}_p} + \frac{W_p}{2B_{\mathrm{ref}}\mathrm{OSNR}_p}\right)^{-1} \tag{6}$$

where $P_s$, $\mathrm{OSNR}_s$ and $\mathrm{OSNR}_p$ are the average signal power, the signal OSNR and the PT OSNR, respectively. $B_{\mathrm{ref}}$ is the reference bandwidth used for the OSNR measurement. Note also that the SNR for an equivalent ID detection system with identical signal OSNR, $\mathrm{OSNR}_{ID} = \mathrm{OSNR}_s$, is given by [23]:

$$\mathrm{SNR}_{ID} = \mathrm{OSNR}_{ID}\frac{2B_{\mathrm{ref}}}{pR_s} \tag{7}$$

From Equations (6) and (7), one may easily compute the OSNR penalty of SHD systems with respect to ID for QPSK signals as the ratio between $\mathrm{OSNR}_s$ and $\mathrm{OSNR}_{ID}$ to reach $\mathrm{SNR}_{SH} = \mathrm{SNR}_{ID}$, in linear units, as:

$$\Delta\mathrm{OSNR} = \left.\frac{\mathrm{OSNR}_s}{\mathrm{OSNR}_{ID}}\right|_{\mathrm{SNR}_{SH}=\mathrm{SNR}_{ID}} = 1 + \frac{W_p}{2pR_s}\frac{\mathrm{OSNR}_s}{\mathrm{OSNR}_p} + \frac{W_p}{2B_{\mathrm{ref}}\mathrm{OSNR}_p} \tag{8}$$

Equation (8) provides a general model for the OSNR penalty of SHD systems with respect to ID systems with arbitrary signal and PT power and OSNR and for single and dual-polarization QPSK signals. It has also been shown to provide a reasonable approximation for the case of square QAM signals [22]. For moderate values of $\mathrm{OSNR}_p$, the third term on the right-hand side of Equation (8) may be neglected, in which case, the SNR penalty becomes mostly dependent on the term, $W_p/R_s$, referred to as the normalized PT bandwidth [21], and the ratio between the OSNR of the signal and PT.

To adapt the result in Equation (8) to the case of PDM-SHD systems, one needs to consider that measured OSNR is dependent on the signal and PT average powers as [21]:

$$\mathrm{OSNR}_{SH} = \frac{P_s + P_p}{2B_{\mathrm{ref}}\,N_s} = \mathrm{OSNR}_s + \mathrm{OSNR}_p \tag{9}$$

where $\mathrm{OSNR}_s$ and $\mathrm{OSNR}_p$ now become the OSNR of the signal and PT components separately, with single-polarization ASE noise. Hence, as in [20], Equation (8) may be rewritten as:

$$\Delta\mathrm{OSNR} = 1 + \frac{P_p}{P_s} + \frac{W_p}{2R_s}\left(1 + \frac{P_s}{P_p}\right) \tag{10}$$

Note that Equation (9) replaces the dependence on the ratio between the signal and PT OSNRs by the ratio between the signal and PT average powers. Ref.25 should appear after ref. 24

Equations (8) and (10) have been experimentally verified in [21,22], for PDM-SHD and SDM-SHD systems, respectively. However, it can be shown that the validity of this model is preserved for more general systems. For example, in digital self-homodyne detection [25], described in Section 3.2.2, and shared carrier reception SDM systems [26], described in Section 4.1, digital processing is used to significantly reduce the PT filter bandwidth and, consequently, the penalty of SHD systems.

## 2.3. DSP Savings and Phase Noise Cancellation

A strong advantage of using SHD in optical communication systems is the reduction of the required DSP complexity at the receiver, which potentially enables significant energy savings compared to an equivalent ID transmission system although we note that energy savings are system dependent and depend largely on what percentage of DSP resource energy is required to mitigate other impairments, such as chromatic dispersion [27,28]. The origin of DSP savings from using SHD is two-fold. Firstly, since the LO and signal are derived from the same laser and assuming that any path difference is orders of magnitude less than the laser coherence length, then no LO-transmitter laser frequency offset needs to be estimated and compensated for [12,13,28]; secondly, the reduction of the total system phase noise enables the carrier phase (CP) to be estimated at a significantly reduced frequency than required in an ID receiver. The origin of these savings is evident inFigure 2, which shows the estimated and unwrapped carrier phase evolution for a single polarization QPSK signal at 25 Gbaud, transmitted through a 10.1 km MCF. The figure shows a 200-μs span containing $2 \times 10^6$ symbols using both an external cavity laser (ECL) and distributed-feedback DFB laser transmitters with the linewidth measured to be 730 KHz and 3 MHz at 193.6 THz, respectively [29]. Measurements were taken for SHD with path lengths closely aligned and then with a PT decorrelated from the signal by bypassing the MCF to approximate ID. To estimate the CP, the Viterbi and Viterbi (V&V) algorithm [28] was used, carrying out the carrier-phase estimation on blocks of 128 samples (64 symbols) each.

From Figure 2, it is evident that in the SHD case, even in the presence of the various noise sources, including inter-core crosstalk, only a residual CP correction is required for both laser types. This result qualitatively verifies that the laser phase noise is cancelled when employing SHD with good path alignment at the receiver [30]. With the path length misaligned by the 10.1 km span, the receiver behaves as in the ID case, and phase wanderings of over a ±600 rad range can be observed for both lasers. Furthermore, we note that the phase variations of the narrower linewidth ECL appear smoother than the DFB case, suggesting that their use with ID would reduce the likelihood of phase slips.

The performance penalty of reducing the CP estimation rate was also investigated by BER measurements in [12,13] with 25 Gbaud, QPSK signals using SDM-SHD in a 19 core fiber. In this study, experimental data was received in a sampling oscilloscope and reprocessed with the V&V algorithm [28] used in CP recovery, and the unwrapped CP estimation buffer was updated only once every N blocks. Consequently, the same estimated CP angle is used for N consecutive blocks, as shown in the diagram of Figure 3. All other variable estimators, such as normalization, CD compensation and CMA equalization, were computed on a block-by-block basis.

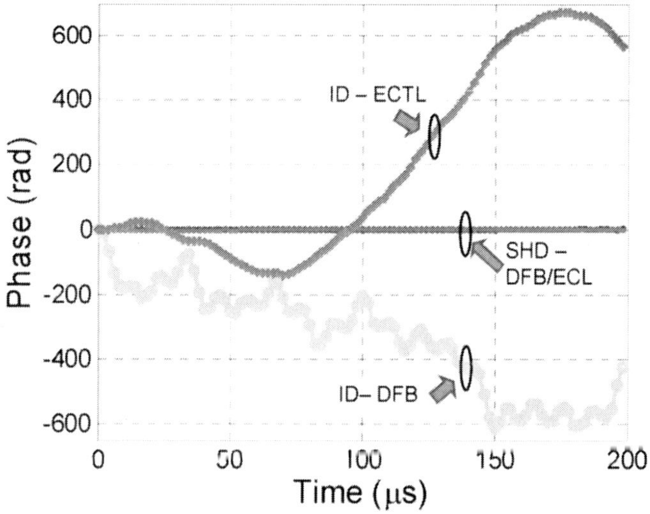


**Figure 2.** Carrier phase evolution over two million symbols for path length aligned and misaligned SDM-SHD for both the external cavity laser (ECL) and DFB laser transmission.

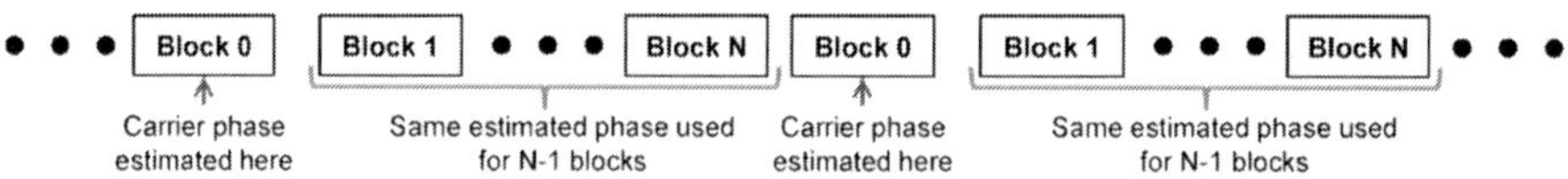


**Figure 3.** Block processing chart for the carrier phase estimation algorithm.

Figure 4 shows the average BER as a function of the number of skipped blocks (N) for V&V estimation block sizes of 32, 64 and 128 samples, for both the SHD (solid lines) and the ID (dashed lines). Figure 4 shows that for low values of N (high CP estimation update rates), longer estimation sequences perform better, since more noise is removed by the longer V&V estimators. As the estimation update rate decreases (larger N), the performance degrades more rapidly for longer estimators. This is due to the fact that a longer estimator is less able to track the phase changes due to the residual linewidth, and thus, the probability of cycle slips increases [31]. However, up to a value of N = $10^3$, the BER penalty is less than 2 × $10^{-6}$ for all the three estimation block sizes considered with SHD. In the ID case, the impact of the laser linewidth is evident with cycle slips after less than 10 skipped blocks, with the more adaptive shorter estimators performing slightly better as N increases. Hence, Figure 4 shows that SHD is able to reduce the required frequency of carrier phase recovery by three orders of magnitude.

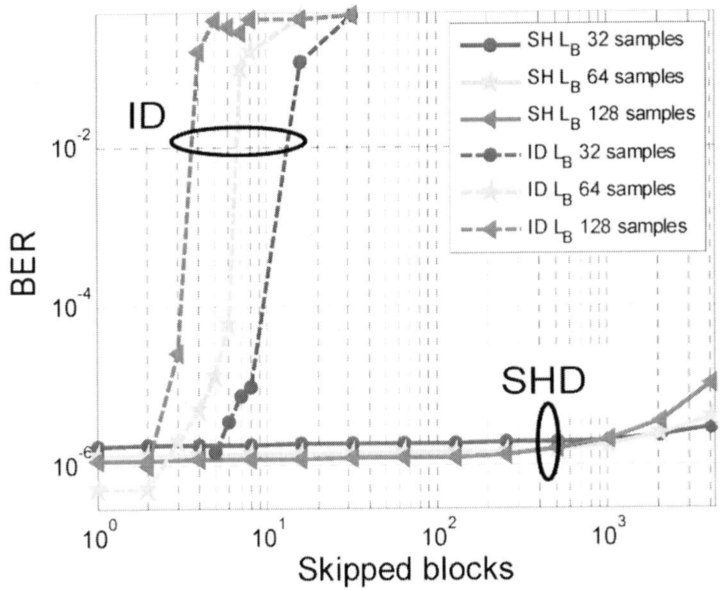


**Figure 4.** Averaged BER as a function of the carrier phase (CP) estimation rate for estimation block sizes of 32, 64 and 128 samples and aligned pilot-tone (PT) and signal paths (solid lines) and misaligned PT and signal paths (dashed lines).

## 2.4. Requirement for Receiver Polarization Control of Pilot-Tone

An implementation problem in SHD systems is the polarization alignment of the LO and signal at the input to the coherent receiver optical signal mixer. In polarization diverse ID systems, the state of polarization (SOP) of the locally-generated LO can be easily aligned to the optical axis of the PBS of the optical hybrid, ensuring data recovery regardless of the incoming signal polarization state. However, for SHD systems, since the SOP of the PT after transmission is unknown and time-varying, this condition must be achieved by tracking and/or controlling the SOP of the PT before the optical hybrid. For PDM-SHD systems, assuming the polarization-orthogonality is maintained after transmission, the signal and PT may be separated by a PBS only if the input SOP is correctly aligned with the PBS optical axis and constant over time. Similarly for SDM-SHD systems, the SOP of the PT must be aligned with the SOP of the signal for single polarization data or at an angle of 45 degrees with respect to the optical axis of the optical hybrid for polarization diverse receivers.

Although not yet demonstrated specifically with SHD systems, active polarization tracking and control systems have been investigated for more than 25 years, and it is envisaged that they could be applied to SHD systems. The first scheme [32] used fiber squeezers and integrated optical components to control the SOP based on feedback from a PC. Subsequently, lithium-niobate polarization transformers [33,34] have been used to improve performance, and experimental demonstrations have reached tracking speeds of up to 50 krad/s for QPSK transmission [35] and with 112Gb/s PDM-RX-DPSK signals [36] and endless polarization tracking at 40 krad/s with 20Gb/s PDM-RZ-DPSK signals [37] and with additional phase control [34]. In addition, it may also be possible to use passive components to split the PT into polarization components using a PBS aligned to the desired SOP and apply a phase shift to one component before recombination. However, such a scheme requires an accurate and constant path length to ensure constructive interference between the two paths and will also not work in the presence of polarization mode-dispersion (PMD) arising from birefringence in the transmission fiber. It is for this reason that SHD systems used for long distance transmission will most likely require active polarization tracking and adjustment before the receiver if such techniques can be sufficiently fast to track SOP fluctuations.

## 2.5. Other Pilot-Tone Transmission Schemes

A variant of SHD, which is compatible with both polarization and space division multiplexing, is the transmission of pilot symbols between data symbols. In these schemes, un-modulated pilot symbols are time-domain multiplexed with data. SHD is achieved by mixing the incoming data signal with versions of itself delayed by one or more bit-slots, to match up data symbols with pilot-symbols at the receiver. Such schemes may be considered analogous to differential detection and have been demonstrated successfully with up to four data symbols for each PT symbol to provide SHD with a 20% overhead [38] and at data rates up to 40 Gb/s [39,40] without the requirement for optical phase locked loops, high speed electronic circuits or complex coding schemes used in other homodyne detection schemes [41].

Although classified as heterodyne detection and, therefore, not the focus of this paper, significant advances in the transmission of high-order QAM formats has been achieved by transmitting data with a PT obtained by frequency shifting the unmodulated carrier [42,43,44,45]. The PT is used as a phase reference to a PLL locking a tunable tracking laser used as the LO at the receiver and allows the use of both polarizations for signal data at the expense of a small increase in the required signal bandwidth. This scheme has allowed the transmission of signals up to 512 QAM [44], as well as 256 QAM-OFDM [45] with high spectral efficiency, but does not allow phase noise cancellation and the linewidth tolerance characteristic of the schemes described in Section 3 and Section 4.

# 3. PDM-SHD: IMPLEMENTATIONS AND DEVELOPMENTS

## 3.1. PDM-SHD Transmission and Advanced Modulation Formats

Following on from the early work of Miyazaki and Kubota [1,2] polarization-multiplexed PTs were used to enable transmission and investigation of advanced modulation formats and led to the development of a new variant of modulator to provide a signal and PT from a single polarization laser input, referred to as a pilot-tone vector modulator (PTVM) [46]. The first transmission results used 20 Gb/s QPSK signals with high phase-noise transmitters, including a spectrum sliced ASE light source [5,6] or a Fabry–Perot diode laser [7]. The PTVM enabled a series of measurements with high-order QAM and multi-level phase modulation formats. Sixteen-QAM modulation was demonstrated up to 40 Gb/s with real-time modulation and de-modulation [47] and with inter-symbol-interference suppression with pre-equalized driving signals [48]. The same technique was used to transit 5 Gbaud 64 QAM over 60 km without dispersion compensation [8], and phase modulated formats included eight-PSK at 30 Gb/s [49,50]. Sixteen-QAM modulation was also used to investigate the tolerance to non-linear impairments in SHD transmission systems over a 116-km link [3].

## 3.2. Developments of PDM-SHD

### 3.2.1. Spectral Interleaving of Signal/PT

A useful technique to reduce the SE cost of PDM-SHD schemes is the spectral interleaving of the signal and PT [14,15]. As shown in schematic form in Figure 5, in this scheme, both polarizations are used to transmit data signals with a frequency shift applied to one polarization to allow the transmission of the spectrally narrow pilot-tone between WDM channels on the orthogonal polarization.

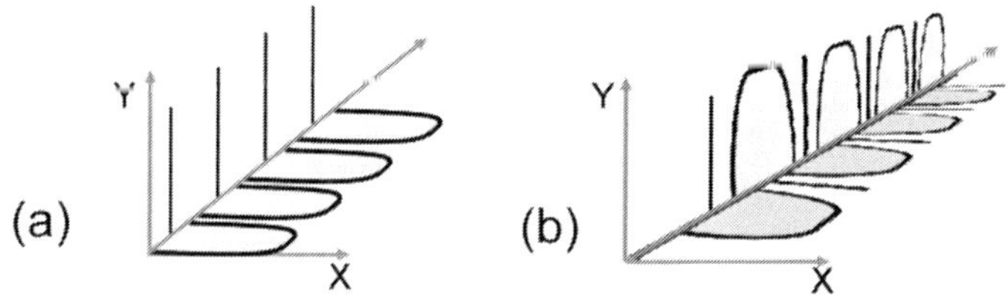


**Figure 5.** PT and signal relationship for PDM-SHD and PT-interleaved PDM-SHD.

Compared to an SHD system transmitting data on a single polarization, this technique has been shown to provide 33% improvement for five wavelength channels carrying 10 Gb/s NRZ signals [15], and simulations of the same set-up suggested that use of a 0.1 GHz PT filter could improve this to 77%, approaching that of ID PDM transmission.

### 3.2.2. Digital Self-Homodyne Detection

As outlined in Section 2.4, coherent detection of polarization multiplexed PT signals requires accurate tracking of the polarization of the PT at the receiver input, which may limit their practicality and increases the complexity of the receiver. As an alternative, digital SHD has been proposed for PDM-SHD [25] and a similar scheme, discussed in Section 4.1, implemented in SDM systems [26]. This approach receives both signal and PT components with a polarization-diverse coherent ID receiver front-end, as shown in Figure 6. After the ID front-end, both components are converted to the digital domain, where they can be digitally aligned. After polarization alignment and assuming perfect dispersion compensation at the ID front-end, the signal and PT components can be approximated by:

$$\begin{aligned} r_s &= e_s \cdot e_{lo}^* \\ r_p &= e_p \cdot e_{lo}^* \end{aligned} \tag{11}$$

where $e_s$, $e_p$ and $e_{lo}$ are the complex envelopes of the signal, PT and LO from the ID front-end. Note that $r_s$ and $r_p$ are intermediate frequency (IF) signals, which are random and time-varying, arising from the frequency offset between the transmitter and LO lasers. From Equation (11), the original signal may be recovered with a single complex multiplication as:

$$r = r_s \cdot r_p^* = e_s \cdot e_p^* \cdot |e_{lo}|^2 \tag{12}$$

Equation (12) shows that the recovered signal no longer includes the LO phase noise, thus preserving the phase noise cancellation characteristic of SHD systems. Consequently, this scheme allows the use of high linewidth lasers for the generation of the signal and PT, as well as the LO.

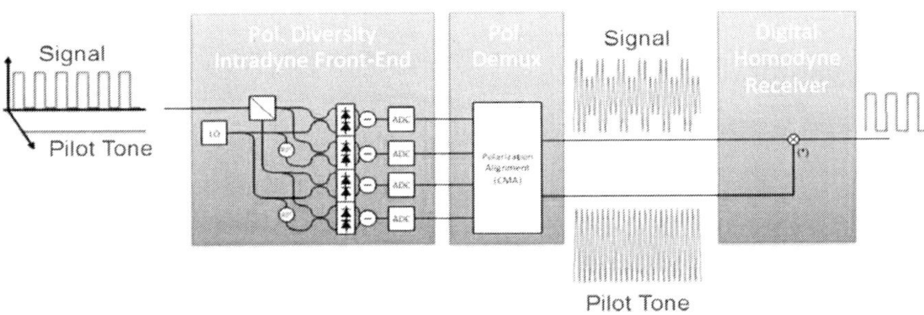


**Figure 6.** Layout for digital self-homodyne detection.

# 4. SDM- SHD: IMPLEMENTATIONS AND DEVELOPMENTS

The recent advancement of SDM technology as a means of meeting the expected growth in the capacity requirement for the next generation of optical networks has also enabled the possibility of sharing the SE cost of SHD between many spatial channels. As described in Section 3.2.1, even with spectral interleaving techniques, using polarization multiplexed PTs still has a strong impact on SE. By transmitting the PT in one SDM channel and using the remaining channels for signal transmission, the spectral efficiency cost is inversely proportional to the number of SDM channels. For example, using a 19-core fiber results in a 5.3% reduction of spectral efficiency compared to an equivalent system with ID, although this figure rises to 14.3% for a seven-core fiber. Although SHD may also be compatible with multi-mode transmission, the impact of modal dispersion has not been investigated experimentally, and PT transmission in multi-mode fiber has only been investigated by simulation [51]. For this reason, all SDM-SHD experiments to date have been performed with MCF with each core diameter being similar to that of a single mode fiber. In an SDM system based on MCF, it is envisaged that different channels experience the same environmental disturbances. This was investigated for spatial super-channels where long-term differential phase drift was shown to vary more slowly with absolute variation 90% lower over 400 us [19]. Such a feature is advantageous for SHD, which relies on matching the path length between pilot-tone and information channels for strong phase noise cancellation. In this section, we outline transmission demonstrations using SHD and shared carrier reception schemes, techniques for in-service path length alignment and approaches to integrate SH transmission with networking and spatial super-channels.

## 4.1. Transmission

Early attempts to apply SHD to SDM systems were performed in a prototype 19-core fiber using a free space coupling system [17,18] with the PT transmitted through one fiber core with data signals transmitted in the remaining cores. In this work [29,52], the feasibility of SHD was determined by noise-loading measurements to determine the penalty of SDM as a function of the WDM channel spacing for a different laser linewidth. An additional signal impairment in SDM systems is crosstalk from other SDM channels, and the impact of this was also studied by measuring the OSNR penalty of transmitting the PT in different fiber cores. As shown in Figure 7, compared to SHD without the presence of dummy signal channels in all cores, using the center core for PT transmission with signals in all cores results in a 1-dB penalty due to the higher crosstalk from the increased number of neighboring cores. The penalty for using the outer core for PT transmission was under 0.5 dB, showing that using a lower crosstalk outer-core for PT transmission is advantageous in an SDM systems.

SDM-SHD systems have been used to demonstrate high capacity transmission using broad linewidth DFB lasers. In [4], a 19-core fiber was used to demonstrate a 105 Tb/s transmission system using low-cost DFB laser transmitters over 10.1 km. This experiment showed the potential for high capacity transmission using SHD by using 125 individual DFB transmitters with frequencies between 189.1 and 195.3 THz on a 50-GHz grid with the linewidth varying from one to 5.3 MHz and each carrying 25 Gbaud QPSK signals. Assuming a 7% FEC overhead, the 105 Tb/s transmission is equivalent to a SE of 16.7 b/s/Hz. The BER was measured for all signal channels and cores and found to vary between $2 \times 10^{-4}$ and $2.7 \times 10^{-6}$, demonstrating error rates well below the $2 \times 10^{-3}$ limit of commercially available FEC systems [53]. The transmission capacity was then doubled to 210 Tb/s (SE = 33.4 b/s/Hz) [54] by adopting polarization-division multiplexing QPSK signals, and the measured BERs for all cores, for five wavelengths, including the edge and center channels, is shown in Figure 8.

A technique similar to the digital self-homodyne detection described in Section 3.2.2 has also been applied to MCF systems. This shared carrier reception (SCR) scheme [26] also relies on the transmission of an unmodulated pilot-tone, tapped from the same laser as signals in a dedicated fiber core, but instead of using this as the LO for coherent reception, the SCR-PT is received by an ID coherent receiver. The precise frequency and carrier phase information extracted from DSP [55] is then used to simplify the DSP of the received signal cores. As with digital SHD, if the LO laser is sufficiently well tuned, then digital domain filtering may be used to improve the transmission characteristics. In [24], a 0.5-dB improvement in the Q-factor was observed and a tolerance of up to 6 GHz to LO/signal detuning with the path length tolerance dependent on the bandwidth of the electrical filter. This technique was used for a long distance transmission of 10 × 128 Gb/s 16 QAM signals with 25 GHz spacing [56] over a distance of 1.815 km using 100 kHz linewidth lasers with phase noise cancellation, allowing a 1,650 km transmission for a 1-MHz linewidth DFB laser, an increase of 165 km from a SCR receiver, with frequency offset correction only.

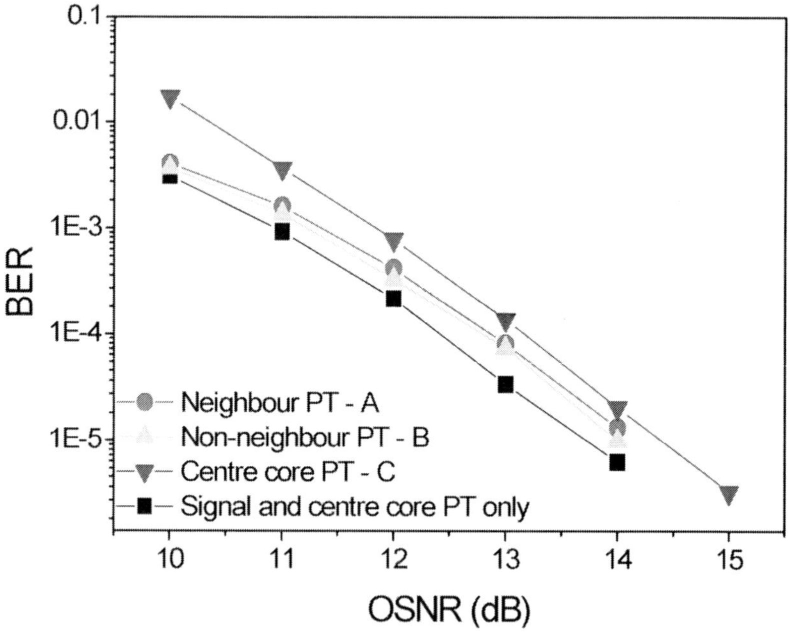


**Figure 7.** Measurement of SDM penalties when transmitting the pilot-tone on different fiber cores compared to SHD transmission without dummy channels. OSNR, optical signal-to-noise ratio.

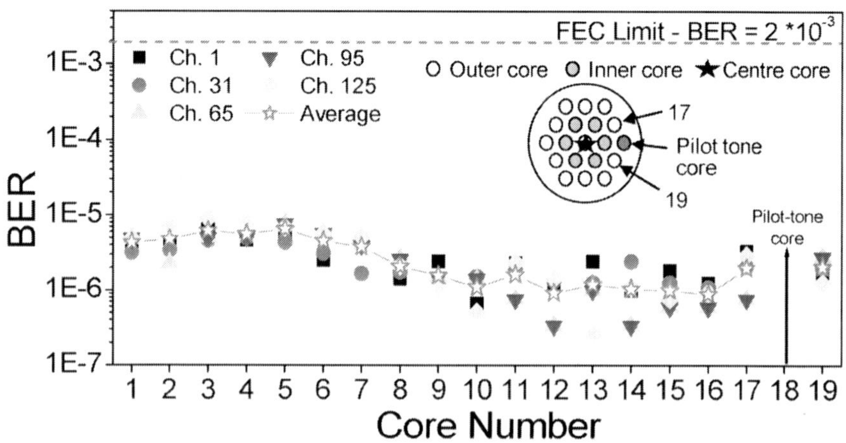


**Figure 8.** Measured BER in each signal core for five channels with equally spaced optical frequencies ranging from 195.3 THz (Ch. 1) to 189.1 THz (Ch. 125).

## 4.2. Moves towards Networking

In addition to a point-to-point transmission demonstration, the first experiments exploring the feasibility of SHD from a networking perspective have also been demonstrated. Previously, architecture-on-demand (AoD) nodes have been widely proposed and also demonstrated to incorporate SDM links [57,58], and reconfigurable add-and-drop-multiplexers (ROADM) designs for the switching of spatial super-channels [19,59] in MCFs have also been proposed. In [60], a reconfigurable SDM-WDM hybrid add-drop node capable of routing both signals and PTs was demonstrated for 100 Gb/s PDM-QPSK signals. Figure 9 shows a schematic of the SDM-WDM add and drop node used in this demonstration.

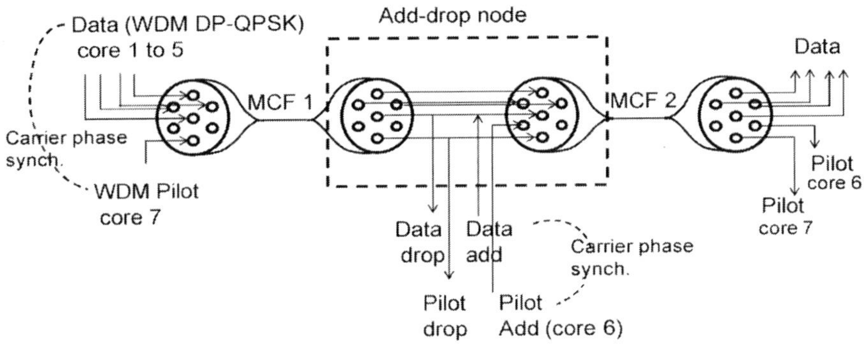


**Figure 9.** Add and drop in the SDM node with PT. MCF, multi-core fiber.

SHD has also been used in a larger network demonstration utilizing AoD nodes and software-defined networking (SDN). In this demonstration, both seven-core and 19-core MCFs were used in a network demonstration using spatial super-channels and a variety of modulation formats with both signal and PTs dynamically assigned by an SDN controller over an open flow interface [61,62]. This demonstration showed that where capacity allows, the advantages of SHD may be exploited even in dynamic networking scenarios by integrating signal and PT management into control plane architectures.

## 4.3. Path Length Alignment Scheme

Phase noise cancellation has been shown in the previous sections to be an essential property of SHD systems. However, optimum performance requires some degree of alignment of the signal and PT path lengths from the point of separation at the transmitter side, up to the balanced photodetectors at the receiver side. The required path length accuracy is currently unclear, being modulation format- and system-dependent. It was estimated to be 10% of the bit-period for PDM-SHD in the presence of non-linearity [3]; yet, measurements with moderate launch power in MCFs [12,13,52] showed that removal of tens of meters of fiber had a minimal impact on BER. Hence, the required path length accuracy for different conditions is the subject of current work, although techniques to minimize it have been investigated, as described below.

The residual phase noise resulting from path-length misalignment in SHD systems may be modeled by extending the complex envelopes of the signal and PT in Equation (1), to give:

$$i_r(t) = 2R\sqrt{P_s\, P_p}\, s(t)\, e^{j\Delta\phi(t)} \tag{13}$$

where $s(t)$, $P_s$ and $P_p$ are the normalized information signal and the average signal and PT powers, respectively. The residual phase noise in SHD is modeled by the term $\Delta\phi(t) = \phi(t) - \phi(t - \tau)$, with $\phi(t)$ as the phase noise of the transmission laser. The term, $\tau$, is the delay between the signal and PT due to path-length misalignment. To model the residual phase noise, one may assume that f(t) can be modeled as a Wiener process with variance increasing linearly with the observation interval. Hence, the variance of $\Delta\phi(t)$ can be obtained by restricting the observation interval to the propagation delay between signal and PT. As such, for small path length delays, $\tau < t$ the residual phase noise has a constant variance [63] given by:

$$\sigma^2_{\Delta\phi} = 2\pi\Delta l\; |\tau| \tag{14}$$

where $\Delta l$ is the linewidth of the transmission laser. Equation (14) shows that the variance of the residual phase noise is directly proportional to the propagation delay between signal and PT. Furthermore, it shows that for very large misalignment, the variance of the phase noise in a SHD system becomes similar to that obtained in an ID system with uncorrelated signal and PT, similar to the experiment described in Section 2.3. Equation (14) provides a basis for the design of DSP for phase noise estimation and compensation in SHD systems, considering a misalignment margin between the signal and PT path lengths.

In addition to laser phase noise, appropriate path-length alignment can also cancel or reduce the impact of nonlinear phase noise introduced by fiber nonlinearities in polarization-multiplexed PT systems [3]. In this case, it has been shown that, in a first order approach, the nonlinear residual nonlinear phase noise in SHD systems over one fiber span is approximated by [3]:

$$\Delta\phi_{\mathrm{NL}}(t) \approx \gamma\,(P_s + P_p)\,L_{\mathrm{eff}}\left[|s(t)|^2 - |s(t-\tau)|^2\right] \tag{15}$$

where $\gamma$ and $L_{\mathrm{eff}}$ are the fiber nonlinearity coefficient and its effective length, respectively. Note that, unlike laser phase noise, the nonlinear phase noise varies at a rate similar to the symbol rate, and its compensation via DSP is generally harder to achieve. Hence, accurate path-length monitoring and alignment in these systems is critical in achieving acceptable performance [3].

In realistic implementations of SDM or PDM-SHD, path length alignment has been achieved using variable delay lines in the signal or PT paths, at the transmitter or receiver sides [4,22,52,54,61,62]. The variable delay lines may be controlled using path-length alignment monitoring subsystems, such as the one proposed in [30]. Figure 10 shows the generic scheme for this subsystem. It is based on modulating the phase of the light wave with a low frequency tone prior to separation of the signal and PT paths. Upon recombination at the at the receiver side, non-zero path-length misalignment will yield a measureable interference signal, as shown in the inset of Figure 10. This signal can be used as feedback to control a variable optical delay line. When perfect path-length alignment is achieved, the interference signal is canceled and has no impact on the system performance.

Figure 11 shows the dependence of the interference signal power on the coordinate of the variable delay line for two cores of a 19-core SDM system using SHD and considering modulation frequencies of 70 MHz and 160 MHz [30]. The path-length monitoring subsystem provides resolutions of 2 dB/cm and 1 dB/cm for 70 MHz and 160 MHz modulation frequencies, respectively, which effectively allow sub-centimeter path-length alignment. Alternatively, an SCR scheme, where signal and PT are independently detected, allows path-length monitoring and alignment fully performed in the digital domain [26,56]. This facilitates the practical implementation of such systems, by avoiding the need for optical delay lines and feedback mechanisms, despite the underlying cost of the shared PT receiver.

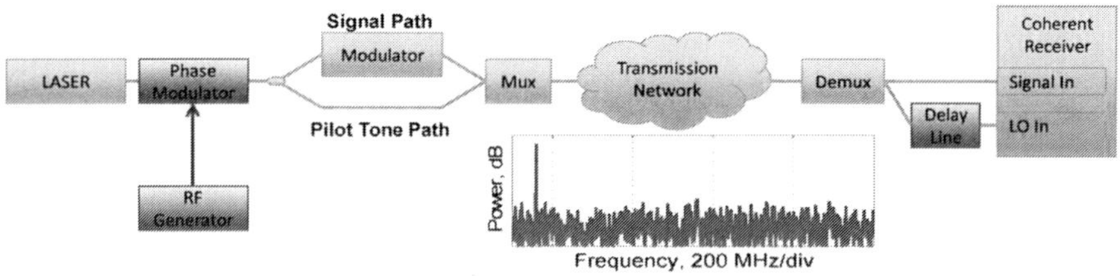


**Figure 10.** Path-length alignment monitoring subsystem. LO, local oscillator.

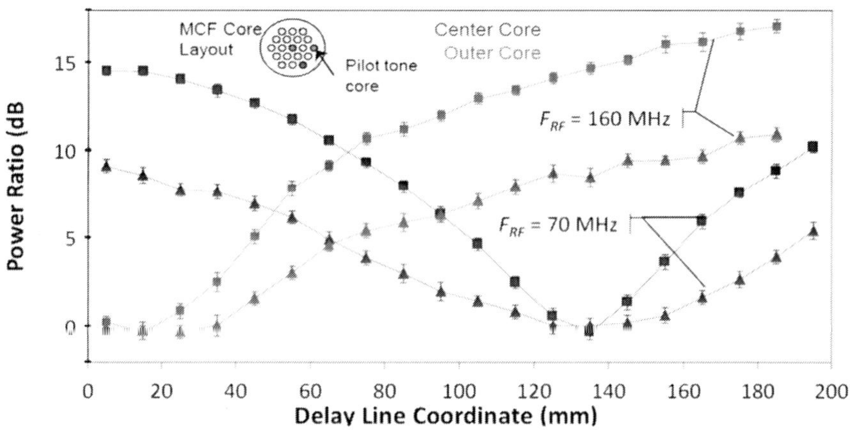


**Figure 11.** The dependence of the interference signal power on the coordinate of the variable delay line for path-length alignment monitoring.

# 5. SUMMARY AND FUTURE CHALLENGES

The work summarized in this paper shows that self-homodyne detection has been well studied in optical transmission links for point-to-point communication in optical fiber systems with clear advantages and disadvantages. Compared to equivalent ID systems, pure SHD, as outlined in Section 2.1 and Section 2.2, results in both a penalty in the required OSNR and a reduction in the achievable spectral efficiency. The spectral efficiency cost is more severe for a polarization multiplexed pilot-tone (PT), where it is typically 50%, reducing to 33% using spectral interleaving techniques. When the PT is space-division multiplexed (SDM), the spectral efficiency reduction is inversely proportional to the number of SDM channels when either SHD or shared carrier reception techniques are used. The OSNR penalty may be well described and depends crucially on the bandwidth of the pilot-tone optical filter, typically varying between 2 dB and 4 dB in experimental investigations.

The advantages of SHD have also been well demonstrated. Phase noise cancellation leads to low susceptibility to phase noise originating from low-cost lasers, which is particularly advantageous when using phase-noise sensitive formats, such as high-order QAM modulation, commonly proposed to improve spectral efficiency, or OFDM. Additionally, we have seen that the energy and complexity cost of receiver digital signal processing can be minimized by using both SDM-SHD and shared carrier reception schemes.

The remaining challenges of SHD research lie in identifying the limits of transmission and in the implementation of the required subsystems, such as polarization alignment schemes. Although SCR schemes have been demonstrated with recirculating transmission, pure SHD has not yet been experimentally demonstrated beyond a point-to-point link. Verifying that phase noise cancellation and its benefits can be maintained in noise limited long-haul transmission links is crucial to proving the practical usefulness of SHD in optical transmission systems. Similarly, the precise level of path length alignment required to obtain the benefits needs to be studied in noise limited links, particularly in the presence of fiber non-linearity and for high-order modulation formats for which SHD is proposed to solve the phase noise dependence. Whether phase noise cancellation can be maintained and whether the required path length accuracy is achievable over the desired transmission lengths will ultimately determine whether self-homodyne detection will be a useful technique for future optical communication systems.

## ACKNOWLEDGMENTS

We acknowledge the support of NICT technical staff for many of the experiments described.

## AUTHORS CONTRIBUTIONS

Benjamin J. Puttnam; Ruben S. Luís and José Manuel Delgado Mendinueta prepared the manuscript with feedback from all authors who also contributed to the work described as reflected in the references

## CONFLICTS OF INTEREST

The authors declare no conflict of interest.

## REFERENCES

1. Miyazaki, T.; Kubota, F. PSK self-homodyne detection using a pilot carrier for multibit/symbol transmission with inverse-RZ signal. *IEEE Photonics Technol. Lett.* **2005**, *17*, 1334–1336.

2. Miyazaki, T. Linewidth-Tolerant QPSK Homodyne Transmission Using a Polarization-Multiplexed Pilot Carrier. *IEEE Photonics Technol. Lett.* **2006**, *18*, 388–390.

3. Johannisson, P.; Sjödin, M.; Karlsson, M.; Tipsuwannakul, E.; Andrekson, P. Cancellation of nonlinear phase distortion in self-homodyne coherent systems. *IEEE Photonics Technol. Lett.* **2010**, *22*, 802–824.

4. Puttnam, B.J.; Delgado-Mendinueta, J.M.; Sakaguchi, J.; Luis, R.S.; Klaus, W.; Awaji, Y; Wada, N.; Kanno, A.; Kawanishi, T. 105 Tb/s Transmission System Using Low-cost, MHz Linewidth DFB Lasers Enabled by Self-Homodyne Coherent Detection and a 19-Core Fiber. In Proceedings of the Optical Fiber Communication Conference and Exposition and the National Fiber Optic Engineers Conference (OFC/NFOEC), Anaheim, CA, USA, 17–21 March 2013.

5. Nakamura, M.; Kamio, Y.; Lu, G.-W.; Miyazaki, T. Ultimately phase-noise tolerant QPSK homodyne using a spectrum-sliced ASE light source. *IEICE Electron. Express* **2007**, *4*, 406–410.

6. Nakamura, M.; Kamio, Y.; Lu, G.-W.; Miyazaki, T. Ultimate Linewidth-Tolerant 20-Gbps QPSK-Homodyne Transmission Using a Spectrum-Sliced ASE Light Source. In Proceedings of the Optical Fiber Communication and the National Fiber Optic Engineers Conference, OFC/NFOEC 2007, Anaheim, CA, USA, 25–29 March 2007.

7. Nakamura, M.; Kamio, Y.; Miyazaki, T. QPSK-Homodyne Transmission using a Multi-Wavelength Fabry-Perot Laser Diode. In Proceedings of the Lasers and Electro-Optics Conference, Baltimore, MD, USA, 6–11 May 2007.

8. Nakamura, M.; Kamio, Y. Linewidth-tolerant, ISI-suppressed 15-Gbit/s 64-QAM transmission over 120-km SSMF. In Proceedings of the 34th European Conference on Optical Communication, ECOC 2008, Brussels, Belgium, 21–25 September 2008.

9. Nakamura, M.; Kamio, Y. 30-Gbps (5-Gsymbol/s) 64-QAM Self-Homodyne Transmission over 60-km SSMF Using Phase-Noise Cancelling Technique and ISI-Suppression Based on Electronic Digital Processing. In Proceedings of the Optical Fiber Communication 2009, San Diego, CA, USA, 22–26 March 2009.

10. Kamio, Y.; Nakamura, M.; Miyazaki, T. 80-Gb/s 256-QAM signals using phase noise and DGD-tolerant pilot-carrier-aided homodyne detection. In Proceedings of the 33rd European Conference and Ehxibition of Optical Communication (ECOC), Berlin, Germany, 16–20 September 2007.

11. Luis, R.S.; Puttnam, B.J.; Mendinueta, J.-M.D.; Sakaguchi, J.; Shinada, S.; Kamio, Y.; Wada, N.; Nakamura, M. Self-homodyne coherent OFDM packet transmission without carrier frequency or common phase error estimation. In Proceedings of the IEEE 4th International Conference on Photonics, Malaka, Malaysia, 28–30 October 2013; pp. 123–125.

12. Delgado Mendinueta, J.M.; Puttnam, B.J.; Sakaguchi, J.; Luis, R.S.; Klaus, W.; Awaji, Y.; Kanno, A.; Kawanishi, T. Investigation of Receiver DSP Carrier Phase Estimation Rate for Self-Homodyne Space-Division Multiplexing Communication Systems. In Proceedings of the Optical Fiber Communication Conference and Exposition and the National Fiber Optic Engineers Conference (OFC/NFOEC), Anaheim, CA, USA, 17–21 March 2013.

13. Delgado Mendinueta, J.M.; Puttnam, B.J.; Sakaguchi, J.; Luis, R.S.; Klaus, W.; Awaji, Y.; Kanno, A.; Kawanishi, T. Energy Efficient Carrier

Phase Recovery for Self-Homodyne Polarization-Multiplexed QPSK. In Proceedings of the 18th OptoElectronics and Communications Conference, Kyoto, Japan, 30 June–4 July 2013.

14. Sjödin, M.; Agrell, E.; Lu, G.-W.; Johannisson, P.; Karlsson, M.; Andrekson, P. Interleaved polarization division multiplexing in self-homodyne coherent WDM systems. In Proceedings of the 36th European Conference and Exhibition on Optical Communication (ECOC), Torino, Italy, 19–23 September 2010.

15. Sjödin, M.; Agrell, E.; Johannisson, P.; Lu, G.-W.; Andrekson, P.A.; Karlsson, M. Filter optimization for self-homodyne coherent WDM systems using interleaved polarization division. *IEEE J. Light. Technol.* **2011**, *29*, 1219–1226.

16. Zhu, B.; Taunay, T.F.; Fishteyn, M.; Liu, X.; Chandrasekhar, S.; Yan, M.F.; Fini, J.M.; Monberg, E.M.; Dimarcello, F.V. 112-Tb/SDM DWDM transmission with 14-b/s/Hz aggregate spectral efficiency over a 76.8-km seven-core fiber. *Opt. Express* **2011**, *19*, 16665–16671.

17. Klaus, W.; Sakaguchi, J.; Puttnam, B.J.; Awaji, Y.; Wada, N.; Kobayashi, T.; Watanabe, M. Free-space coupling optics for multi-core fibers. In Proceedings of the IEEE Photonics Society Summer Topical Meeting Series, Seattle, WA, USA, 9–11 July 2012.

18. Sakaguchi, J.; Puttnam, B.J.; Klaus, W.; Awaji, Y.; Wada, N.; Kanno, A.; Kawanishi, T.; Imamura, K.; Inaba, H.; Mukasa, K.; *et al.* 19-core fiber transmission of 19 × 100 × 172-Gb/s SDM-WDM-PDM-QPSK signals at 305 Tb/s. In Proceedings of the National Fiber Optic Engineers Conference on Optical Fiber Communication Conference and Exposition (OFC/NFOEC), Los Angeles, CA, USA, 4–8 March 2012.

19. Feuer, M.D.; Nelson, L.E.; Zhou, X.; Woodward, S.L.; Isaac, R.; Zhu, B.; Taunay, T.F.; Fishteyn, M.; Fini, J.M.; Yan, M.F. Joint Digital Signal Processing Receivers for Spatial Superchannels. *IEEE Photonics Technol. Lett.* **2012**, *24*, 1957–1960.

20. Camatel, S.; Ferrero, V. Homodyne coherent detection of ASK and PSK signals performed by a subcarrier optical phase-locked loop. *IEEE Photonics Technol. Lett.* **2006**, *18*, 142–144.

21. Sjödin, M.; Johannisson, P.; Karlsson, M.; Tong, Z.; Andrekson, P.A. OSNR requirements for self-homodyne coherent systems. *IEEE Photonics Technol. Lett.* **2010**, *22*, 91–93.

22. Luis, R.; Puttnam, B.; Mendinueta, J.; Klaus, W.; Sakaguchi, J.; Awaji, Y.; Kawanishi, T.; Kanno, A.; Wada, N. OSNR Penalty of Self-Homodyne Coherent Detection in Spatial-Division-Multiplexing Systems. *IEEE Photonics Technol. Lett.***2014**, *26*, 477–479.

23. Essiambre, R.; Kramer, G.; Winzer, P.J.; Foschini, G.J.; Goebel, B. Capacity Limits of Optical Fiber Networks. *IEEE J. Light. Technol.* **2010**, *28*, 662–701.

24. Rebola, J.; Cartaxo, A. Q-factor estimation and impact of 248 spontaneous-spontaneous beat noise on the performance of optically 249 preamplified systems with arbitrary optical filtering. *IEEE J. Light. Technol.* **2003**, *21*, 87–95.

25. Luis, R.S.; Puttnam, B.J.; Mendinueta, J.-M.D.; Sakaguchi, J.; Shinada, S.; Nakamura, M.; Kamio, Y.; Wada, N. Self-homodyne detection of polarization-multiplexed pilot tone signals using a polarization diversity coherent receiver. In Proceedings of the 39th European Conference and Exhibition on Optical Communication (ECOC 2013), London, UK, 22–26 September 2013.

26. Le Taillandier de Gabory, E.; Arikawa, M.; Hashimoto, Y.; Ito, T.; Fukuchi, K. A shared carrier reception and processing scheme for compensating frequency offset and phase noise of space-division multiplexed signals over multicore fibers. In Proceedings of the Optical Fiber Communication Conference and Exposition and the National Fiber Optic Engineers Conference (OFC/NFOEC), Anaheim, CA, USA, 17–21 March 2013.

27. Kuschnerov, M.; Bex, T.; Kainzmaier, P. Energy Efficient Digital Signal Processing. In Proceedings of the Optical Fiber Communication Conference and Exposition and the National Fiber Optic Engineers Conference (OFC/NFOEC), San Francisco, CA, USA, 9–13 March 2013.

28. Savory, S.J. Digital Coherent Optical Receivers: Algorithms and Subsystems. *IEEE J. Sel. Top. Quantum Electron.* **2010**,*16*, 1164–1179.

29. Puttnam, B.J.; Sakaguchi, J.; Delgado-Mendinueta, J.M.; Klaus, W.; Awaji, Y; Wada, N.; Kanno, A.; Kawanishi, T. Investigating self-homodyne coherent detection in a 19 channel space-division-multiplexed transmission link. *Opt. Express* **2013**, *21*, 1561–1566.

30. Luis, R.S.; Puttnam, B.J.; Mendinueta, J.M.D.; Sakaguchi, J.; Klaus, W.; Awaji, Y.; Wada, N.; Kanno, A.; Kawanishi, T. In-service method of path length alignment in SDM systems with self-homodyne detection. In Proceedings of the 18th OptoElectronics and Communications Conference, Kyoto, Japan, 30 June–4 July 2013.

31. Meyr, H.; Moeneclaey, M.; Fechtel, S.A. *Digital Communication Receivers: Synchronization, Channel Estimation, and Signal Processing*; John Wiley & Sons, Inc.: New York, NY, USA, 1997.

32. Noé, R.; Heidrich, H.; Hoffmann, D. Endless polarization control systems for coherent optics. *IEEE J. Light. Technol.* **1988**, *6*, 1199–1207.

33. Hidayat, A.; Koch, B.; Mirvoda, V.; Zhang, H.; Bhandare, S.; Ibrahim, S.K.; Sandel, D.; Noe, R. Fast Optical Endless Polarization Tracking with LiNbO3 Component. In Proceedings of the Optical Fiber communication/National Fiber Optic Engineers Conference, San Diego, CA, USA, 24–28 February 2008.

34. Koch, B.; Noé, R.; Mirvoda, V.; Sandel, D. 20 krad/s Endless Optical Polarisation and Phase Control. *Electron. Lett.* **2013**, *49*, 483–485.

35. Pfau, T.; Peveling, R.; Hoffinann, S.; Bhandare, S.; Ibrahim, S.; Sandel, D.; Adamczyk, O.; Porrmann, M.; Noe, R.; Achiam, Y.; *et al.* PDL-Tolerant Real-time Polarization-Multiplexed QPSK Transmission with Digital Coherent Polarization Diversity Receiver. In Proceedings of the Digest of the IEEE/LEOS Summer Topical Meetings, Portland, OR, USA, 23–25 July 2007.

36. Koch, B.; Noé, R.; Mirvoda, V.; Griesser, H.; Bayer, S.; Wernz, H. Record 59-krad/s Polarization Tracking in 112-Gb/s640-km PDM-RZ-DQPSK Transmission. *IEEE Photonics Technol. Lett.* **2010**, *22*, 1407–1409.

37. Koch, B.; Noé, R.; Mirvoda, V.; Sandel, D.; Jan, O.; Puntsri, K. 20-Gb/s PDM-RZ-DPSK Transmission with 40 krad/s Endless Optical Polarization Tracking. *IEEE Photonics Technol. Lett.* **2013**, *25*, 798–801.

38. Kamio, Y.; Miyazaki, T. Pilot-Symbol-Aided Self-Homodyne Detection for High-Efficiency Optical-Fiber Transmission System. In Proceedings of the Pacific Rim Conference on Lasers and Electro-Optics, CLEO/Pacific Rim 2005, Tokyo, Japan, 30 July–2 August 2005.

39. Lu, G.-W.; Nakamura, M.; Kamio, Y.; Miyazaki, T. 40-Gb/s QPSK with Inserted Pilot Symbols Using Self-Homodyne Detection. In Proceedings of the Optical Fiber Communication and the National Fiber Optic Engineers Conference, OFC/NFOEC 2007, Anaheim, CA, USA, 25–29 March 2007.

40. Lu, G.-W.; Nakamura, M.; Kamio, Y.; Miyazaki, T. 40-Gb/s QPSK and 20-Gb/s PSK with inserted pilot symbols using self-homodyne detection. *Opt. Express* **2007**, *15*, 7660–7666.

41. Kikuchi, K. Phase-diversity homodyne detection of multilevel optical modulation with digital carrier phase estimation. *IEEE J. Sel. Top. Quantum Electron.* **2006**, *12*, 563–570.

42. Nakazawa, M.; Yoshida, M.; Kasai, K.; Hongou, J. 20 msymbol/s, 64 and 128 QAM coherent optical transmission over 525 km using heterodyne detection with frequency-stabilised laser. *Electron. Lett.* **2006**, *42*, 710–712.

43. Nakazawa, M.; Hirooka, T.; Yoshida, M.; Kasai, K. Ultrafast Coherent Optical Transmission. *IEEE J. Sel. Top. Quantum Electron.* **2012**, *18*, 363–376.

44. Okamoto, S.; Toyoda, K.; Omiya, T.; Kasai, K.; Yoshida, M.; Nakazawa, M. 512 QAM (54 Gbit/s) coherent optical transmission over 150 km with an optical bandwidth of 4.1 GHz. In Proceedings of the 36th European Conference and Exhibition on Optical Communication, Torino, Italy, 19–23 September 2010.

45. Omiya, T.; Yoshida, M.; Nakazawa, M. 1 Tbit/s 256 QAM-OFDM transmission over 560 km with 14.3 bit/s/Hz spectral efficiency. In Proceedings of the 18th OptoElectronics and Communications Conference held jointly with 2013 International Conference on Photonics in Switching (OECC/PS), Kyoto, Japan, 30 June–4 July 2013.

46. Nakamura, M.; Kamio, Y.; Miyazaki, T. PMD- and Dispersion-Tolerance of QPSK Homodyne Detection using a Polarization-Multiplexed Pilot Carrier. In Proceedings of the European Conference on Optical Communications, ECOC 2006, Cannes, France, 24–28 September 2006.

47. Nakamura, M.; Kamio, Y.; Miyazaki, T. Real-time 40-Gbit/s 16-QAM self-homodyne using a polarization-multiplexed pilot-carrier. In

Proceedings of the Digest of the IEEE/LEOS Summer Topical Meetings, Acapulco, Mexico, 21–23 July 2008.

48. Nakamura, M.; Kamio, Y.; Miyazaki, T. Linewidth-tolerant real-time 40-Gbit/s 16-QAM self-homodyne detection using a pilot carrier and ISI suppression based on electronic digital processing. *Opt. Lett.* **2010**, *35*, 13–15.

49. Kamio, Y.; Nakamura, M.; Miyazaki, T. Pilot-Carrier Based Linewidth-Tolerant 8PSK Self-homodyne using Only One Modulator. In Proceedings of the 33rd European Conference and Ehxibition of Optical Communication (ECOC), Berlin, Germany, 16–20 September 2007.

50. Nakamura, M.; Kamio, Y.; Miyazaki, T. Linewidth-tolerant 30 Gbit/s 8-PSK self-homodyne using single modulator and phase-noise cancelling technique. *Electron. Lett.* **2009**, *45*, 368–369.

51. Qu, Z.; Fu, S.; Zhang, M; Tang, M; Shum, P.; Liu, D. Analytical Investigation on Self-Homodyne Coherent System Based on Few-Mode Fiber. *IEEE Photonics Technol. Lett.* **2014**, *26*, 74–77.

52. Puttnam, B.J.; Sakaguchi, J; Delgado Mendinueta, J.M., Klaus, W.; Awaji, Y; Wada, N.; Kanno, A.; Kawanishi, T. Investigating self-homodyne coherent detection in a 19-core spatial-division-multiplexed transmission link. In Proceedings of the 38th European Conference and Exhibition on Optical Communications (ECOC), Amsterdam, The Netherlands, 16–20 September 2012.

53. ITU. Recommendation G.975 forward Error Correction for Submarine Systems. In Proceedings of the World Telecommunications Standardization Assembly, Montreal, QC, Canada, 27 September–6 October 2000.

54. Puttnam, B.J.; Delgado-Mendinueta, J.M.; Sakaguchi, J.; Luis, R.S.; Klaus, W.; Awaji, Y.; Wada, N.; Kanno, A.; Kawanishi, T. 210Tb/s Self-Homodyne PDM-WDM-SDM Transmission with DFB lasers in a 19-Core Fiber. In Proceedings of the IEEE Photonics Society Summer Topical Meeting Series, Waikoloa, HI, USA, 8–10 July 2013.

55. Maher, R.; Thomsen, B. Dynamic linewidth measurement technique using digital intradyne coherent receivers. *Opt. Express* **2011**, *19*, B313–B322.

56. Le Taillandier de Gabory, E.; Arikawa, M.; Ito, T.; Fukuchi, K. DWDM transmission of 128 Gb/s PM-16QAM signal over 1815km of 7-core MCF using shared carrier reception for improving the received signal quality. In Proceedings of the 39th European Conference and Exhibition on Optical Communication (ECOC 2013), London, UK, 22–26 September 2013.

57. Amaya, N.; Irfan, M.; Zervas, G.; Nejabati, R.; Simeonidou, D.; Sakaguchi, J.; Klaus, W.; Puttnam, B.J.; Miyazawa, T.; Awaji, Y.; *et al.* First Fully-Elastic Multi-Granular Network with Space/ Frequency/Time Switching Using Multi-Core Fibres and Programmable Optical Nodes. In Proceedings of the European Conference and Exhibition on Optical Communication, London, UK, 16–20 September 2012.

58. Amaya, N.; Irfan, M.; Zervas, G.; Nejabati, R.; Simeonidou, D.; Sakaguchi, J.; Klaus, W.; Puttnam, B.J.; Miyazawa, T.; Awaji, Y.; *et al.* Fully-elastic multi-granular network with space/frequency/time switching using multi-core fibres and programmable optical nodes. *Opt. Express* **2013**, *21*, 8658–8872.

59. Feuer, M.D.; Nelson, L.E.; Abedin, K.; Zhou, X.; Taunay, T.F.; Fini, J.F.; Zhu, B.; Isaac, R.; Harel, R.; Cohen, G.; *et al.* ROADM System for Space Division Multiplexing with Spatial Superchannels. In Proceedings of the Optical Fiber Communication Conference and Exposition and the National Fiber Optic Engineers Conference (OFC/NFOEC), Anaheim, CA, USA, 17–21 March 2013.

60. Sakaguchi, J; Klaus, W.; Puttnam, B.J.; Miyazawa, T.; Awaji, Y; Delgado-Mendinueta, J.M.; Luis, R.S.; Wada, N. SDM-WDM hybrid reconfigurable add-drop nodes for self-homodyne photonic networks. In Proceedings of the IEEE Photonics Society Summer Topical Meeting Series, Waikoloa, HI, USA, 8–10 July 2013.

61. Amaya, N.; Yan, S.; Channegowda, M.; Rofoee, B.R.; Shu, Y.; Rashidi, M.; Ou, Y.; Zervas, G.; Nejabati, R.; Simeonidou, D.; *et al.* First Demonstration of Software Defined Networking (SDN) over Space Division Multiplexing (SDM) Optical Networks. In Proceedings of the 39th European Conference and Exhibition on Optical Communication (ECOC 2013), London, UK, 22–26 September 2013.

62. Amaya, N.; Yan, S.; Channegowda, M.; Rofoee, B.R.; Shu, Y.; Rashidi, M.; Ou, Y.; Zervas, G.; Nejabati, R.; Simeonidou, D.; *et al.* Software defined networking (SDN) over space division multiplexing (SDM)

optical networks: Features, benefits and experimental demonstration. *Opt. Express* **2014**, *22*, 3638–3647.

63. Zan, Z.; Lowery, A.J. Experimental demonstration of a flexible and stable semiconductor laser linewidth emulator.*Opt. Express* 2014, *18*, 13880–13885.

10

# ENERGY-EFFICIENCY VERSUS RESILIENCE: RISK AWARENESS VIEW ON DIMENSIONING OF OPTICAL NETWORKS WITH A SLEEP MODE

**Piotr Chołda , Piotr Jaglarz**

AGH University of Science and Technology, Kraków, Poland

## ABSTRACT

This article proposes to solve the trade-off between energy-efficiency and resilience with a focus on business mechanisms. Risk engineering is used as a foundation. Financial impact (penalty) quantification with various compensation policies is applied, and business-relevant risk measures are used during the risk assessment. Then, risk mitigation strategies are evaluated to select the appropriate risk response. The approach is presented in networks with energy profiles supporting a sleep mode. An effective heuristic is used to assign flows, and it is shown that the energy-efficiency performance is substantially independent of the recovery methods selected for risk mitigation. It is also demonstrated that backup resources can be switched off in the normal state without having a considerable impact from a financial viewpoint.

## KEYWORDS

Energy-efficiency, Green networks, Network recovery, Reliability, Risk engineering

# 1. INTRODUCTION

Failures result in either node or link outages. In effect, some services are interrupted, which in turn is responsible for financial losses for operators. To counter these losses, automatic *recovery methods* are designed. They operate in various ways (with different levels of sharing, scope, etc. [20]) and result in different parameters, such as *resilience* (i.e., survivability to random failures) related to client needs. They also incur increased capital expenditure (CAPEX) and operational expenditure (OPEX) costs, since in order to bypass fault-affected elements, it is necessary to use *backup* resources. Here, we focus on one of the most prominent OPEX costs: the cost of energy usage. Idzikowski et al. [34] rate resilience (and reliability-related parameters) among the main criteria of assessing energy-aware design. The reason is that large optical pipes carry a lot of traffic which should not be left interrupted without a response.

In this paper, we deal with dimensioning of energy-efficient resilient backbone optical networks. Such networks aim to minimize energy usage or prevent the consumption of non-renewable energy (e.g., produced by traditional coal plants). From the business viewpoint, there are at least three aspects encouraging energy-efficiency [2]: (a) high costs of energy; (b) if energy usage is too high, its supply can simply be cut off, which can replace capacity as a bottleneck in network management and operation [10]; (c) pressures on the industry to protect the environment may force regulators to introduce energy-saving policies. All these aspects are important for the operators rather than the clients, who simply require uninterrupted services. Wiatr et al. [64] elaborate on the existence of the trade-off between energy-efficiency and quality performance, demonstrating that power minimization cannot always be the main driver in network design. Likewise, resilience involves a trade-off, since its introduction applies increased energy usage due to the introduction of backup resources. Here, we propose to approach this trade-off with the application of some elements of *risk engineering* [21] and to look at this problem from the business perspective. Therefore, we shift the focus from reducing energy usage (while satisfying various demands) to assessing the potential impact of *making networks environmentally friendly* (*green*) on the resilience perceived by a network client. Then, we are able to deal with the mitigation of recognized risks (failures) combined with energy-efficiency. We are not aware of any other approach of combining the problems of energy-efficient recovery methods with their selection based on risk awareness.

Precisely, we deal with network dimensioning problems (in uncapacitated networks) using Yaged's heuristic approach providing fast solutions that can be obtained due to the assumed energy profiles supporting *sleep modes*, typical for elements of optical networks. Our optimization bears in mind that during a normal, failure-free network state (more than 99 % of the time), there is no need to consume energy. Thus, by shifting the interest from capacity (no longer important due to the pervasive overprovisioning of optical resources) to the energy usage viewpoint, we are able to show that the most attractive options for the abovementioned trade-off solution are quite surprising. Dedicated protection provides the most benefits; however, using a sleep mode with spare (backup) resources with re-routing appears to be an interesting option.

The remainder of the paper is organized as follows. First, to show the context of our results, we discuss related work in Sect. 2. Section 3 presents energy profiles that are assumed in our optimization and performance evaluation studies. The optimization for network dimensioning is based on a fast design method and provides energy-efficient routing with various recovery methods. Section 4 outlines risk management: the framework we use to deal with the intrinsic trade-off between energy-efficiency and resilience provisioning (selected risk engineering aspects). After presenting its organization (risk management cycle), we focus on various methods of expressing the monetary impact of failures (compensation policies) and methods for meaningful quantification of the predicted financial losses (risk measures), as well as business-relevant selection of countermeasures (risk mitigation). Section 5 shows the impact of the selected risk mitigation strategies on various recovery settings. As the results assume one recovery method for all demands, in Sect. 6 we propose a simple optimization method to find a combination of recovery methods for the assumed mitigation strategy. In Sect. 7, we formulate general conclusions on how the assignment will look like and should be carried out. Finally, we provide a summary with a view on future work in the closing Sect. 8.

## 2. RELATED WORK

### 2.1 Energy-efficiency in resilient networks

While the first papers on greening (improving environmental credentials) of networks appeared in the early 2000s, here we limit our presentation to relatively recent relevant literature. The field of network energy-efficiency is described in many excellent surveys [7, 9, 49, 67]. They present the justification for dealing with energy-efficiency in optical backbone networks. They give details on the main methods to provide it and present different approaches.

Sleep mode is one of the methods of consolidating resources [7], where traffic is consolidated in some network elements while the other elements are put in sleep mode (i.e., some links are made unavailable). Wiatr et al. [63] discuss various types of sleep (in passive optical networks): power shedding, deep/fast sleep, and dozing, with all their advantages and disadvantages. However, we deal with the problem from a more abstract perspective. Network sleep mode is typically assumed with the design of energy-aware networks, where the diminishing return effect (also known as bandwidth discount [12]) typically takes place. It is suggested that the mode is used in a twofold manner: (a) on a short time scale (traffic engineering), where it is adjusted to short-term load variability (e.g., over a day), and (b) on a longer time scale (network dimensioning), where it is used for spare resources to be switched on when they are needed after failures occur. The latter approach is relevant to this paper. While generally the sleep mode mechanisms are not present in contemporary network devices, some attempts to include them in protocol suites exist. For instance, Morea et al. [44] propose enhancements in the Generalized MultiProtocol Label Switching (GMPLS) control plane protocols. Additionally, documents have been produced in the Internet Engineering Task Force by Eman (a working group on energy management, established in 2009) to provide the management plane with the necessary ontology to represent data useful in energy-aware networks, taking into account the sleep mode [22, 52]. Energy profiles supporting our approach and adapted in the design and optimization of optical networks, sometimes with resilience provisioning, are reported in [5, 9, 13, 14, 36, 44,56, 67]. While assuming the most common fixed + proportional ($f + p$) energy profile for single elements, the total energy usage for the whole network has the concave character as a function of the summarized network load [36, 67]. An in-depth study of energy usage is also given in [59], where a very simple rule of thumb for energy usage is proposed as a universal value for the electrical layer (1 Gb/s costs 10 W for a MPLS router); in the optical layer, it is necessary to consider optical switching and calculate the number of transponders and regenerators on links. It is shown that power efficiency is obtained by reducing marginal energy usage per unit capacity of a card port, thus justifying using of concave energy profiles. Chiaraviglio et al. [14] consider some risks related to using the sleep mode based on the $f + p$ energy profile. Using elegant theoretical modeling, they show that sleep is advantageous unless the fixed cost of switching on components is an order of magnitude lower than the proportional energy cost.

It should also be noted that sometimes reported energy profiles are different than those relevant to this work. For instance, Kilper et al. [37] take into account fixed switching on costs, but they also assume that the unit cost of variable energy usage increases with capacity; as such they consider convex characteristics, at least in access networks. They claim that for capacities lower than 10 Gb/s, the power of line cards is not dependent on capacity, but that for larger capacities, there is some dependency. Convexity also concerns the electronic level: there is a square or cubic relationship with the frequency or voltage [8, 28, 53]. However, at the network element level, it seems that the effect is reduced to $f + p$.

Restrepo et al. [53] analyze various energy profiles: on–off (step function), logarithmic-like functions (representants of concave functions whose usage justifies the application of the sleep mode), cubic-like (convex) characteristics responsible for representing increases in energy usage with voltage or frequency in electronic circuits, and purely linear functions (some switching devices are said to have this property). They use all these energy profiles in the minimized goal functions and present how to route traffic accordingly. They present linear optimization problems applying a very coarse linearization (with two segments only) in order to not increase the complexity of the linear problem too much. To some extent, we refine their approach with a more sophisticated optimization of concave functions and additionally propose an efficient heuristic to obtain solutions of a high quality (a very small optimization gap).

While literature regarding resilience and energy-efficiency issues separately is extensive, papers discussing both together also exist. However, they typically deal with optimization problems from the energy-efficiency minimization viewpoint, considering resilience needs as additional constraints [47]. From this perspective, it has been shown that energy-efficiency approaches (mainly using sleep modes for spare resources) enable operators to save a significant percentage of energy usage, even with protection methods applied, but the reliability performance is put to jeopardy with the increased sleep rate. While it has also been noted that energy minimization and resilience provisioning are opposing goals and some trade-offs are needed (e.g., by using money saved by the decreased energy usage to finance the increased costs of recovery as proposed by Wiatr et. al. [62]), no proposals for countering this have been provided. We attempt to fill this gap by using the risk-aware approach by mapping both energy and reliability performance into one monetary scale. Many papers provide optimization formulations with $f + p$ energy profiles; however, to our knowledge, none of them propose the use of Yaged's heuristic. A typical approach in the context of resilient energy-efficient networks is to minimize energy usage with mixed integer linear programming (MILP) formulations such as in [44], where routing and wavelength assignment (RWA) is optimized jointly with 1:1 dedicated protection, and then the optimization formulation is solved by a commercial solver. For instance, Liu et al. [38] present algorithms for energy-efficient routing of connections protected with shared backup paths, so that the shared spare resources are put to sleep. They also take into account shared-risk link groups (SRLG), a notion important in multilayer networks, not to be confused with our risk management approach (two links are elements of the same SRLG if they fail together). Lopez et al. [39] work on the fact that provisioning various service classes to clients provides some gains to operators (e.g., in energy savings or capacity costs). They focus on energy usage and deal with various recovery classes. Nevertheless, the authors use fixed scenarios with given percentages of clients assigned various service classes. In contrast, we consider how to provide a given client, defining their needs with a specific compensation policy, with a recovery method conforming to the operator's risk mitigation strategy. Muhammad et al. [45] investigate switching off devices that are used for recovery purposes only. We follow this approach in relation to re-routing, but in the case of protection methods, we assume that sleeping elements cannot be woken on demand. Francois et al. [29] propose a new method that provides resilience while also aiming to use energy more efficiently. They propose the use of green backup paths to carry some traffic (so that other links can go sleep); they are used again as real backup paths when a failure happens (this resembles a preemptive mode in 1:1 automatic protection switching standardized for transport networks). They also present a two-goal optimization problem (by minimizing the maximum link usage and maximizing the total amount of energy saved), as well as solving the RWA problem. The fact that energy-efficiency as each cost-focused goal must be traded off with desirable non-functional properties has been noted before. For instance, Cavdar et al. [13] pay attention to the resilience-energy trade-off by noting that the use of sleep modes and the focus on reducing the number of elements used increase the number of single points of failure. Additionally, as traffic carried via some elements is greater than if the sleep mode is not applied, an average failure affects the operation in a more significant way. Jirattigalachote et al. [36] consider a dedicated path protection 1:1, where links carrying backup resources are switched off while they are not used. To ensure that it is possible to put a considerable number of links to sleep, they aim to separate links with working and backup resources. They note the trade-off between energy minimization and performance (the probability of blocking) in their dynamic case. Wiatr et al. [64] note the fact that energy-aware network design mainly focuses on power minimization, while other aspects should be traded off with energy considerations. They also focus on quality issues, such as request blocking. Finally, Vizcaino et al. [61] also note the trade-off between energy-efficiency and resilience constraints (represented as service availability). They analyze various approaches to the design of path protection mechanisms (both dedicated and shared) in elastic optical networks. Similar to [50], they skip re-routing due to its long operation following a failure. It is worth noting that the papers recognizing the trade-offs do not consider the financial aspect of transferring the opposing aspects into monetary units to express the goal functions in a unified way and thus supporting the decision.

## 2.2 Risk awareness in resilient networks design

Franke [30] notes that generally the discussion of the relationship between the technical aspect and the business context related to the management of telecommunication networks is poorly developed. While the methods and protocols for network resilience, described for instance in [20, 55, 60], are not a new topic, a business-oriented approach to resilient network design has not been studied in depth, and many problems remain. Risk had originally been dealt with in two types of industries: financial (where investment portfolios are being selected) and technological (where failures jeopardize human safety or societal welfare). The two approaches have been combined in the telecommunications sector (selection of new investments [26], or in security against faults generated by malicious behavior [1]). We focus on resilience provisioning, a topic covered in general in [21]. However, in the context of risk management in resilient networks, risk assessment is the most popular topic. While the value-at-risk measure was postulated to be used in communications networks for resilience quantification [1, 21, 43], this has little reference to optimization methods used in the financial sector, where value-at-risk is basic [41].

A proposal of linear optimization of risk mitigation strategies is outlined in [18]. Vajanapoom and Tipper [58] present a set of papers on the topic, with the most comprehensive presenting linear programming-based models applying the risk exposure measure. A similar average-based risk measure in network design is proposed in [48]. Dikbiyik et al. [25] present an integer linear program that also bases risk response on risk exposure, where the consequences are based on the cumulative downtime exceeding the assumed threshold. Gonzalez and Helvik [32] provide a set of optimization approaches using two-stage stochastic programs to increase the provider's gain (to minimize recovered connection costs and penalties paid for outages incurred). As mentioned before, these approaches are typically not based on an exact analysis of quantile risk measures relevant to business. An approach somewhat similar to ours is given in [24], which considers large-scale failures (disasters) in cloud environments. It presents various approaches to optimized design of disaster-resilient cloud infrastructures, when one of the approaches is based on risk minimization.

To the authors' knowledge, the only work that combines energy-efficiency problems with optimization methods specific for risk management has been presented by Cano et al. [11]. However, the approach is totally different to ours. Apart from the fact that the authors deal with energy-efficient planning of heating/cooling systems in buildings rather than with the networking problems, risk management is related to the investment process (the risks are related to a lack of energy sources, etc.) and not to adverse events such as failures. In contrast to our approach, they develop a two-state stochastic optimization problem usually encountered in risk-based investment planning, for instance in the chemical industry [6, 68].

# 3. DESIGN OF RESILIENT ENERGY-EFFICIENT NETWORKS

We take a typical assumption that energy can be saved by using a *sleep mode*. There are three options for the operation of devices with respect to power management procedures investigated by researchers [50]: (a) active mode (fully used): at each time point, the device can operate at full capacity; (b) sleep mode (low power/standby/idle/hibernation): the device is using some energy and can switch to active mode almost instantaneously; (c) switched off mode (inactive): the device is not using any energy, and it takes some considerable time for the device to be woken up. While the difference between the last two modes is important in traffic engineering, here we are interested in network dimensioning (long-term behavior), and we identify the sleep mode with switching off [36]. Perello et al. [50] deal with two basic types of sleep modes: link sleeping mode (LSM) and optoelectronic device sleep mode (OESM). The latter assumes switching off transponders or regenerators, while the former mainly concerns optical amplifiers. Since the switching on process in OESM can be fast (order of milliseconds), it may be suitable for protections. LSM (with wake-up time taking seconds) is too slow for protection methods, and we take it into account for re-routing methods only. Later, we study the influence of this assumption on risk assessment. Another aspect relating the sleep mode to resilience is stressed by Caria et al. [12], who emphasize that due to switching off, traffic paths are longer, increasing delays, susceptibility to failures, and degrading connectivity.

From the optimization viewpoint, the sleep mode is feasible only with selected *energy profiles*, i.e., energy usage characteristics or functions of traffic load being carried through a component. Ricciardi et al. [54] partition energy profiles into three groups: (a) experimental: presenting real-world data taken from device vendors (declarations) or operators (measurements), etc.; (b) analytical: models focused on an operating environment, omitting many details and configurations while still presenting exact characteristics; and (c) theoretical: the simplest abstractions that grasp the most general behavior of the devices. Here, we use the last type, since we want to show a new methodology of heuristic-based optimization and the risk mitigation-based approach. We posit that dealing with more complex energy profiles may hinder the presentation and unnecessarily lengthen the paper, and the influence on the general results will be low. While it is assumed that the desirable energy profile should be linear (i.e., with a zero cost at origin and the slope proportional to the load), in practice other cases are encountered that ensure the feasibility of the sleep mode. For almost all devices [49], energy profiles consist of the static (fixed) part, active only when a link is used, and the dynamic part dependent on various parameters (frequency, voltage, or traffic load). The simplest model adequate for the sleep mode assumes that when the link is not used, the energy usage *EU* is zero; then, if a link is used, the fixed cost E0E0 is significant and the dynamic part is proportional with a constant EpEp to the link load LL. We call this model $f + p$, i.e., fixed + proportional. The mathematical definition of $f + p$ is as follows (see Fig. 1a):

$$EU = \begin{cases} 0 & \text{if } L = 0 \\ E_0 + E_p \times L & \text{if } L > 0 \end{cases} \tag{1}$$

Generally, each model with a fixed cost or *concave* dynamic part favors the sleep mode, since in this case, it is energy-efficient to increase the loads in some elements and switch off others (by decreasing traffic flows carried via them to zero). A concave function is characterized by decreasing marginal energy usage per traffic capacity unit, being responsible for the economical effect known as the *decreasing return to scale*(see Fig. 1b). Here, we assume that the energy profile of all links in networks we study is simply an increasing concave function, since it not only approximates $f + p$ well, but also is justified for many other important cases. This type of energy usage is reported in some transport technologies (as mentioned in Sect. 2), and it also approximates the behavior of energy profiles characteristic for adaptive/multiple line rate (ALR) approaches [37]. The latter are proposed as Ethernet 802.11az, or they are relevant for optical networks with wavelength division multiplexing (WDM), where some wavelengths and related transponders can be switched off; additionally the wavelengths themselves can have various rates. The relevant models are illustrated in Fig. 1c–d.

We assume the static network dimensioning scenario in a transparent optical network, where for a given network we look for routing lightpaths for a long horizon, and where the traffic demand matrix does not change. The aim is to minimize total energy usage. We do not want to focus on RWA aspects; therefore, we assume that all flows can attain continuous values, and routing can be described with linear constraints. Nevertheless, as the goal is to minimize energy usage constructed out of concave energy profiles, the optimization problem is known as a difficult *concave optimization* task. In such a case, it can be linearized by providing additional constraints and binary variables that are used to model the concave function with linear segments. The problem can be represented in the form of an *NP*-hard MILP optimization task. The general methodology is described in [51], ch. 5]. The inclusion of many binary variables is prohibitive for major problems, since it is not possible to obtain the exact solution in practice. Therefore, to perform the numerical studies, we decided to use a modified Yaged's method [51], ch. 5.6], a heuristic polynomial algorithm with rapid convergence, designed to solve problems where the concave goal function is minimized. The method is outlined in Fig. 2. The algorithm allocates flows iteratively using shortest paths. We use Dijkstra's algorithm for single paths and Suurballe's algorithm for the shortest cycles (necessary to find backup paths for protections). Here, "shortest" means "minimizing energy usage", although for comparison we also show results based on distance and hop-count minimization. Yaged's method is based on the observation that concave energy costs mean that it is better to use loaded links rather than opening new ones. As a result, the demand flows are not bifurcated, i.e., they are routed with single paths, a feature highly desirable in optical networks. The complete solution provides a set of links that do not carry traffic (they can be switched off).

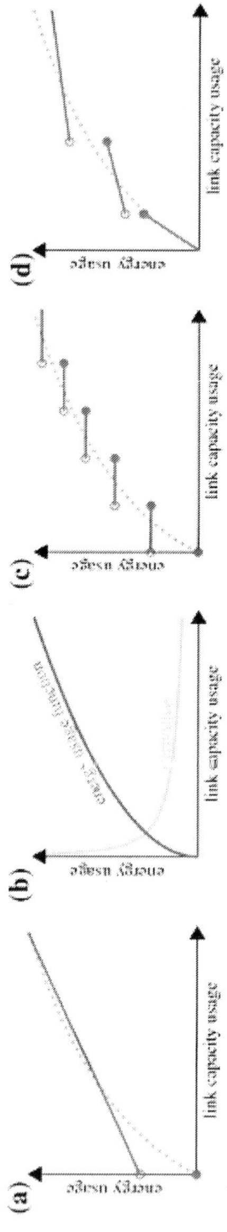


**Figure. 1** Various energy profiles supporting sleep modes from the optimization viewpoint. *Dotted curves* represent the concave approximations. **a** Fixed + proportional (*f* + *p*) energy profile (with an approximating concave *curve*).**b** Concave energy profile with its derivative, illustrating the *decreasing return to scale.* **c** Energy profile for ALR: step function (with an approximating concave *curve*). **d** Energy profile for ALR: linear stepwise function or piecewise concave cost (with an approximating *curve*)

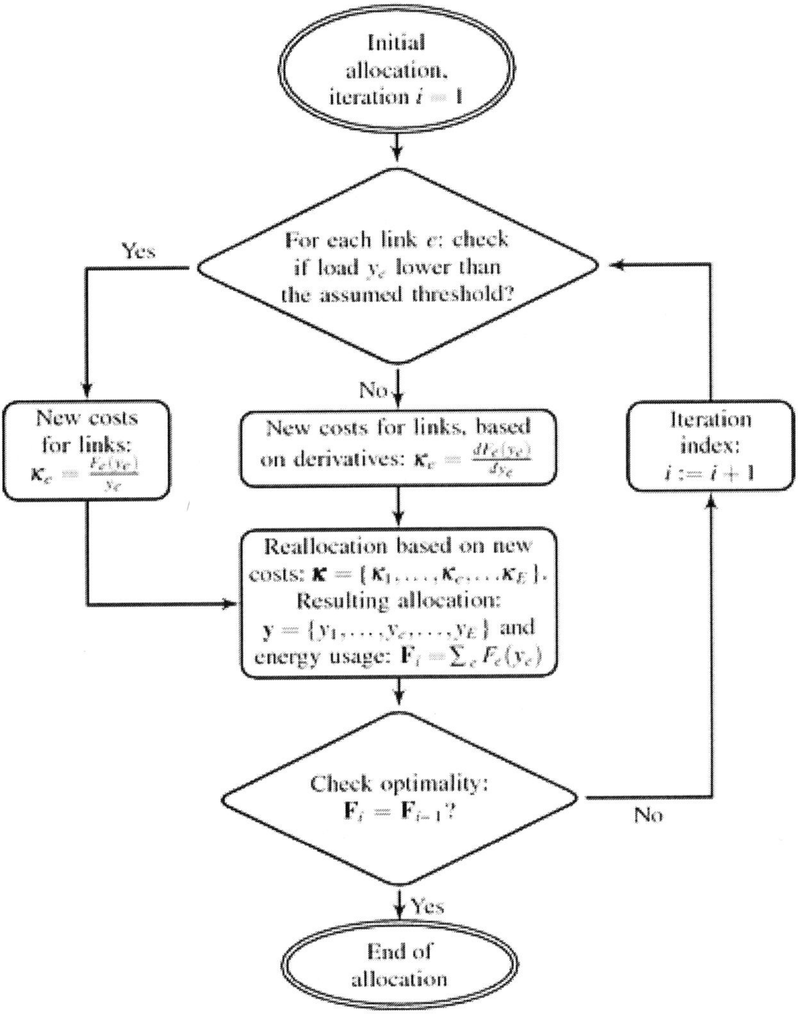


**Figure. 2** A sketch of routing optimization with a modified Yaged's method

Each link ee (in a network consisting of EE links), carrying traffic load yeye, has its own concave energy profile Fe(ye)Fe(ye). First, the algorithm calculates the initial flow allocation y⋆={y⋆1,...,y⋆e,...,y⋆E} (e.g., simply with the shortest hop routing), and then the output energy usage: F1=∑eFe(y⋆e). Then, the weights of links κκκκ are calculated as derivatives of energy usage found for the current link loads: $\kappa_e = \frac{dF_e(y_e)}{dy_e}|_{y_e=y_e^\star}$. The new output energy usage found on the basis of shortest routing with the set of link weights κκκκ, i.e., F2=∑eFe(y⋆⋆e) will be lower than F1F1. The algorithm iterates this way until energy usage ceases to decrease. To make the solutions independent of the initial allocation, we use a modified algorithm, where to calculate a new weight for a link with a low traffic load (below an assumed threshold), we do not use the derivative, but rather the tangent of the secant between the origin and point $(y_e^\star, F_e(y_e^\star))$: $\kappa_e = \frac{F_e(y_e^\star)}{y_e^\star}$.

Yaged's method performs well not only for concave cost functions, but for all cost functions that have non-increasing derivatives of increasing costs. Therefore, it can be used not only for simple concave approximates of the $f + p$ profile, but directly for such a profile. This profile has a derivative that is constant (Ep) except for the Dirac delta function in 0, which can be modeled as a rapidly decreasing monotonous function dependent on the relation between E0 and Ep. The character of this decrease is not a problem from the perspective of calculation complexity, as application of Yaged's method performs a simple numerical calculation where the derivative can be given as a set of numbers, and symbolic calculations are not necessary. We apply Yaged's method to the $f + p$ profile in Sect. 5.

# 4. RISK MANAGEMENT IN RESILIENT NETWORK DESIGN

Network designers are experienced in recovery methods and mathematical tools to describe and optimize their operations. However, this specialist work is influenced by business management decisions on how money should be spent. The interface between the technological and business levels is handled by *risk management*, providing design results useful for the main business, i.e., ensuring the continuity of the operator's mission in the presence of adverse events. One of the relevant aspects is to match recovery methods to client needs defined in the service level agreement (SLA). From this perspective, it may be unreasonable not to provide recovery to clients whose SLAs predict a high level of penalties if the connections are interrupted. Similarly, it does not make sense to provide 1+11+1 protection (the most reliable option from the engineering viewpoint) to all clients, since the fees charged to many clients do not balance the cost for such protection, making it an unnecessary cost for operators [19, 39].

Dealing with client needs is discussed during the *risk management cycle*. It follows the plan–do–check–act philosophy of the Deming cycle [18, 65] and organizes how businesses approach the analysis and optimization of responses (recovery methods in our case) to adverse random events (failures), which are described by risk engineering. Risk is understood as "an uncertain event or condition that, if it occurs, has a positive or negative effect on an objective" [35], where each operator's objective is to increase profits and minimize losses. As both are dependent on random events, the risk is described with two basic parameters: by *probability* and *impact*, the latter best expressed in monetary (financial) units. Therefore, the first stage of the risk management cycle is *risk assessment* to identify types of risks (e.g., node/link failures) and evaluate their parameters (i.e., using statistical methods) with the aim of expressing the impact as penalties imposed on the operator with the relevant measures. After recognizing and prioritizing risks, it is possible to control them using *risk response*, first choosing feasible methods (e.g., preselecting specific recovery methods relevant to a given technology) and then deciding how to combine them to conform to policies assumed by the network operator while assessing the predicted decrease in risk and costs involved. Finally, the countermeasures should be *deployed* (involving the re-configuration of resources, testing, etc.) and*monitored* (to check whether the response implemented meets the intended goals). These two closing stages are less interesting from the network dimensioning viewpoint; therefore, we do not discuss them.

## 4.1 Risk assessment

Energy is one of the costs, and its minimization is the concern of the operator, while high levels of resilience are in the client's interest. During risk management, it is possible to trade off these opposing aspects. If the client is risk-neutral [31], the best way is to express them with the same monetary measure. If there is a known energy profile, the energy cost can be easily expressed monetarily, and the way to express the monetary impact of not conforming to the client's needs is to use penalties. A model that transforms technical loss (as perceived by the client) into its monetary equivalent is defined by *compensation policies* [43]. A penalty is paid to the clients affected by failures if such a case is predicted in their SLA. The most typical compensation policy is to assume that the penalty is proportional to the cumulative downtime over a given interval. As this policy can be expressed by *interval availability*, we call it Av. An opposite approach is to base the policy on the number of all outages perceptible at the service level over a given interval. This means that the interest is shifted to *continuity* rather than availability; this applies to highly demanding services which are rendered useless by a failure no matter how fast the recovery method (a case for real-time control [15]). We call this policy Co. These two extreme policies can be combined. For instance, it can be assumed that the penalty is based on the number of outages exceeding a selected downtime threshold [25] or that the penalty is not scaled proportionally over an interval [23]. As this is the first approach to energy-efficient design with risk awareness, we stick to the basic policies in order not to overcomplicate the view.

Failure is a random event; therefore, the value of a penalty is a random variable. Risk measures are the means for predicting the behavior of network connections before and after deploying the selected risk response. Various known risk measures can be chosen [21, 66]. The simplest to interpret is the mean penalty per interval, known as *risk exposure* (RE): it is the average amount of penalties that reduce the profit. It is also *coherent* [4], a useful feature in the context of convex optimization [17]. Nevertheless, it does not grasp the variability of the impact or extreme values (i.e., two totally different distributions may have the same mean value [31]). Hence, quantile risk measures have been proposed. The most fundamental is value-at-risk (VaR), the maximum penalty with a given confidence interval. VaRVaR is commonly accepted in investment management for which many optimization methods elaborated in *modern portfolio theory* exist. They have also been proposed for networking [1, 21, 33, 43]. Let $\xi$ be the level of penalties, pertaining to a single connection or a whole network. If $P\xi(x)=Pr\{\xi\leq x\}$ is the cumulative distribution function of $\xi\xi$, VaRVaR is defined as the maximum penalty with a given confidence interval $\eta\eta$: $VaR\eta=P-1\xi(\eta)=inf\{x:Pr\{\xi\leq x\}\geq\eta.\}$ Although VaR is theoretically not a coherent risk measure [1, 4, 21], another study [17] shows that from a practical perspective, it behaves as a coherent measure. While the value of RERE can be treated as a mean measure of operator's profit reduction over a given interval, the value of a quantile risk measure can be treated as a suggestion how much money should be buffered in the worst case to deal with the results of adverse events over an interval. The latter approach is typically assumed in the risk-based project management [27].

## 4.2 Risk response

An *informed business decision* on risk response is to invest in recovery if the penalties are high or recovery costs are low. From the business viewpoint, basic responses (e.g., risk avoidance, transfer, or acceptance [16, 18]) can be very different, but a crucial option in the design of resilient networks is *risk mitigation*. It involves recovery methods to decrease the risk RR by reducing impact (i.e., penalties) of failures. However, other than from the operator's viewpoint, both the levels of RR and BB (the energy budget/cost for risk mitigation/recovery provisioning) are taken into account as an important goal. These two objectives are contrary to each other, and there are two basic methods to deal with such cases [70]: either (a) constructing an aggregate objective function (with weights assigned to various goals) or (b) modeling one objective as a constraint in the optimization of another. Risk mitigation uses the first approach by weighing both goals with monetary values. We follow this path since energy profiles and compensation policies enable us to define monetary equivalents for both of them. This approach is also more flexible, since in the risk-neutral contexts, it enables us to relax the requirements to some extent. The operator still needs to pay penalties, but it is not necessary to define restrictive thresholds on resilience that cannot be exceeded.

The aggregation approach finds a convex characteristic, representing decreasing Pareto-optimal (non-dominated) solutions. Formally, option (r1,b1, with risk level r1r1 and related budget b1b1, dominates the option (r2,b2) if and only if: $(r1,b1) \succeq (r2,b2) \Leftrightarrow r1<r2 \wedge b1 \leq b2$. The Pareto front, i.e., a theoretical curve presenting non-dominated options, is shown in the budget-risk (B,R) space in Fig. 3. The level of risk before deploying the response is known as *baseline risk* Rbase, while the level below which it is not possible to drop (independently of the budget, since failures will happen) is known as *residual risk* Rres. The following *business strategies* to select risk mitigation are elaborated in the IT security field (they are summarized in Table 1). (a) *Risk minimization*, RM [3]: the aim is to downsize risk as much as possible (minRminR) for the lowest possible cost: argminB(dR(B)dB=0). From the theoretical viewpoint, if treated as absolute minimization, this option is perceived as extremely costly and not useful in a typical carrier's practice (except for operating critical infrastructures). However, in Sect. 5, we show that this is not true for the specific networks with concave energy costs. (b) *Total benefit coverage*, TC [3]: the cost of organizing the strategy is equal to the risk reduction: D=Rbase−R(B). This means we look for the budget so that argminB≠0|D=B|, and as a result, uncertain risk is exchanged for a known (certain) cost. This can also be presented with linear

constraints using properties of non-dominated solutions, since we are interested only in values of DD better than the budget BB (which means the solution will be feasible if the curve is below the (Rbase−R)=B line in Fig. 3, which is not always the case), and we select the option with the highestDD level. (c) *Cost balance*, CB [57]: the aim is to find a point where the risk level and the budget involved are the same: |R(B)=B|. This is recommended as a strategy for finding business continuity for US federal institutions; again, we can present this strategy with linear constraints ensuring that the risk is greater than the budget B≤R and finding the minimum risk level. (d) *Profit maximization*, PM [3]: the aim is to maximize the total monetary gain. Looking for such a solution is a well-known economical problem of *net utility maximization* [40], where utility is the risk reduction and the price is the budget involved. Mathematically, the solution is the point where the marginal risk reduction is balanced by the marginal budget increase or min(R(B)+B). We also show *risk acceptance*, RA, which is not a risk mitigation strategy; however, we use it as a benchmark, since it is a budget optimization solution.

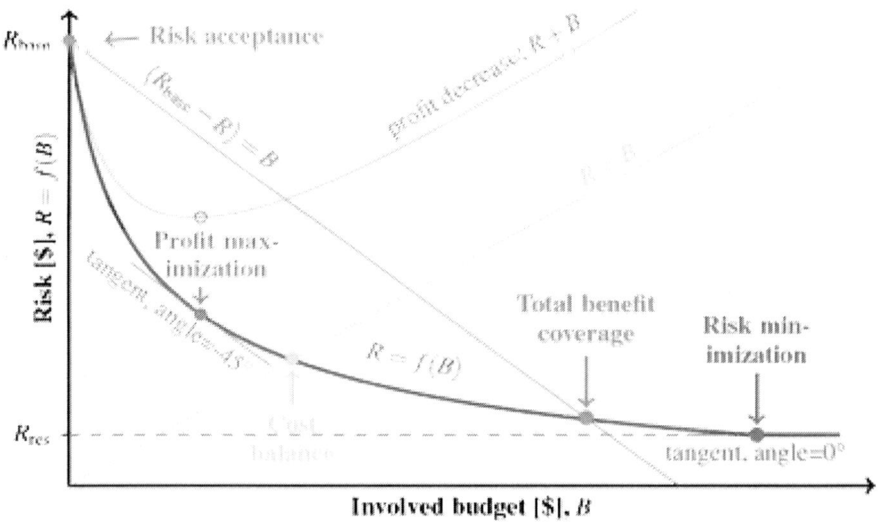


**Figure. 3** Basic risk mitigation strategies shown in the (B,R)(B,R) space. The *blue line marks* the non-dominated solutions. Line R=BR=B is the oblique asymptote for the profit decrease *curve*

**Table 1** Summary of the risk mitigation optimization approaches

| Strategy | Additional constraints | Goal function |
|---|---|---|
| Risk acceptance | (none) | minBminB |
| Risk minimizatio n | (none) | minRminR |
| Total benefit coverage | B≤D=Rbase−RB≤D=Rbase −R | maxDmaxD |
| Cost balance | B≤RB≤R | minRminR |
| Profit maximizatio n | (none) | min(R+B)min(R+ B) |

# 5. NUMERICAL EXAMPLE, PART 1: COARSE-GRAINED OPTIMIZATION

First, we optimize the use of resources from the viewpoint of energy usage. We demonstrate that the modified Yaged's heuristic performs well in comparison with the exact optimization. Second, we show simulation results of real energy usage and risk values and look for solutions for the assumed risk mitigation strategies. While these results are being discussed for protections, we also check how re-routing with switching off backup resources performs from the risk awareness viewpoint. It appears that re-routing outperforms protection methods if only the amount of backup resources in switched on links exceeds a specific threshold.

The network and connections within it are modeled with a very simple approach, i.e., a network is represented by a graph with a set of perfectly reliable nodes and a set of unreliable links connecting the nodes. A demand is defined as a requirement to carry a volume of data between two nodes. It should be fully satisfied with a single non-bifurcated connection. Network topologies and demands are retrieved from the SNDlib library (http://sndlib.zib.de) as models of two different networks: PL, the Poland Network (polska.xml) representing a sparse topology; and Ge, the German Research Network (nobel-germany.xml) representing a dense topology.1 Each demand is assigned a single recovery option out of the following set. (a) *No recovery* (NR): all demands are carried in a way optimizing energy usage, which is treated as basic cost B0; if failures affect connections, they are interrupted. (b) *Dedicated path protection 1:1* (DP): the connections are routed with two disjoint paths, where the whole set of connections optimizes energy usage; in the normal state, the shorter of the disjoint paths is used as a working path, and the other (a backup path) is not used. This means that although the backup capacity must be reserved, it does not use energy if not needed [36]. (c) *Dedicated link protection 1:1* (DL): the working paths are routed as for NR, but backup segments for protecting working capacity in links are added (also assuming energy usage optimization). (d)*Shared backup path protection* (SP): a pair of paths for the connections is found as for DP; however, energy usage optimization takes into account that backup resources are shared among working paths. To calculate the usage, a single failure assumption is taken; this is challenged in simulations where multiple failures can be present. (e) *Shared backup link protection* (SL): it differs from DL in that backup capacity is shared among segments protecting the working capacity of various links (again, an absence of multiple failures is assumed during optimization). (f) *Re-routing (restoration) with x % backup capacity reserved in working links and an option to switch on sleeping links*: working paths are routed as in NR, then the links not being used are put to sleep; they can be woken up if it is not possible to re-route the affected connections using the backup capacity reserved in switched on links (the amount of the reserved backup capacity equals to xx % of the working capacity in a link). As for protections, the re-routing is studied in path (RPx) and link (RLx) versions. We do not present results for dedicated path/link protections 1+1, set analogously to DP/DL, but rather with traffic steadily repeated to backup resources (also in the normal state). From the risk perspective, these methods offer little improvement (with the existing technological solutions, the switching time is negligible); therefore, they are clearly dominated as being the most costly. Other methods can be applied in practice [20]; however, here we focus only on those that can be modeled in a compact way, and we show that even with this small set of options, non-trivial results can be obtained.

We assume that the energy profile is modeled with the concave *square root function* of a load on a link. The profile is linearized with linear segments, where each has the length of two in abscissa (representing the traffic load in a link). Table 2 compares results of the optimization of energy usage when calculated with the exact MILP formulation with the optimization software CPLEX 12.5 (on a server with an Intel Xeon E5-2680@2.93 MHz processor, 24 GB of RAM, and 12 cores) with those obtained by the MATLAB application of the Yaged's method. It was not possible to solve the exact model in any case (memory overflow due to many branching processes); therefore, a level of relaxation gap for branch and bound was assumed to not exceed 10 % and set prior to conducting the calculations. The optimization process took several hours (typically around 24 h), but it was able to provide results either up to 4 % better than with the heuristic Yaged's method or even worse. It has been demonstrated that Yaged's method converges after a finite number of steps; in our case, this is fewer than ten, and the whole process takes no more than 0.6 s.

**Table 2** Comparison of exact versus heuristic optimization for the PL network

| **Recovery option** | **Goal: CPLEX [CC]** | **Goal: modified Yaged's heuristic [YY]** | **Relative difference:** Y−Cmin{Y,C}Y−Cmin{Y,C} **(%)** | **Number of iterations of the Yaged's heuristic** |
|---|---|---|---|---|
| NR | 528.04 | 522.06 | -1.15 | 6 |
| DP | 971.77 | 972.09 | 0.03 | 4 |
| DL | 1186.40 | 1174.04 | -1.05 | 2 |

| Recove ry option | Goal: CPLE X [CC] | Goal: modified Yaged's heuristic [YY] | Relative difference: Y−Cmin{Y,C }Y−Cmin{Y, C} (%) | Number of iterations of the Yaged's heuristic |
|---|---|---|---|---|
| SP | 820.6 2 | 801.50 | -2.38 | 3 |
| SL | 900.6 2 | 936.37 | 3.97 | 1 |

Next, MATLAB was used to develop simulations for finding dynamic energy usage and risk values, assuming that the paths are assigned as those found during the optimization stage. Two distributions are associated with each link: the first describes exponential time between failures, while downtimes are modeled with the Pareto distribution. The distribution parameters are proportional to link lengths [60], and their values are retrieved from [42]. The independent failure and repair interchanging process is simulated in each link, and then the effect on connections (with the assumed recovery option in operation) is calculated with both compensation policies. Each connection has its own parameters necessary to find the exact value of the penalty, which is assumed to be the product of downtime hours (Av) or number of outages (Co) and the demand volume given with network models (in Mb/s). When risk values are calculated for the whole network, the distributions of the penalties estimated for individual connections are summed. For each simulation scenario, we assume that all the connections use the same recovery option; we held 1000 simulations over 100,000 h, enabling us to find the distributions of penalties and the risk measures (RE andVaR0.95). The penalty distributions do not have heavy tails, a phenomenon generally found in resilient networks [17]. For each simulation time point, we calculate the link loads and transform them with assumed square root energy profiles to the temporary energy usage. The risk measures and mean energy usage are expressed for an interval of 10 years.

A difficult problem is finding meaningful monetary coefficients to obtain a financial value of a unit of energy cost and penalty. Penalty coefficients depend on the market, SLAs in place, and technical network conditions (topologies, failure rates, etc). For the authors, as researchers working at academia, the challenge of accessing such data has been too great. Therefore, we use values found online. Taking into account that the presented estimation may be inexact, we also study results obtained when the orders of these coefficients are modified. The base values of coefficients are obtained as follows. We assume that an approximate cost of one kilowatt-hour in the USA is 0.2 USD,2 and following [46], we take the approximate energy for sending 1 Gb/s as being around 100 W on average, taking into account all the optical components. Carrying all the demands with NR generates a load of 26,243 Mb/s. During 10 years of operation (approximately 100,000 h), the total energy cost is 0.27119 million USD, which gives the energy unit coefficient of 520 USD, since optimization results give a value of 521.5 energy units. We also assume that the rental cost of 100 Mb/s optical fiber is approximately 1500 USD per month,3 which gives a total cost per hour of work of 1 Gb/s as approximately 20.5 USD. According to the given penalty policy,4 we assume that the penalty for 1 h of downtime is equal to the reduction in the payments of one day, which means the risk coefficient for Av is 24 times the cost of the rental per hour that is approximately 500 USD. To find the risk coefficient for Co, we apply the rule presented in [43], where the reimbursement for a single outage is the product of the reimbursement for a unit of downtime multiplied by the average duration of an outage. In the case of NR, the latter is 10 h, which means that the risk coefficient for Co is 5000 USD. That is, we assume that approximately the risk to energy unit coefficient is equal to 1 in the case of Av, and 10 in the case of Co.

Simulation results related to protections are shown in Fig. 4 (note that the abscissa gives the total energy used, i.e., the involved risk mitigation budget can be found by subtracting the amount of energy necessary to provide RA). The first surprising result is that for various risk measures, the relative position of solutions to various strategies differs very little, an effect quite opposite to the results obtained for the budgets based on capacity reserved reported in [16]. As for protections we assume to use OESM that switches on very fast, we do not penalize the cases when it is necessary to increase link capacities, and in this case, Av is a more adequate compensation policy. The results basing budgets on energy usage appear to be almost invariant of the optimization method (note that energy-efficiency promotes much larger paths, which are more prone to failures), and some of the optimum points overlap. This stems from the fact that with concave energy profiles, it is easy to provide a risk minimization strategy (in all cases, it is DL) that surpasses other strategies, since the difference in risk of various recovery options (except for NR) is very low, and the most reliable one is preferred. Even if the initial optimization is conducted with a classic capacity-oriented approach (minimization of the summarized number of hops) or reliability-oriented approach (minimization of the summarized physical distances reducing failure rates), the results show almost no differences from the solution perspective, although the numbers differ. It should be noted that although optimization changes, mitigation is still based on energy usage. Another surprising fact is that in such a coarse-grained approach to risk mitigation (i.e., one recovery option for all), the theoretical characteristic shown in Fig. 3 is highly degenerated. We show in Sect. 7 that this effect is less pronounced if the optimization takes into account reserved capacity (i.e., an approach taken in classic network dimensioning) rather than energy usage. It is necessary to emphasize that the degeneration effect is not related to the used monetary coefficients, but to the character of the budget consumption, where in our case, the energy is consumed only when the links are used. Due to very large penalty in the NR case, the monetary coefficients do not have the impact on the solutions. It is also notable that although local protections are more costly during capacity usage optimization, they are less costly from the dynamic energy usage perspective than their path counterparts. Additionally, local versions provide a higher reliability (i.e., decrease the risk), since they are more resilient to multiple failures. The path methods (DP and SP) use very long working and backup paths, since this is more advantageous from the energy usage perspective (more links are loaded); this effect also appears in the case of the backup segments protecting the links, although it is not as strong. Therefore, the link methods dominate.

As the use of the concave energy profile may seem controversial, we decided to provide the results also for a plain *f* + *p* profile. We used two profiles of this kind. They are reported in [59]. The first energy profile (known as OTN-100 Gb/s) has a very big fixed cost (E0=360 W) and a small proportional cost ($E_p = 3.6\,\mathrm{W/Gb/s}$). The second profile (known as OTN-10 Gb/s) has almost ten times smaller fixed cost (E0=34 W), and a very similar proportional cost ($E_p = 3.4\,\mathrm{W/Gb/s}$). Some results related to the use of these profiles are shown in Fig. 5. We can see that the energy values are different (we do not rescale the energy coefficients), but the qualitative character of the results is analogous to ones presented in Fig. 4, although it is possible to see a modest influence of the risk measure on the final results (domination and a solution for the strictly business-related PM strategy). This stems from the fact that the effect of scale is again present and that the energy is consumed mostly in the failure-free situations. Therefore, we can assume that the presented results are valid for a very broad range of energy profiles supporting the sleep mode.

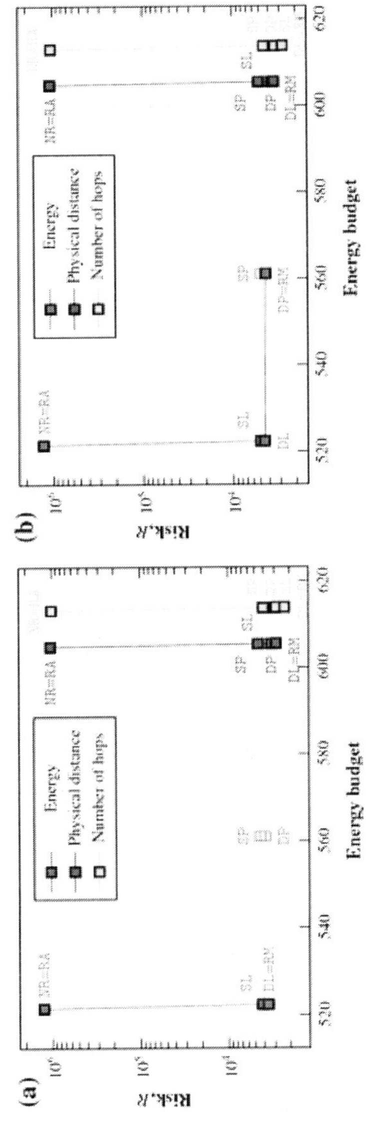


**Figure. 4** Summary of simulation results of various protection methods (the Av compensation policy is used to find the penalties) in the PL network with concave energy profiles. *Filled squares* denote the non-dominated solutions (they are joined by *lines*), while the *empty squares* represent dominated solutions. The risk/energy coefficients do not affect domination. In all the cases: RM = PM = CB = TC. **a** Risk measure: *RE*. **b** Risk measure: *VaR*

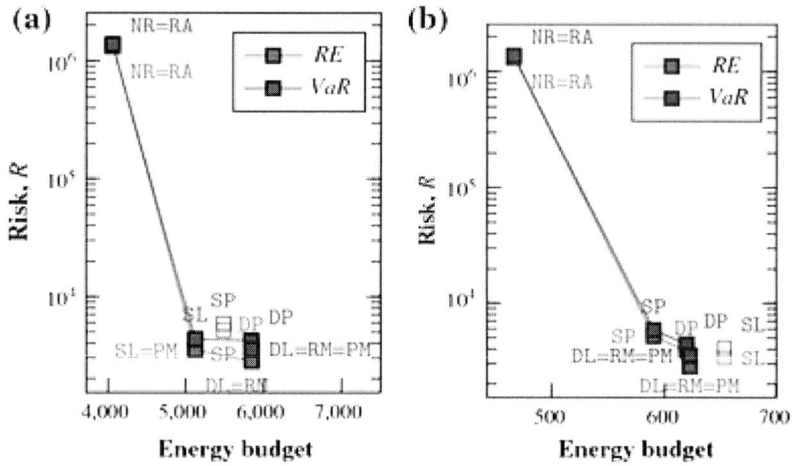


**Figure. 5** Summary of simulation results of various protection methods (the Av compensation policy is used to find the penalties) in the PL network with two $f + p$ energy profiles. *Filled squares* denote the non-dominated solutions, while the *empty squares* represent dominated solutions. In all the cases: RM = CB = TC. **a** OTN-100 Gb/s profile. **b** OTN-10 Gb/s profile

We assess protections separately from the re-routing options since their establishment and operation philosophies are different: for protections, we find paths that use the shortest cycles which makes the working paths longer and increases energy usage, while re-routing uses the shortest paths. This also involves various protocols, signaling procedures, and switching times. However, some general comparisons are also instructive. In the case of re-routing, we check the fact that they may use resources in switched off links. Contrary to [13, 36] which claim the usefulness of switching off backup resources, Perello et al. [50,61] state that switching off the backup resources fully is not acceptable from the reliability viewpoint. Here we investigate the real impact of the opposite assumption. We assume that switching on the sleep resources is triggered when backup resources are necessary in the failure state, since other backup resources are consumed in fully utilized links or there is no physical connectivity to redirect a connection for an affected demand. In our case, whole WDM links are put to sleep, and contrary to the protection scenario, we assume to use the LSM mode. Therefore, switching on a WDM link takes a lot of time and considerably interrupts the connection. We consider it as an outage and quantify the related penalty with the Cocompensation policy.

The results are shown in Fig. 6 (while the results are presented for RE, the character of results for VaR does not change). An interesting fact about the re-routing options RPx/RLx can be observed: the greater the capacity reserved as backup in switched on links (i.e., ones in a non-sleep mode in the normal state), the lower the probability that it will be necessary to switch on sleeping links and therefore the total energy usage is lower. Hence, from the energy viewpoint, a good option is to switch off as many links as possible and switch them on only if it is not possible to route the connections that should be restored with non-sleeping links. This effect is shown as decreasing energy usage with increasing backup rate xx. The risk also decreases simultaneously, which is a phenomenon more obviously significant in the case of the Co policy. In fact, the best option is to use RL∞ which needs to wake up additional links only when there is no physical connectivity in the basic set of links due to multiple failures. Additionally, we can see that there is a threshold x (here, x=60%) when the re-routing starts to dominate the protection methods and is most advantageous for an operator from the risk mitigation viewpoint. However, the latter finding is valid for long agreements, which may be contrary to the assumptions supporting the Co policy [15]. We can generalize these results to the two following extreme cases: if reservation of capacity is costly (e.g., it must be leased), then protection still makes much sense. If, on the other hand, it is possible to overprovision network capacity and the operator's cost is related mainly to the energy usage (e.g., since the network is owned by this operator), re-routing is the preferred solution. Again, the results stem from the scale effects due to concavity of energy profiles.

# 6. RISK MITIGATION: FINE-GRAINED OPTIMIZATION IN ENERGY-EFFICIENT NETWORKS

Here, we propose a simplified approach to finding the optimal combination of recovery options, assuming that each connection can use its own option, contrary to the coarse-grained approach presented so far. Since in the PL network, there are 66 connections (121 for the Ge network), the number of all possible combinations, where each connection may have a different recovery option (skipping re-routing), is 566566. They cannot all be analyzed by simulations to obtain the exact risk and energy values. This is why we decided to base the calculations on linear programming with risk and energy values estimated in five simulation scenarios presented in Sect. 5, where all the connections apply a single recovery option. Output numbers will not be precise, since risk measures are nonadditive for quantile measures (even though they may be subadditive in practice [17]) and for shared protections, the resulting values are pessimistic (as the combination reduces the sharing of backup resources). Another simplification is that we assume linear aggregation of energy costs. We do not have an easy method for including concave energy costs in the constraints taking into account goal functions containing risk (unlikely to be the case when the concave energy usage function is the only element of the goal function as relevant in the linearization methods mentioned in Sect. 3). Although the approach is simplistic, a somewhat similar approach is successfully used in the project risk management field [69]. A more precise approach based on modeling of the total reserved capacity is outlined in [16]. Demands are denoted with d (there are D demand overall), and the recovery options used are denoted with t (so that t∈{NR,DP,DL,SP,SL}. The constant rdt gives the value of the risk measure obtained for demand d if it applies recovery option t, while the constant edtedt gives the energy consumption of demand dd when it uses recovery option tt. The overall energy usage if all demands are assigned NR (risk acceptance) is denoted as B0, while the baseline risk in this case is Rbase. All the constants are given in monetary units (thus, we do not disturb the formulation with energy/risk monetary coefficients). The binary variable xdt equals 1 if demand d is assigned recovery method t, and 0 otherwise. The constraints common to all the optimization programs to find solutions for different mitigation strategies are as follows:

$$\sum_t x_{dt} = 1 \qquad d = 1, \ldots, D \tag{2}$$

$$\sum_d \sum_t e_{dt} x_{dt} - B_0 = B \tag{3}$$

$$\sum_d \sum_t r_{dt} x_{dt} = R \tag{4}$$

$$R_{\text{base}} - R = D \tag{5}$$

Equation (2) enforces that each demand is assigned only one recovery method. Equation (3) defines the budget involved as the total cost of the enforcement of the combination of various recovery methods minus the cost of the risk acceptance case. Equation (4) determines the value of the risk incurred by the combination of various recovery options assigned to the connections, while Eq. (5) gives the value of the risk decrease. To finalize the formulation for an assumed mitigation strategy, it is necessary to add the goal function and additional constraints describing an assumed risk mitigation strategy, as enumerated in Table 1.

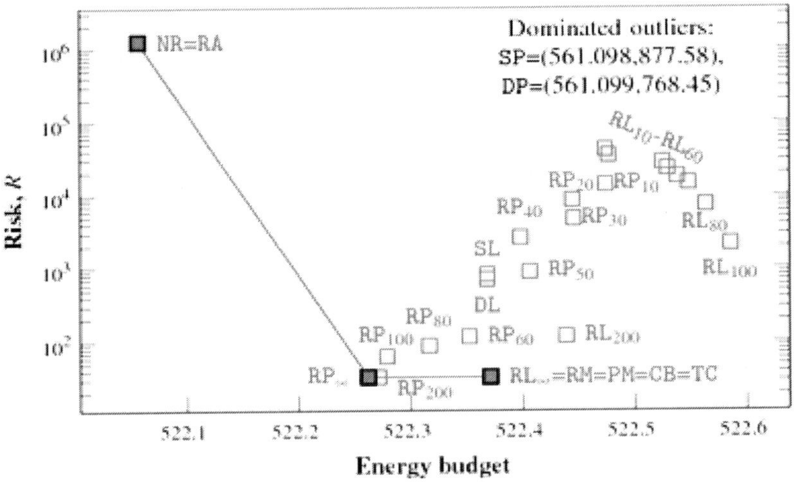


**Figure. 6** Summary of simulation results of various re-routing and protection methods (the Co compensation policy is used to find the penalties) in the PL network with concave energy profiles. Seven of 18 links are switched off in the normal state for re-routing methods. *Filled squares* denote the non-dominated solutions, while the*empty squares* represent dominated solutions. Routing is based on energy usage. The risk measure applied is RE

# 7. NUMERICAL EXAMPLE, PART 2: FINE-GRAINED OPTIMIZATION

CPLEX was used to develop optimization problems presented in the previous section to obtain solutions with mixed recovery options assigned to various demands. Although the problem assumes that MILP is used, the calculations are very fast (each takes less than 2 s) as there are only up to a few hundred binary variables. On the basis of the five basic configurations, where all the demands are assigned only one recovery option (NR, DP, DL, SP, and SL), we find risk values rdtrdt for all connections and the relative input to the total energy usage, edtedt. The latter is calculated as a sum over all links used by a demand, and proportional to the total energy used in the link and the demand volume.

We assume that although the demands can use various recovery options, they all apply the same compensation policy and risk measure. In this case, the optimization results presented in Fig. 7 are based on the Av policy (since the combined recovery options are limited only to protections, and for them, we assume application of this policy since the recovery switching is fast) and RERE (again, the results obtained with VaR are similar). We show the results for various values of the monetary coefficients by which we multiply a unit of energy used, capacity reserved, or penalty. The upper part shows the results of optimization if the budget is based on the energy used. For comparison, the lower part relates to optimization when it is conducted on the basis of the reserved capacity. Note that in the case of risk minimization, the method of budget calculation (energy or capacity) and the monetary budget coefficient do not matter. There are three basic facts seen if we optimize on the basis of the actual energy used. (1) The optimal solutions for all the strategies and coefficient combinations are very similar (the distributions of recovery options do not change, and the assignment of different options to connections is steady). For various coefficients, the resulting (B,R) values differ, so that if we decrease the cost coefficient 10 times, then BB decreases approximately 10 times, too. (2) The optimal solutions mainly combine two recovery options (both dedicated protections, excluding the shared link protection which seemed to be attractive due to the coarse-grained optimization; see Fig. 4). For a few demands, shared protection performs better than dedicated protection. The differences are very small and stem from more advantageous sharing of resources in backup links. This is again due to the fact the energy costs are not proportional to the capacities used. (3) The combination of the recovery options is also very similar across the coefficients representing monetary cost. Therefore, to find the fine-grained optimum, the best option is to look for profit maximization, which is almost identical to solutions to other strategies. It is based on the greatest ratio of the unit energy budget per unit decrease in risk found among various recovery options for each connection separately. This approach follows theoretical results in the net utility maximization theory as mentioned in Sect. 4. In the case of fine-grained optimization based on capacity reservation, the dependence on the monetary coefficients is more important, and there is no easy way to find a single solution for all the risk mitigation strategies or risk measures applied (a fact reported also in [16]).

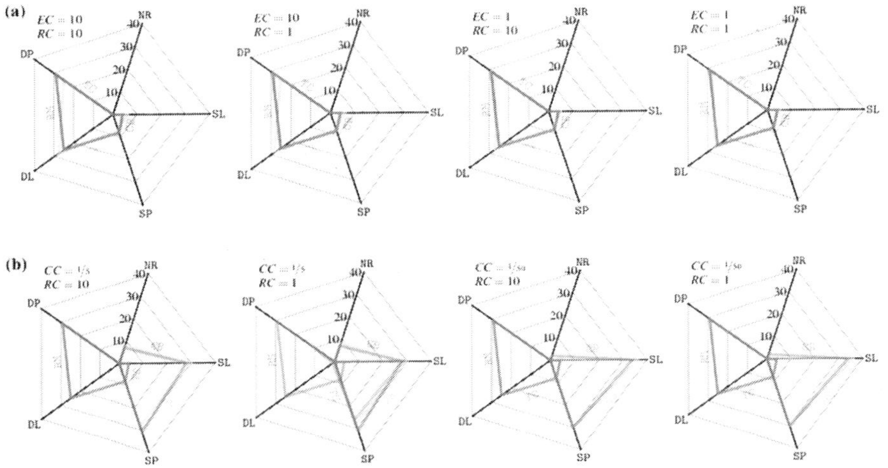


**Figure. 7** Summary of results for various risk mitigation strategies found with the fine-grained optimization based on various monetary coefficients. The RA strategy always applies NR to all the 66 demands in the PL network.**a** Involved recovery budget is based on energy usage. EC monetary energy coefficient. RC monetary risk coefficient. **b** Involved recovery budget is based on the amount of reserved capacity. CC monetary capacity coefficient. RC monetary risk coefficient

In Fig. 8, diagrams in the (B,R) space show selected results, again confirming the overshadowing options if we base the budget on energy cost (the effect is also shown with the coarse-grained approach described in Sect. 5). Again, this stems from the fact that additional energy usage over NR for providing recovery is low and similar across various recovery options due to the relatively low occurrence of failures when backup resources need to be used. This effect is additionally made stronger by the concavity of the energy profiles. The situation would be different if the budget were based on monetary costs proportional to reserved capacity, as typically done in network dimensioning. Related results are shown in Fig. 8 and in the lower part of Fig. 7, where more variability is seen for the capacity-based solutions. To make both situations comparable, the capacity and energy monetary coefficients are selected so that the total cost for non-recovered connections is of the same order as the cost when the budget is based on energy usage. Although risk parameters do not change, the budget structure is quite different and changes the character of the results. We can see that the solutions differ, and even NR appears in the combinations responsible for risk mitigation. There is no simple method of using profit maximization as a universal basis for finding solution to risk mitigation. In Fig. 8b, it can even be seen that we are able to find solutions for cost balance and total benefit coverage by looking for non-dominated solutions and characteristic lines shown in Fig. 3. Therefore, when the budget is based on the costs proportional to reserved capacities, the theoretical characteristic assumed in Sect. 4 is not degenerated. However, as can be seen, the absence of degradation is in fact a negative effect from the practical perspective.

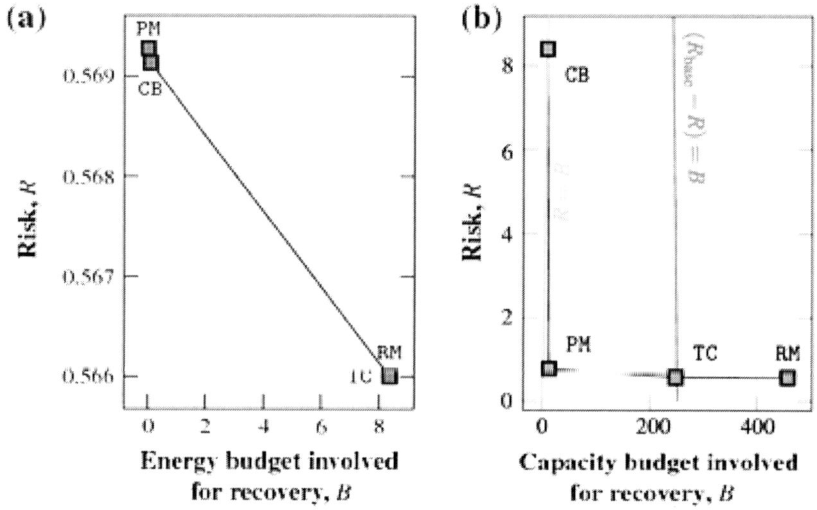


**Figure. 8** Results presenting various response strategies with fine optimization. None of the optimal solutions are dominated by another. The energy and risk monetary coefficients equal 1. The capacity coefficient equals $1/50$. **a** Budget based on energy usage. **b** Budget based on capacity reservation

# 8. CONCLUSIONS AND CHALLENGES

We present a framework for dealing with the assignment of recovery methods in energy-efficient resilient networks using the sleep mode. As energy-efficiency is meant to be traded off with resiliency, we propose to overtake this trade-off by using concepts elaborated in risk engineering. First, to find the optimal flows for recovery methods, we use the properties of typical energy profiles (concave and $f + p$) and then we are able to apply an effective Yaged's heuristic. Then, after expressing energy and risk with monetary equivalents, we show the properties of various solutions of respected risk mitigation strategies used in business management. Surprisingly, in the case of networks dealing with sleep modes, the solutions to these strategies are typically reduced to almost the same configurations, providing network designers with simple rules of thumb independently of the assumed compensation policies, risk measures, topologies assumed, and monetary relationships. Since it is sometimes claimed that the sleep mode is not relevant for backup resources, such an option is also checked from the risk engineering viewpoint. It is shown that under some conditions, this is a feasible and advantageous option. We show that the intuitions related to the recovery costs based on capacity reservation are not necessarily valid when these costs are based on energy usage. For instance, as the lengths of paths are typically longer in the energy-efficient networks, the approach where shorter connections are recovered with re-routing, and longer ones with protections, is not necessarily useful in general. It is simply better to be directed with the type of reservation costs: when they are non-negligible, a more reasonable decision is to apply protections for all the connections. On the other hand, if the capacity reservation is free of charge or its cost is not significant, then it may be reasonable to concentrate working connections in a small number of links to be able to put others to sleep and save energy. This will cost a negligible increase in the total penalty risk.

We perceive the following problems as our further work: (a) we currently assume that in each scenario, all the demands apply the same compensation policy, although these differ in real situations; it is necessary to elaborate the modeling that allows us to calculate risk measures for other connections differentiated from this perspective; (b) the optimization model proposed for combining recovery options is naïve (although it appears to be useful); we predict that the iterative method (cyclically repeating the risk assessment and risk response stages as in the risk management cycle) that combines interchanging optimization and simulation with updates of the risk measures and the recovery assignment can be convergent and will provide exact risk and energy usage data; (c) here, there is only one budget dimension taken into account; however, it may be also important to take into account the combination of the monetary cost of energy usage and leasing the capacity, as it is crucial in the protection versus re-routing decision; (d) energy costs and profiles may change hourly or daily; multi-time-period design can be used to model more exactly the aim to provide energy-efficiency; (e) another option for improving the precision of the models presented is to add more exact energy profiles describing various elements in the optical network, also involving the opaque model, multiple layers, RWA constraints, or limited amount of capacity installed in the links.

## FOOTNOTES

1 Due to a lack of space, we are not able to present all the results here; therefore, others are given here:http://home.agh.edu.pl/~cholda/research/energy-efficiency-vs-resilience/. The results obtained for PL and Genetworks are similar in nature; therefore, we do not show those relating to Ge in the paper.

2 Electric power monthly: http://www.eia.gov/electricity/monthly/epm_table_grapher.cfm?t=epmt_5_6_a.

3 BTnet Leased Line: http://business.bt.com/broadband-and-internet/leased-lines/.

4 BTnet Service Level Agreement: http://business.bt.com/assets/pdf/broadband-and-internet/datasheet/BTnet_leased_line_service_level_agreement.pdf.

## ACKNOWLEDGMENTS

The authors would like to thank P. Boryło, P. Guzik, A. Kamisiński, A. Lasoń, K. Rusek, J. Rzasa, and A. Szymański for useful discussions related to the topic of this paper.

## REFERENCES

1. Ackermann, T.: IT Security Risk Management. Perceived IT Security Risks in the Context of Cloud Computing. Springer, Berlin (2013)

2. Aleksić, S.: Energy-efficient communication networks for improved global energy productivity. Telecommun. Syst. **54**(2), 183–199 (2013)

3. Anderson, E.E.: Firm objectives, IT alignment, and information security. IBM J. Res. Dev. **54**(3), 5 (2010)

4. Artzner, P., et al.: Coherent measures of risk. Math. Financ. **9**(3), 203–228 (1999)

5. Bao, N.H., et al.: Power-aware provisioning strategy with shared path protection in optical WDM networks. Opt. Fiber Technol. **18**(2), 81–87 (2012)

6. Barbaro, A., Bagajewicz, M.J.: Managing financial risk in planning under uncertainty. AIChE J. **50**(5), 963–989 (2004)

7. Bianzino, A.P., et al.: A survey of green networking research. IEEE Commun. Surv. Tutor. **14**(1), 3–20 (2012)

8. Bolla, R., et al.: Designing optimal energy profiles for network hardware. In: Proceedings of GLOBECOM'12. Anaheim, CA (2012)

9. Bolla, R., et al.: Energy efficiency in the future Internet: a survey of existing approaches and trends in energy-aware fixed network infrastructures. IEEE Commun. Surv. Tutor. **12**(2), 223–244 (2011)

10. Bolla, R., et al.: Enabling backbone networks to sleep. IEEE Netw. **25**(2), 26–31 (2011)

11. Cano, E.L., et al.: Energy efficiency and risk management in public buildings: strategic model for robust planning. Comput. Manag. Sci. **11**(1–2), 25–44 (2014)

12. Caria, M., et al.: A comparative performance study of load adaptive energy saving schemes for IP-over-WDM networks. IEEE/OSA J. Opt. Commun. Netw. **4**(3), 152–164 (2012)

13. Cavdar, C., et al.: Energy-efficient design of survivable WDM networks with shared backup. In: Proceedings of GLOBECOM'10. Miami, FL (2010)

14. Chiaraviglio, L., et al.: Modeling sleep mode gains in energy-aware networks. Comput. Netw. **57**(15), 3051–3066 (2013)

15. Chołda, P., et al.: Continuity-based resilient communication. In: Proceedings DRCN 2009. Washington, D.C. (2009)

16. Chołda, P., et al.: Risk mitigation in resilient networks. In: Proceedings of RNDM 2014. Barcelona, Spain (2014)

17. Chołda, P., et al.: Risk-awareness in resilient networks design: value-at-risk is enough. In: Proceedings of NETWORKS 2014. Funchal, Portugal (2014)

18. Chołda, P.: Risk-aware design and management of resilient networks. In: Proceedings of RISI 2014. Fribourg, Switzerland (2014)

19. Chołda, P., et al.: A survey of resilience differentiation frameworks in communication networks. IEEE Commun. Surv. Tutor. **9**(4), 32–55 (2007)

20. Chołda, P., Jajszczyk, A.: Recovery and its quality in multilayer networks. IEEE/OSA J. Light. Technol.**28**(4), 372–389 (2010)

21. Chołda, P., et al.: Towards risk-aware communications networking. Reliab. Eng. Syst. Saf. **109**, 160–174 (2013)

22. Claise, B., et al.: Energy object context MIB. IETF Internet-draft (work in progress) (2014)

23. Clemente, R., et al.: Risk management in availability SLA. In: Proceedings of DRCN 2005. Lacco Ameno, Island of Ischia, Italy (2005)

24. Colman-Meixner, C., et al.: Disaster-survivable cloud-network mapping. Photonic Netw. Commun. **27**(1), 141–153 (2014)

25. Dikbiyik, F., et al.: Minimizing the disaster risk in optical telecom networks. In: Proceedings of OFC/NFOEC 2012. Los Angeles, CA (2012)

26. Elnegaard, N.K., Stordahl, K.: Modelling uncertainty and risk in telecommunication investment project. Telektronikk **104**(3/4), 119–135 (2008)

27. Fan, M., et al.: Choosing a project risk-handling strategy: an analytical model. Int. J. Prod. Econ. **112**(2), 700–713 (2008)

28. Francini, A., Stiliadis, D.: Rate adaptation for energy-efficiency in packet networks. Bell Labs Tech. J.**15**(2), 131–146 (2010)

29. Francois, F., et al.: Leveraging MPLS backup paths for distributed energy-aware traffic engineering. IEEE Trans. Netw. Serv. Manag. **11**(2), 235–249 (2014)

30. Franke, U.: Optimal IT service availability: shorter outages, or fewer? IEEE Trans. Netw. Serv. Manag.**9**(1), 22–33 (2012)

31. Garvey, P.R.: Analytical Methods for Risk Assessment. A Systems Engineering Perspective. CRC Press, Boca Raton (2009)

32. Gonzalez, A.J., Helvik, B.E.: Guaranteeing SLA availability in telecommunications networks. In: Proceedings of NETWORKS 2012. Rome, Italy (2012)

33. Gonzalez, A.J., Helvik, B.E.: SLA success probability assessment in networks with correlated failures. Comput. Comm. **36**(6), 708–717 (2013)

34. Idzikowski, F., et al.: TREND in energy-aware adaptive routing solutions. IEEE Commun. Mag. **51**(11), 94–104 (2013)

35. IEEE Std 1490TM1490TM-2003: IEEE guide adoption of PMI standard. A guide to the Project Management Body of Knowledge (2004)

36. Jirattigalachote, A., et al.: Dynamic provisioning strategies for energy efficient WDM networks with dedicated path protection. Opt. Switch. Netw. **8**(3), 201–2013 (2011)

37. Kilper, D., et al.: Energy challenges in current and future optical transmission networks. Proc. IEEE**100**(5), 1168–1187 (2012)

38. Liu, Y., et al.: Delay-SRLG constrained, backup-shared path protection in WDM networks with sleep scheduling. Comput. Comm. **36**(2), 211–222 (2013)

39. López, J., et al.: Differentiated quality of protection to improve energy efficiency of survivable optical transport networks. In: Proceedings of OFC/NFOEC 2013. Anaheim, CA (2013)

40. Maillé, P., Tuffin, B.: Telecommunication Network Economics. From Theory to Applications. Cambridge University Press, Cambridge (2014)

41. Mansini, R., et al.: Twenty years of linear programming based portfolio optimization. Eur. J. Oper. Res.**234**(2), 518–535 (2014)

42. Markopoulou, A., et al.: Characterization of failures in an operational IP backbone network. IEEE/ACM Trans. Netw. **16**(4), 749–762 (2008)

43. Mastroeni, L., Naldi, M.: Compensation policies and risk in service level agreements: a value-at-risk approach under the ON-OFF service model. In: Proceedings of ICQT 2011. Paris, France (2011)

44. Morea, A., et al.: Protocol enhancements for "greening" optical networks. Bell Labs Tech. J. **18**(3), 211–230 (2013)\

45. Muhammad, A., et al.: Energy-efficient WDM network planning with dedicated protection resources in sleep mode. In: Proceedings of GLOBECOM'10. Miami, FL (2010)

46. Musumeci, F., et al.: A power consumption analysis for IP-over-WDM core network architectures. IEEE/OSA J. Opt. Commun. Netw. **4**(2), 108–117 (2012)

47. Musumeci, F., et al.: Energy-efficiency of protected IP-over-WDM networks with sleep-mode devices. J. High Speed Netw. **19**(1), 19–32 (2013)

48. Ogino, N., Nakamura, H.: Telecommunications network planning method based on probabilistic risk assessment. IEICE Trans. Commun. **E94–B**(12), 3459–3470 (2011)

49. Orgerie, A.C., et al.: A survey on techniques for improving the energy efficiency of large-scale distributed systems. ACM Comput. Surv. **46**(4), 47:1–47:31 (2014)

50. Perelló, J., et al.: Link vs. opto-electronic device sleep mode approaches in survivable green optical networks. In: Proceedings of OFC/NFOEC 2013. Anaheim, CA (2013)

51. Pióro, M., Medhi, D.: Routing, Flow and Capacity Design in Communication and Computer Networks. Morgan Kaufmann Publishers, Los Altos (2004)

52. Quittek, J., et al.: Requirements for energy management. IETF RFC 6988 (2013)

53. Restrepo, J.C.C., et al.: Energy profile aware routing. In: Proceedings of GreenComm'09. Dresden, Germany (2009)

54. Ricciardi, S., et al.: Modeling energy consumption in next-generation wireless access-over-WDM networks with hybrid power sources. Math. Comput. Model. **58**(5–6), 1389–1404 (2013)

55. Schupke, D.A.: Multilayer and multidomain resilience in optical networks. Proc. IEEE **100**(5), 1140–1148 (2012)

56. Shen, G., Tucker, R.S.: Energy-minimized design for IP over WDM networks. IEEE/OSA J. Opt. Commun. Netw. **1**(1), 176–186 (2009)

57. Swanson, M., et al.: Contingency planning guide for federal information systems. NIST SP 800–34 Rev. 1. National Institute of Standards and Technology, Gaithersburg (2010)

58. Vajanapoom, K., et al.: Risk based resilient network design. Telecommun. Syst. **52**(2), 799–811 (2013)

59. Van Heddeghem, W., et al.: Power consumption modeling in optical multilayer networks. Photonic Netw. Commun. **24**(2), 86–102 (2012)

60. Vasseur, J.P., et al.: Network Recovery, Protection and Restoration of Optical, SONET-SDH, IP, and MPLS. Morgan Kaufmann Publishers, Los Altos (2004)

61. Vizcaíno, J.L., et al.: Protection in optical transport networks with fixed and flexible grid: cost and energy efficiency evaluation. Opt. Switch. Netw. **11**(part B), 55–71 (2014)

62. Wiatr, P., et al.: Energy efficiency and reliability tradeoff in optical core networks. In: Proceedings of OFC 2014. San Francisco, CA (2014)

63. Wiatr, P., et al.: Green WDM-PONs: exploiting traffic diversity to guarantee packet delay limitation. In: Proceedings of ONDM 2013. Brest, France (2013)

64. Wiatr, P., et al.: Power savings versus network performance in dynamically provisioned WDM networks. IEEE Commun. Mag. **43**(5), 48–55 (2012)

65. Wickboldt, J.A., et al.: A framework for risk assessment based on analysis of historical information of workflow execution in IT systems. Comput. Netw. **55**(13), 2954–2975 (2011)

66. Xia, M., et al.: Risk-aware routing for optical transport networks. In: Proceedings of INFOCOM 2010. San Diego, CA (2010)

67. Xia, M., et al.: Green provisioning for optical WDM networks. IEEE J. Sel. Top. Quantum Electron. **17**(2), 437–445 (2011)

68. You, F., et al.: Risk management for a global supply chain planning under uncertainty: models and algorithms. AIChE J. **55**(4), 931–946 (2009)

69. Zhang, Y., Fan, Z.P.: An optimization method for selecting project risk response strategies. Int. J. Qual. Proj. Manag. **32**(3), 412–422 (2014)

70. Zhao, Y., et al.: Load balance vs. energy efficiency in traffic engineering: a game theoretical perspective. In: Proceedings of INFOCOM 2013. Turin, Italy (2013)

11

# COMMUNICATION NETWORK ARCHITECTURES BASED ON ETHERNET PASSIVE OPTICAL NETWORK FOR OFFSHORE WIND POWER FARMS


**Mohamed A. Ahmed [1], Jae-Kyung Pan [2], Minho Song [3] and Young-Chon Kim [4,*]**

[1] Division of Electronics and Information Engineering, Chonbuk National University, Jeonju 561-756, Korea

[2] Department of Electrical Engineering, Wind Energy Grid-Adaptive Technology Research Center, Smart Grid Research Center, Chonbuk National University, Jeonju 561-756, Korea

[3] Division of Electronics and Information Engineering, Wind Energy Grid-Adaptive Technology Research Center, Chonbuk National University, Jeonju 561-756, Korea

[4] Department of Computer Engineering, Wind Energy Grid-Adaptive Technology Research Center, Smart Grid Research Center, Chonbuk National University, Jeonju 561-756, Korea



## ABSTRACT

Nowadays, with large-scale offshore wind power farms (WPFs) becoming a reality, more efforts are needed to maintain a reliable communication network for WPF monitoring. Deployment topologies, redundancy, and network availability are the main items to enhance the communication reliability between wind turbines (WTs) and control centers. Traditional communication networks for monitoring and control (*i.e.*, supervisory control and data acquisition (SCADA) systems) using switched gigabit Ethernet will not be sufficient for the huge amount of data passing through the network. In this paper, the optical power budget, optical path loss, reliability, and network cost of the proposed Ethernet Passive Optical Network (EPON)-based communication network for small-size offshore WPFs have been evaluated for five different network architectures. The proposed network model consists of an optical network unit device (ONU) deployed on the WT side for collecting data from different internal networks. All ONUs from different WTs are connected to a central optical line terminal (OLT), placed in the control center. There are no active electronic elements used between the ONUs and the OLT, which reduces the costs and complexity of maintenance and deployment. As fiber access networks without any protection are characterized by poor reliability, three different protection schemes have been configured, explained, and discussed. Considering the cost of network components, the total implementation expense of different architectures with, or without, protection have been calculated and compared. The proposed network model can significantly contribute to the communication network architecture for next generation WPFs.

## KEYWORDS

Ethernet passive optical network; switched Ethernet; communication network; wind power farm; reliability; optical power budget; optical path loss; network cost

## 1. INTRODUCTION

In the near future, our dependence on conventional fossil fuels as the main energy source will need to be replaced by the use of renewable sources of energy. The wind power industry is becoming a mature technology, with many countries planning and constructing large-scale wind power farms (WPFs), with a huge number of wind turbines (WTs). The question arises of how to monitor and control these platforms. Traditional communication networks for monitoring and control using switched gigabit Ethernet will not be adequate for the huge amount of data passing through the network. The need for a new technology that supports high data traffic, satisfies system requirements, and matches the required quality of service (QoS) is needed. In a conventional switched-based architecture, industrial switches are active elements which require a power supply because of powered electronics. Therefore, the network configuration requires an electric power supply as well as backup power in the field. The failure of one switch will disrupt the communication network and may affect the rest of the network. Compared to a normal Ethernet switch, the prices of industrial switches are much more expensive.

This paper proposes an EPON (Ethernet Passive Optical Network) technology as one of the promising candidates for next generation WPFs. The topologies used for offshore WPF are based on an electrical collector system (power cables). EPON technology uses optical fiber to transport data. There are no active elements between WTs and their control center, which significantly reduces the network cost. In this paper, we describe the communication network design for the newly-planned WPFs. Many researchers have suggested the application of EPON technology to electric power communication networks. Zaker *et al.* [1] propose a gateway design for the fiber-wireless sensor network (Fi-WSN) architecture employed in the smart grid environment. The proposed architecture combines the advantage of high speed and low latency for EPON with the flexibility and low cost of WSNs. Raza *et al.* [2] propose EPON communication architecture for connecting substations in a power system. The simulation result shows the average packet end-to-end delays and packet loss ratio in both the upstream and downstream. Jiang *et al.* [3] present a variety of communication technologies for construction of a subscriber power consumption information acquisition system. The paper analyzes different communication technologies, such as optical fiber network (EPON and industrial Ethernet), and wireless networks. Yao *et al.* [4] present an integrated communication platform of a photovoltaic (PV) power station, which integrates EPON, power line carrier, and wireless communication. The authors in [5] and [6] present the reliability and availability analysis of Ethernet communication architecture for a transmission substation in a power system. The authors used a reliability block diagram (RDB) approach to compare different network architectures (cascade, star, and ring). Franken *et al.* [7] present a method for reliability calculations of the electrical system (sub-sea cables, switchgear, transformers, and WTs) within offshore wind farms. Results of energy not supplied, using different WPF sizes and different architectures, were calculated and compared.

In this paper, we propose different communication network architectures based on the electrical topology for small-scale WPFs. The reliability block diagram approach is used to calculate the availability of the proposed EPON model between optical line terminal (OLT) and WT-ONUs, with different protection schemes. The main objective of this work is to:

- Propose communication network architectures for WPF monitoring based on EPON technology;
- Configure the network model of small-scale WPFs, based on electric power topologies;
- Analyze the optical power budget and path loss to ensure that the received signal power at the wind turbines side is sufficient to maintain acceptable performance;
- Make a reliability analysis of the communication network, consisting of OLT, feeder fiber (FF), passive optical splitter (POS), distributed fiber (DF), and ONU;
- Compare the connection availability between OLT and WT-ONUs with different protection schemes; and
- Compare the network costs for 20 different architectures with, and without, protection schemes.

This paper is structured as follows. Section 2 provides related work. In Section 3, the proposed EPON-based wind power farm topologies are described. Section 4 provides the performance evaluation, results, and discussion. Finally, Section 5presents the conclusion and the direction of future work.

# 2. RELATED WORK

## 2.1. Communication Network for Wind Power Farms (WPFs)

The supervisory control and data acquisition (SCADA) system enables operators to monitor, control and record data of a wind power plant, from a remote place called a central control station [8,9,10]. It consists of three main components as shown in Figure 1.

- Inside the turbine tower, including the wind turbine controller (WTC), remote terminal units (RTU), intelligent electronic devices (IEDs), and sensors. The WTC collects all data using short communication links, and makes them available for processing and transferring to the control center. The closed-circuit television (CCTV) system and the Internet connection may share or use an independent communication network between the wind turbine and the control center.

- Communication network based on Ethernet equipment (gigabit Ethernet switch) is considered, to transfer data between the WTs and the control center. Most WPFs use the same power cable routes for the SCADA communication network as for the power distribution.

- Control center connects individual WTs and meteorological stations to the control center. It enables the operators to manage the behavior of all WTs as a whole. It requires a long distance for data transmission.

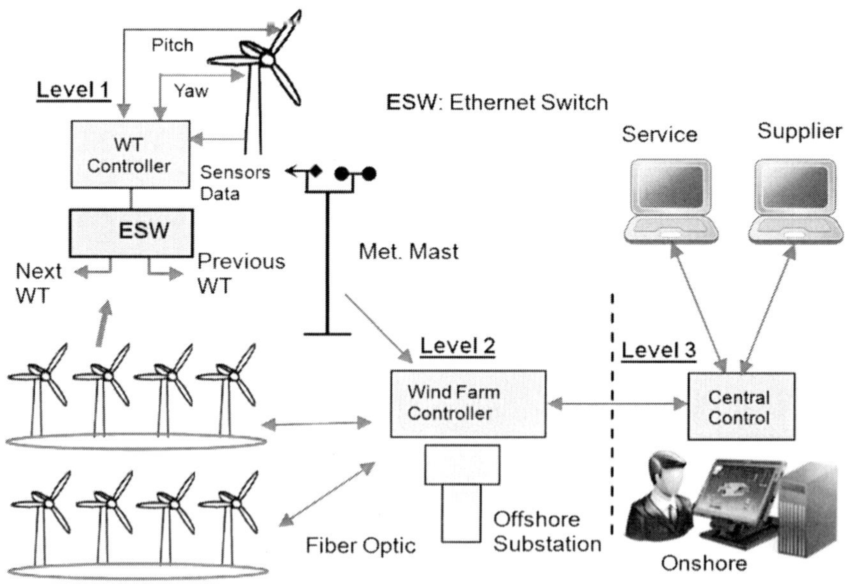


**Figure 1.** Schematic view of conventional wind power farms (WPF) network.

## 2.2. Ethernet Passive Optical Network (EPON)-Based Communication Network for WPFs

EPON consists of an OLT located in the control center, multiple ONUs, and POS. In the downstream direction, EPON is a point-to-multipoint network; the OLT broadcasts control messages and data packets to all ONUs through a passive splitter. In the upstream direction, EPON is a multipoint-to-point network; multiple ONUs transmit data to the OLT through a passive combiner [11].

Figure 2 shows a schematic view of the proposed EPON-based communication network for WPF [12]. The proposed network consists of an optical network device (ONU) deployed on the WT's side for collecting different data, including wind turbine operation, meteorological data, fault parameters, and security from different internal networks. All ONUs from different WTs are connected to an OLT unit, placed on the control center side. The path between WTs and the control center does not contain any active elements, which saves costs and reduces the complexity of maintenance and deployment, compared with current switched gigabit Ethernet.

In Figure 2, the proposed architectures (star, cascade, *etc.*) represent the structure design of new planned WPFs. The hybrid configuration (Ethernet switch and ONU) represents modifying existing WPFs to support EPON technology, which is out of the scope of this work. In a star configuration, four wind turbines are connected with distributed fibers to the control center using one POS (1 × 4). In the cascade configuration, four POS (1 × 2) are connected in cascade; one port is connected to the next WT, while the other port is connected to the WT-ONU unit.

In this work, EPON uses a 1-Gbps single fiber, with a transmission range of 20 km. Two different wavelengths are used to support downstream and upstream transmission, 1490 nm and 1310 nm, respectively. Each WT transmits data in its time slots, to avoid data collisions. EPON must employ a medium access control (MAC) mechanism, to arbitrate access to the shared medium, to prevent different ONU's data from colliding in the upstream direction. OLT efficiently shares the upstream transmission bandwidth among all turbines' ONUs. Failure to gain access to this shared channel in a timely manner could negatively impact the communication and should be considered as an aspect of reliable communication. There have been many studies and frameworks for managing the medium access control in [13,14,15,16].

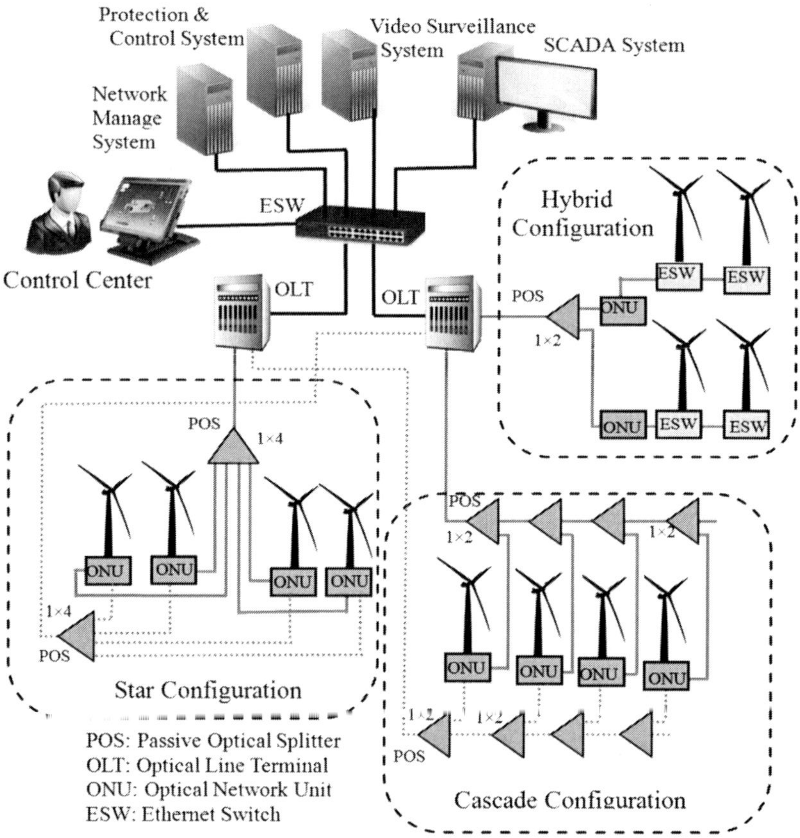


**Figure 2.** Schematic view of proposed Ethernet Passive Optical Network (EPON) communication network for WPF.

# 3. PROPOSED EPON-BASED WPF TOPOLOGIES

This section considers a small scale WPF, consisting of 12 WTs, with a generated power of 60 MW, close to the grid (5 km). All WTs are connected together in a radial topology (WT1 → WT6 and WT7 → WT12). Five different architectures (A, B, C, D, and E) are proposed. Among the five mentioned architectures, any other WPF layout can be constructed. We assumed that:

- An OLT is installed in the control center;
- Each WT has only one ONU device;

- All internal networks inside a WT (which include monitoring, protection, and security) may be connected with one or more ONU buffers; and
- Different network architectures are designed, using FF, DF, and POS.

## 3.1. Architecture (A)

This begins with an OLT unit located in the control center, connected with a 5 km feeder fiber to the first wind turbine (WT1). Cascade splitters are used to reduce the amount of deployed fiber in the network. WT1 has two POS (primary and secondary); both are (1 × 2). The primary POS has two output ports; one is connected to WT1, while the other is connected to WT7. For all other WTs (WT2 → WT12), each WT has only one POS (1 × 2); one port is connected using DF to the next WT, while the other port is connected to the WT-ONU unit. All DFs in this architecture are of 1 km length (the distance between WTs). Note that, POS (1 × 2) is used in both WT6 and WT12, in order to help extend the network, in case of installing new WTs (one port is used, while the other is left empty). Figure 3 shows the network elements of architecture (A).

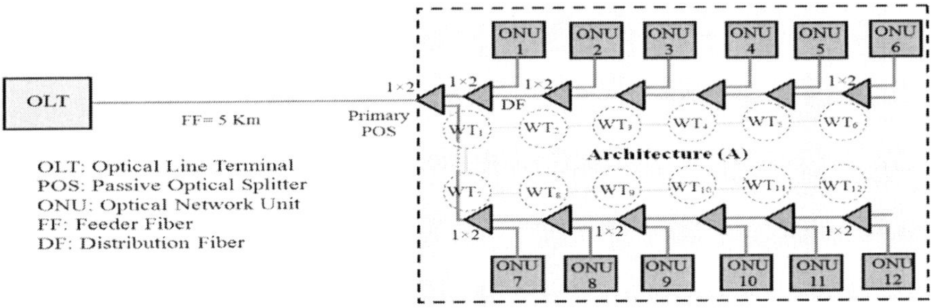


**Figure 3.** EPON-based architecture (A).

## 3.2. Architecture (B)

This differs from architecture (A), as it has three POS (one primary POS and two secondary POS). The primary POS is (1 × 2), the same as architecture (A). The first secondary POS is (1 × 6) in WT1, with six outputs. Each WT from (WT1 WT6) has dedicated optical fiber cable from secondary POSs located in WT1. Additionally, the secondary POS in WT7 is (1 × 6), and connects via a dedicated path to (WT7 ➔ WT12). Figure 4 shows the network elements of architecture (B).

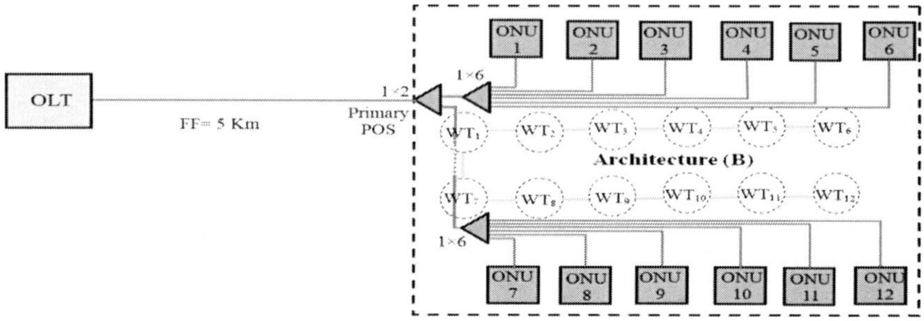


**Figure 4.** EPON-based architecture (B).

## 3.3. Architecture (C)

This combines the advantage of both configurations (A) and (B) of low fiber deployment and less number of POS. It differs from architecture (B), as it has five POS (one primary POS and four secondary POS). All secondary POS are (1 × 4), in WT1, WT4, WT7, and WT10. In WT1, three output ports from secondary POS are connected to (WT1–WT3), the remaining port is connected to another POS (1 × 4) exist in WT4 with four output ports. The remaining ports of POS in WT4 and WT10 are left for future extensions. Figure 5 shows the network elements of architecture (C).

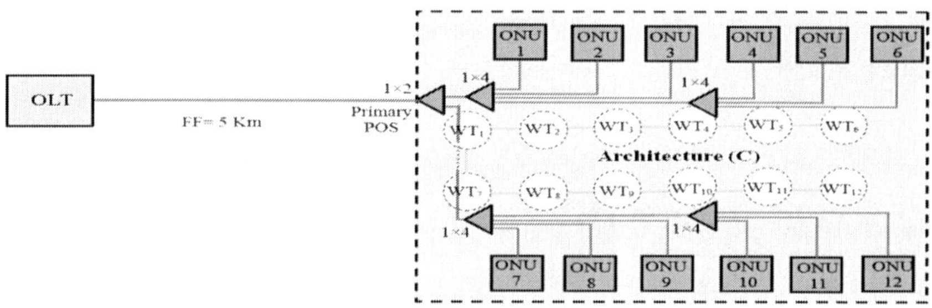


**Figure 5.** EPON-based architecture (C).

## 3.4. Architecture (D)

It is a modification of architecture (C). Each WT from (WT4–WT6) and (WT10–WT12) has only one POS (1 × 2); one port is connected using DF to the next WT, while the other port is connected to the WT-ONU unit. Both architecture (C) and architecture (D) can be expandable to include more WTs. Architecture (D) has more POSs than architecture (C). Howerver, architecture (D) has less fiber deployment. Figure 6 shows the network elements of architecture (D).

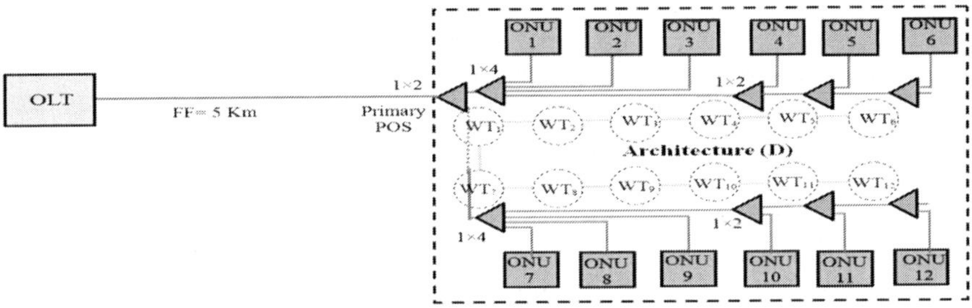


**Figure 6.** EPON-based architecture (D).

## 3.5. Architecture (E)

This architecture can be applied to small-scale onshore WPF, with WTs distributed around the control center, as shown inFigure 7. It is similar to architecture (B), with the only difference being that only one primary POS (1 × 12) is used, and it resides in the control center. The length of the feeder fiber to WT1 is neglected, where WT1 is nearby the control center. The advantage of this architecture is that the serviceability of the network is improved, as all network hardware is located in the control center. Additionally, this architecture reduces the costs and complexity of maintenance and deployment.

A disadvantage of architecture (E) is the large amount of distributed optical fibers is used. Table 1 shows the details of network elements for the proposed EPON-based network architectures, where an OLT unit is located at the control center and 12 ONUs are deployed on the WTs side.

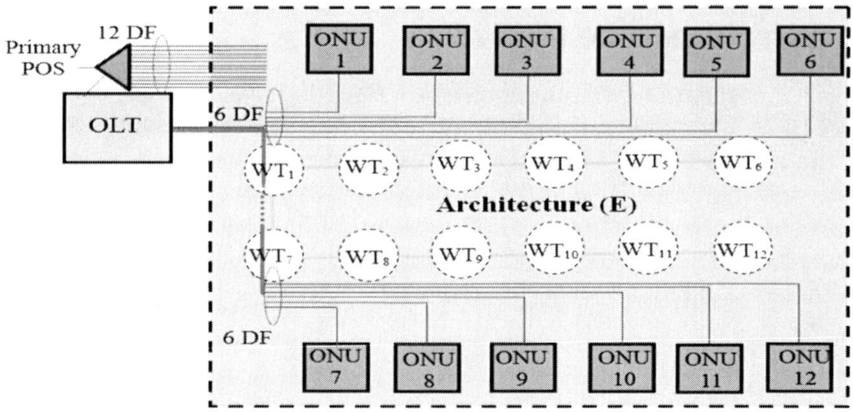


**Figure 7.** EPON-based architecture (E).

**Table 1.** Network elements of small wind power farms (WPF) (unprotected).

| EPON | OLT | FF (km) | Number Primary POS | Number Secondary POS | Number Other POS | DF (km) |
|---|---|---|---|---|---|---|
| A | 1 | 5 | 1 (1 × 2) | 12 (1 × 2) | - | 11 |
| B | 1 | 5 | 1 (1 × 2) | 2 (1 × 6) | - | 31 |
| C | 1 | 5 | 1 (1 × 2) | 4 (1 × 4) | - | 19 |
| D | 1 | 5 | 1 (1 × 2) | 2 (1 × 4) | 6 (1 × 2) | 17 |
| E | 1 | - | 1 (1 × 12) | - | - | 36 |

WPF: wind power farm; EPON: Ethernet Passive Optical Network; OLT: optical line terminal; FF: feeder fiber; POS: passive optical splitter; DF: distributed fiber.

# 4. PERFORMANCE EVALUATION

## 4.1. Optical Power Budget

The optical power budget is analyzed to ensure that the received signal power is enough to maintain acceptable performance even though the lengths of communication links between the control center and WTs are different. The power budget for EPON specified in IEEE 802.3ah standard is 26.0 dB in the case of 1000 Base-PX20, for both upstream and downstream traffic. The optical budget [dB] is defined as the difference between the minimum transmitter launch power ($P_{tx}$, dBm) at the input of the optical link, and the minimum sensitivity of the receiver ($P_{rx}$, dBm) at the output of optical links [17,18,19] as shown in Equation (1):

$$\text{Power Budget} = P_{tx} - P_{rx} \tag{1}$$

There are many sources of attenuation including splitters ($Loss_{POS}$), connectors ($Loss_{conn}$), and the fiber cable itself ($Loss_{fiber}$). The total optical power loss ($P_{Loss}$) of optical link given in [dB] is calculated as shown in Equation (2):

$$P_{Loss} = \sum Loss_{POS} + \sum Loss_{Conn} + \sum Loss_{Fiber} \tag{2}$$

Table 2 shows the EPON specification according to IEEE 802.3ah standard. The component insertion loss is shown in Table 3. We considered the fiber attenuation is 0.4 dB/km, and the connector loss is 0.2 dB. Passive optical splitter (1 × 2) insertion loss is 0.4 dB for a split ratio of 5:95, while the insertion loss is 3 dB for split ratio of 50:50.

**Table 2.** IEEE STD 802.3 EPON specifications.

| Parameter | Data Rate | Power Budget |
|---|---|---|
| EPON | 1.25 Gbps<br>(D and U)<br>1000Base-PX20 | PX-20U 26 dB<br>PX-20D 26 dB |

Figure 8 shows the total optical path loss calculation for the five different configurations. The highest optical path loss value represents the farthest turbine, while the lowest value represents the nearest turbine. For example, the highest optical path loss value is about 20.2 dB for WT12 in architecture (C), while the lowest value is about 6.6 dB for WT1 in architecture (A). In architecture (A), with 1 km between WTs, fiber attenuation is 0.4 dB, connector loss is 0.4 dB and POS (1 × 2) has insertion loss of 0.4 dB. This is the reason why the total optical path loss is growing linearly.

The analysis of optical power budget shows that architecture (A) offers a lower optical path loss among other architectures due to a small insertion loss of POS (1 × 2). However, architecture (B) has a higher optical path loss due to a higher splitting ratio and length of fiber cables. Architecture (C) and architecture (D) combine the advantage of both configuration (A) and configuration (B).

**Table 3.** Component insertion loss.

| Component | Attenuation |
|---|---|
| Fiber | 0.4 dB/km |
| Connector | 0.2 dB |
| Splitter | 1 × 2 (5%:95%) → 0.4 dB |
| | 1 × 2 (50%:50%) → 3.0 dB |
| | 1 × 4 → 6.0 dB |
| | 1 × 8 → 9.0 dB |
| | 1 × 16 → 12.0 dB |

## 4.2. Reliability

In this section, a reliability analysis of EPON-based communication network between OLT and WT-ONUs for a small-size offshore WPF is given. Five basic topologies without any redundancy (architecture A, B, C, D, and E) are constructed. Since redundancy for collector systems in WPFs considers additional cables for protection, we consider additional FF, POS, and DF as network protection components. Failure rate and mean time to repair (MTTR) are considered, based on network components from reference [20,21,22].

The following assumptions are considered in our study:

- The WTs used in this study are 5 MW. The distance between two WTs is 1 km.
- Reliability analysis considers only the communication network between the OLT and ONUs. It consists of OLT, FF, POS, DF, and ONU.

- We assume that the failure rate of the OLT and ONU is small, compared with the number of failures in cables. Therefore, the equipment failures of OLT and ONUs are neglected in our analysis.
- The communication network equipment inside each WT is not considered, and only the ONU is considered in our calculation.

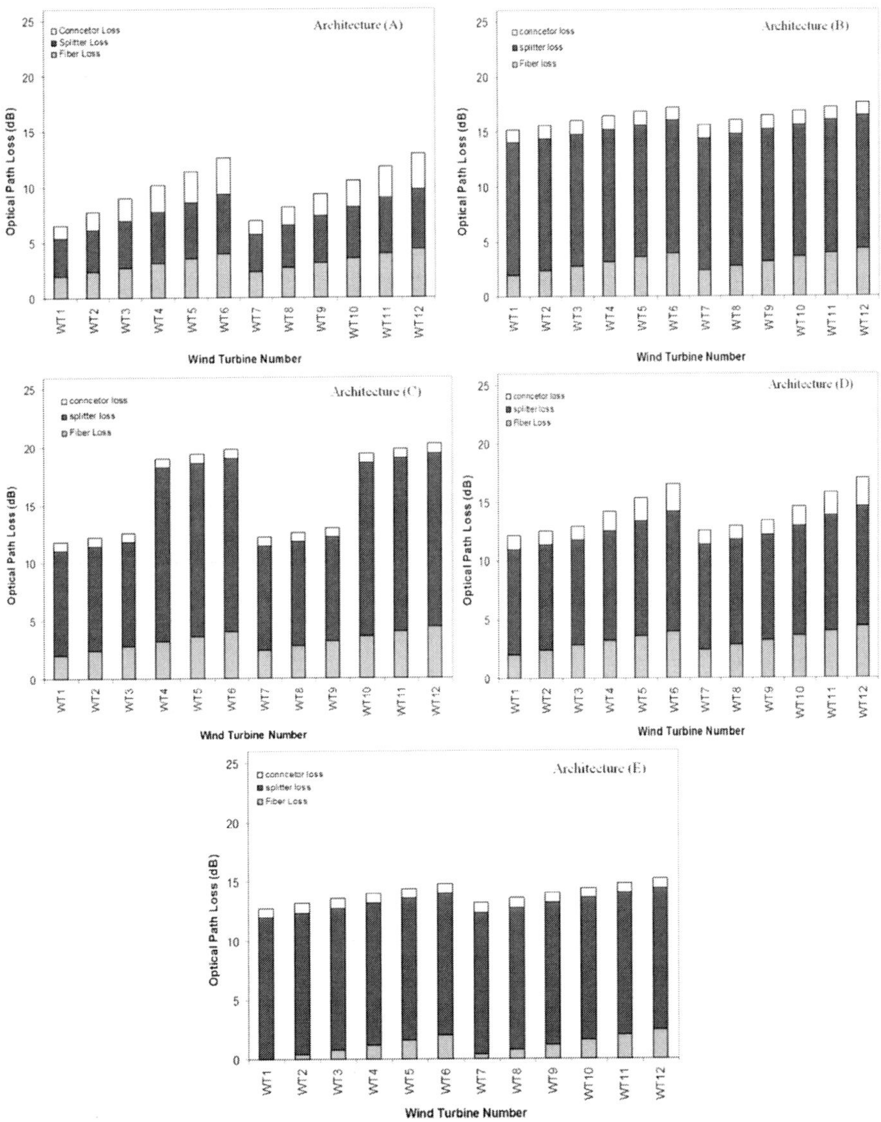


**Figure 8.** Total optical path loss for different architectures.

### 4.2.1. Reliability Block Diagram

We consider time division multiplexing passive optical network (TDM-PON) architecture defined by international telecommunication union-telecommunication sector (ITU-T), with basic architecture and protection schemes [23]. The reliability block diagram for basic PON without protection is shown in Figure 9, where each block in the diagram represents either a component or a fiber link. For each block, a characteristic parameter of unavailability is given in Table 4.

**Figure 9.** Reliability block diagram for unprotected EPON architecture.

**Table 4.** Reliability data [22].

| Components | Failure Rate (FIT) * | MTTR (h) | Unavailability |
|---|---|---|---|
| OLT | 256 | 4 | $1.024 \times 10^{-6}$ |
| ONU | 256 | 24 | $6.144 \times 10^{-6}$ |
| Switch | 1250 | 24 | $3.00 \times 10^{-5}$ |
| 1:2 Splitter | 50 | 24 | $1.20 \times 10^{-6}$ |
| 1:N (2:N) Splitter | 120 | 24<br>6 | $2.88 \times 10^{-6}$<br>$7.20 \times 10^{-7}$ |
| Fiber (per km) | 570/km | 24 | $1.368 \times 10^{-5}$ |

MTTR: mean time to repair; * 1 FIT = 1 failure/$10^9$ h.

The series configuration, from a reliability point of view, means that a system fails if one or more of its components fail [24,25]. Using the failure rate of a communication network component (OLT, POS, Fiber, and ONU), unavailability of a component ($U_x$) is derived from its failure rate in FIT (1 FIT = 1 failure/$10^9$ h), and the mean repair time (MTTR) in hours, as shown in Equation (3). Equation (4) shows the mean down time (MDT) in (min/year) for component *x* ($MDT_x$):

$$U_X = (\text{Failure rate} \times \text{MTTR}_X)/10^9 \qquad (3)$$

$$MDT_X = (U_X \times 365 \times 24 \times 60) \quad (4)$$

The expression for the connection unavailability ($U_{EPON}$) for EPON-based architecture without protection is given in Equation (5), where, $U_{OLT}$, $U_{FF}$, $U_{POS}$, $U_{DF}$, and $U_{ONU}$ are the unavailability of OLT, FF, POS, DF, and ONU, respectively.

$$U_{EPON} = U_{OLT} + U_{FF} + U_{POS} + U_{DF} + U_{ONU} \quad (5)$$

For example, architecture (A) has a minimum number of fiber deployments, by using cascaded splitters with FF = 5 km and DF = 11 km. The expected downtime for architecture (A) is 163 min/year.

### 4.2.2. EPON-based Protection Schemes

Redundancy is a major way to enhance the reliability of WPF communication network. There are two types of redundancy, device redundancy (OLT and ONU), and fiber redundancy (FF and DF). We assume that the failure rate of the OLT and ONU is small, compared with the number of failures in cables. Therefore, protection of both OLT and ONUs are neglected in our analysis. In feeder fiber protection architecture, only the FF is protected as shown in Figure 10. In this case, the feeder fiber is normally working, while the spare fiber is in standby. In the case of FF failure, the OLT is responsible for switching to the spare fiber. In a full protection architecture, all network components are fully duplicated.

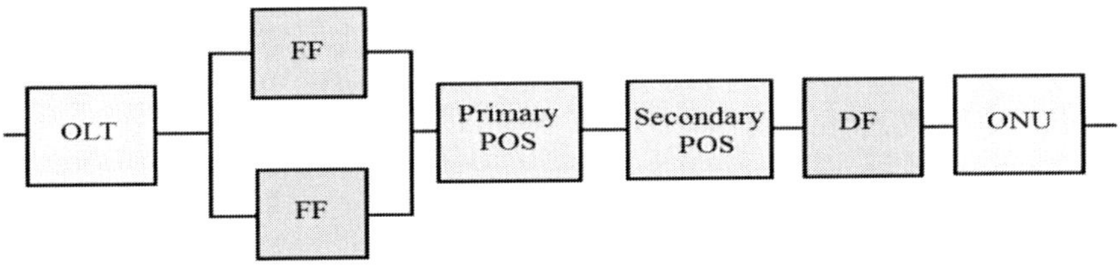


**Figure 10.** Reliability block diagram for feeder fiber (FF) protection architecture.

There are three different protection architectures; category (1) represents architectures with feeder fiber protection; one is working, while the other is in standby. In category (2), DFs are duplicated, and also secondary POS are changed to support the redundancy in the DF, as shown in Figure 11. In the last category (3), all architectures enjoy full protection, with all network components duplicated as shown in Figure 12.

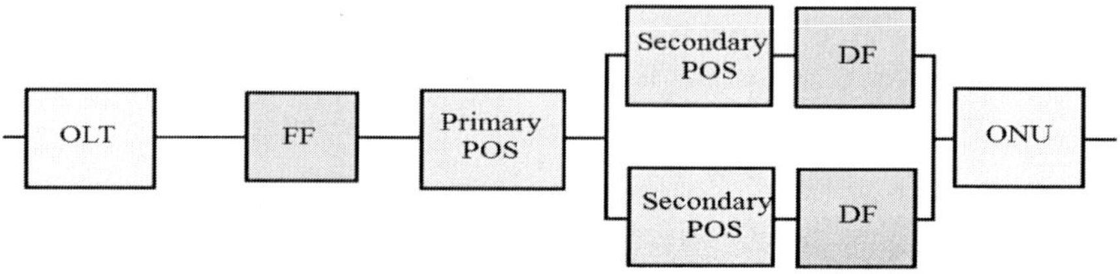


**Figure 11.** Reliability block diagram for (distributed fiber) DF protection architecture.

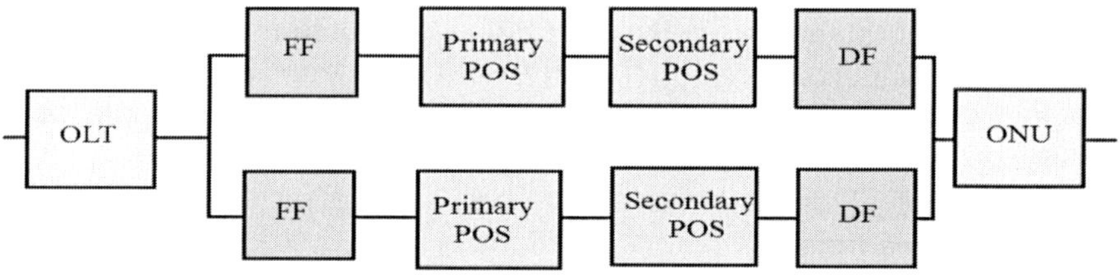


**Figure 12.** Reliability block diagram for Full protection architecture.

We calculated the unavailability of 20 different network architectures, using the component reliability data of Table 4 and Equations (6–8). Note that, the MTTR for the OLT is considered as only 4 h, as it is located in the control center. All other network components located on the WT side are considered to take 24 h to repair, in case of a fault.

$$U_{EPON(FF)} = U_{OLT} + (U_{FF})^2 + U_{POS} + U_{DF} + U_{ONU} \quad (6)$$

$$U_{EPON(DF)} = U_{OLT} + U_{FF} + U_{POS1} + (U_{POS2} + U_{DF})^2 + U_{ONU} \quad (7)$$

$$U_{EPON(DF)} = U_{OLT} + (U_{FF} + U_{POS1} + U_{POS2} + U_{DF})^2 + U_{ONU} \quad (8)$$

Figure 13 shows the expected downtime for different network architectures. In the case of an unprotected network, architecture (A) has the lowest MDT of about 163 min/year, while architecture (E) has the highest MDT, of about 300 min/year, as shown in Figure 13. In the case of full protection, all three architectures have the same MDT, of about 39 min/year. As we can see, conventional switched-based architecture has the highest downtime of about 320 min/year for an unprotected network configuration, while about 205 min/year downtime for full protection configuration.

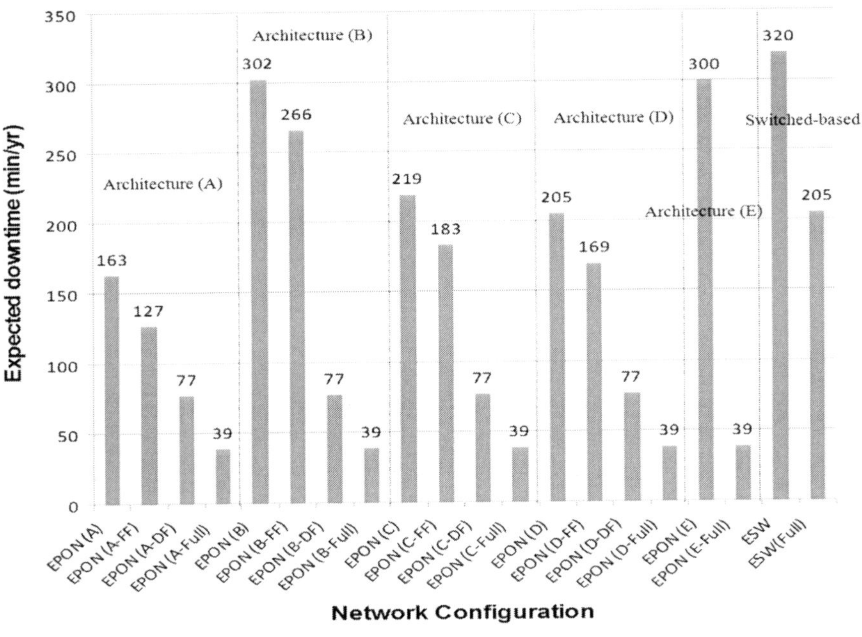


**Figure 13.** Mean down time for different architectures for a small (wind power farms) WPF.

## 4.3. Network Cost

Capital expenditure (CAPEX) consists of both initial network equipment and network installation costs [25,26]. In this work, the power cable line comprises the electrical collection cables and optical fiber cables. Since the communication network part is integrated with the electrical system, we consider only the network equipment cost to evaluate and compare different network architectures.

The communication network cost can be divided into active device costs (OLT, ONU, and Ethernet switch) and passive component costs (POS and fiber). The total network cost for the EPON-based and switch-based architectures can be represented as follows:

$$C_{EPON} = C_{OLT} + C_{FF} + C_{POS} + C_{DF} + C_{ONU} \times N_{ONU} \quad (9)$$

$$C_{Ethernet} = C_{ESW} + C_{FF} + C_{DF} \quad (10)$$

where, $C_{OLT}$, $C_{POS}$, $C_{ESW}$, and $C_{ONU}$ represent the equipment of the OLT, Ethernet switch, POS, and ONU, respectively. CFF and CDF represent the costs of optical fiber cable of the feeder fiber and distributed fiber, respectively. $N_{ONU}$represents the number of WT-ONUs.

The equipment costs used in our model are detailed in Table 5. We consider that the OLT cost is $12,100, and the ONU cost is $350. As expected, unprotected EPON is the most economic deployment solution for EPON-based architectures, compared with the conventional switched-based architectures.

**Table 5.** Component cost (US $) [22].

| Parameter | Cost (US $) |
|---|---|
| OLT | 12100 |
| ONU | 350 |
| ESW | 1800 |
| Splitter (1 × 2) | 50 |
| Splitter (1 × N) | (50 per port) |
| Fiber (per km) | 160 |

In Figure 14, results show that the network cost of architectures (A, C, and D) with full protection is lower than unprotected architecture for conventional switched-based architectures. Using the results obtained of reliability analysis and network cost, an effective and highly reliable WPF architecture can be chosen and deployed.

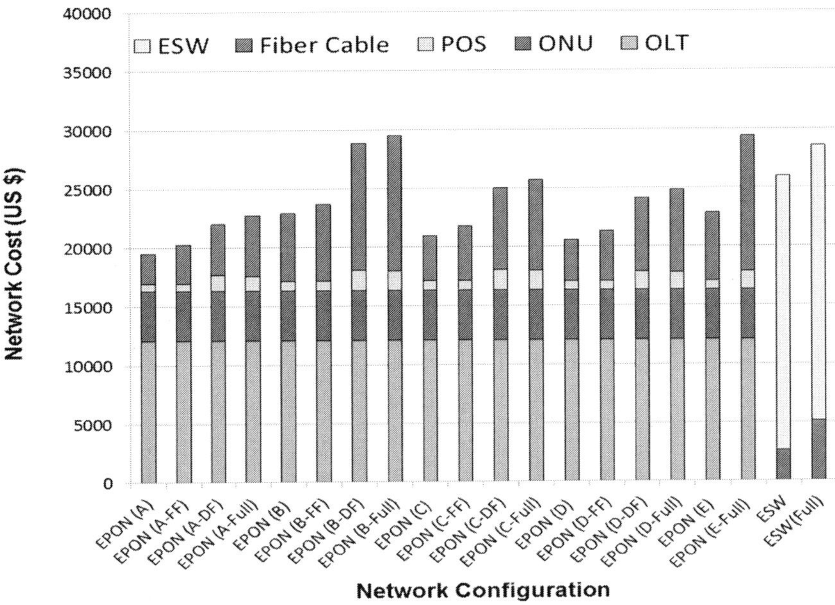


**Figure 14.** Network cost for WPF architectures.

# 5. CONCLUSIONS

In this paper, we propose an EPON-based communication network architecture for a small-size offshore WPF. The main objective aims to deploy ONUs on the WT side, for collecting data from different internal networks. The path between WTs and the control center does not contain any active elements, compared with current switched gigabit Ethernet. The reliability block diagram approach is used to calculate the availability between the OLT and WT-ONUs, for five different architectures (A, B, C, D, and E). Based on ITU-T, three different protection schemes (FF, DF, and Full) are configured, explained, and discussed. In view of network availability, results show that the unprotected architecture (A) has the lowest MDT of about 163 min/year; while in the case of full protection, the three architectures have similar MDT of about 39 min/year. In view of network cost, architecture (A) has the lowest network cost. In addition, architectures (A, C, and D) with full protection are lower in network cost than unprotected architectures for conventional switched-based architecture. We can conclude that architecture (A) has the best performance in view of both network availability and network cost. An effective and highly reliable WPF architecture can be chosen to be deployed by considering the numerical results of optical path loss, reliability analysis, and network cost.

## ACKNOWLEDGMENTS

This work was supported by the Brain Korea 21 PLUS and the National Research Foundation of Korea (NRF) grant funded by the Korea government (MSIP) (2010-0028509).

## AUTHOR CONTRIBUTIONS

Mohamed A. Ahmed and Young-Chon Kim proposed the communication network architectures of this paper; Jae-Kyung Pan and Minho Song helped in analysis of the results; Young-Chon Kim provided guidance during the whole research.

## CONFLICTS OF INTEREST

The authors declare no conflict of interest.

## REFERENCES

1. Zaker, N.; Kantarci, B.; Erol-Kantarci, M.; Mouftah, H.T. Smart grid monitoring with service differentiation via EPON and wireless sensor network convergence. *Opt. Switch. Netw.* **2014**, *14*, 53–68.

2. Raza, A.; Ullah, K.; Ahmed, S.; Hyuk-Soo, J.; Hyo-Sik, Y. Baseline Efficiency in EPON Based Architecture to Inter-Network Substations. In Proceedings of the IC4 2009: 2nd International Conference on Computer, Control and Communication, Karachi, Pakistan, 17–18 February 2009; pp. 1–6.

3. Jiang, H.; Yao, J.; Zhang, G. The construction scheme analysis of distant communication systems for subscriber power consumption information acquisition systems. *J. Int. Counc. Electr. Eng.* **2011**, *1*, 482–485.

4. Yao, H.; Peng, Q.; He, W.; Zhang, X. Integrated Communication Technology for Supervisory Control and Data Acquisition System of PV Power Station. In Proceedings of the ISDEA 2012: 2nd International Conference on Intelligent System Design and Engineering Application, Sanya, China, 6–7 January 2012; pp. 1277–1280.

5. Kanabar, M.; Sidhu, T.S. Reliability and Availability Analysis of IEC 61850 Based Substation Communication Architectures. In Proceedings of the IEEE PES Power and Energy Society General Meeting 2009, Calgary, AB, Canada, 26–30 July 2009; pp. 1–8.

6. Ali, I.; Thomas, M.S.; Gupta, S.; Hussain, S.M.S. IEC 61850 substation communication network architecture for efficient energy system automation. *Energy Technol. Policy* **2015**, *2*, 82–91.

7. Franken, B. *Reliability Study. Analysis of Electrical Systems within Offshore Wind Parks, Elforsk Report 07:65*; Elforsk: Stockholm, Sweden; November; 2007; Available online: http://www. elforsk. se/Global/Vindforsk/Rapporter%20fran%20Vindforsk%20II/V-118%20R_07_65.pdf (accessed on 10 January 2016).

8. Gardner, P.; Craig, L.M.; Smith, G.J. Electrical Systems for Offshore Wind Farms. In Proceedings of the European Wind Energy, 20th BWEA Conference, Cardiff, UK, 2–4 September 1998; pp. 1–9.

9. Pettener, A.L. SCADA and Communication Networks for Large Scale Offshore Wind Power Systems. In Proceedings of the RPG 2011: IET Conference on Renewable Power Generation, Edinburgh, UK, 6–8 September 2011; pp. 1–6.

10. Ahmed, M.A.; Kim, Y.C. Communication network architectures for smart-wind power farms. *Energies* **2014**, *7*, 3900–3921.

11. Maier, M.; Ghazisaidi, N. QoS provisioning techniques for future fiber-wireless (FiWi) access networks. *Future Internet***2010**, *2*, 126–155.

12. Ahmed, M.A.; Yang, W.H.; Kim, Y.C. Communication Network Architectures Based on EPON for Offshore Wind Power Farm. In Proceedings of the ENERGY 2013: 3rd International Conference on

Smart Grids, Green Communications and IT Energy-aware Technologies, Lisbon, Portugal, 24–29 March 2013; pp. 15–20.

13. Kramer, G.; Pesavento, G. Ethernet passive optical networks (EPON): Building a next-generation optical access network. *IEEE Commun. Mag.* **2002**, *40*, 66–73.

14. Kantarci, B.; Mouftah, H.T. Bandwidth distribution solutions for performance enhancement in long-reach passive optical networks. *IEEE Commun. Surv. Tutor.* **2012**, *14*, 714–733.

15. McGarry, M.P.; Reisslein, M. Investigation of the DBA algorithm design space for EPONs. *J. Lightwave Technol.* **2012**,*30*, 2271–2280.

16. Li, X.; Dan, L.; Wu, Q. Adaptive dynamic bandwidth allocation algorithm supporting multi-services over Ethernet passive optical networks. *Opt. Int. J. Light Electron Opt.* **2013**, *124*, 287–291.

17. IEEE Standard 802.3ah, IEEE-SA Standard Board, June 2004. Available online: http://www.ieee802.org/21/doctree/2006_Meeting_Docs/2006-11_meeting_docs/802.3ah-2004.pdf (accessed on 10 January 2016).

18. Ng, B.; Syuhaimi, M.; Premadi, A.; Jumari, K. Optical power budget and cost analysis in PON-based I-FTTH. *Res. J. Inf. Technol.* **2010**, *2*, 127–138.

19. Huang, H.; Zhang, H. Application and analysis of long distance EPON in transmission lines monitoring system. *Adv. Mater. Res.* **2011**, *317–319*, 1583–1589.

20. Chen, J.; Kantor, M.; Wajda, K.; Wosinska, L. *Next-Generation FTTH Passive Optical Networks: Research towards Unlimited Bandwidth Access*; Prat, J., Ed.; Springer: Dordrecht, Netherlands, 2008.

21. Kourtessis, P; Almeida, C.; Chang, C.H.; Chen, J.; Di Bartolo, S.; Fasser, P.; Gagnaire, M.; Leitgeb, E.; Lima, M.; Löschnigg, M.; *et al.* Evolution of Optical Access Networks. In *Towards Digital Optical Networks*; Springer: Berlin, Germany, 2009; Volume 5412, pp. 97–131.

22. Ahmed, M.A.; Yang, W.H.; Kim, Y.C. Performance Evaluation of EPON-Based Communication Network Architectures for Large-Scale Offshore Wind Power Farms. In *Multimedia and Ubiquitous Engineering*; Springer: Dordrecht, Netherlands, 2013; Volume 240, pp. 841–848.

23. ITU-T G.983.1, Broadband Optical Access Systems based on Passive Optical Networks (PON), October 1998. Available online: https://www.itu.int/rec/T-REC-G.983.1-199810-S/en (accessed on 10 January 2016).

24. Machuca, C.M.; Wosinska, L.; Chen, J. Assessment methodology of protection schemes for next generation optical access networks. *Opt. Fiber Technol.* **2015**, *26*, 82–93.

25. Wosinska, L.; Chen, J.; Larsen, C.P. Fiber access networks: Reliability analysis and Swedish broadband market. *IEICE Trans. Commun.* **2009**, *E92-B*, 3006–3014.

**26.** Fernández, Á.; Stol, N. CAPEX and OPEX simulation study of cost-efficient protection mechanisms in passive optical networks. *Opt. Switch. Netw.* **2015**, *17*, 14–24.

12

# ENERGY-SAVING MECHANISM IN WDM/TDM-PON BASED ON UPSTREAM NETWORK TRAFFIC

**Paola Garfias [1,*], Marilet De Andrade [2], Massimo Tornatore [2], Anna Buttaboni [2], Sebastià Sallent [1] and Lluís Gutiérrez [1]**

[1] Universitat Politècnica de Catalunya, Esteve Terradas 7, 08860 Barcelona, Spain

[2] Politecnico di Milano, Via Ponzio 34/5, 20133 Milan, Italy

## ABSTRACT

One of the main challenges of Passive Optical Networks (PONs) is the resource (bandwidth and wavelength) management. Since it has been shown that access networks consume a significant part of the overall energy of the telecom networks, the resource management schemes should also consider energy minimization strategies. To sustain the increased bandwidth demand of emerging applications in the access section of the network, it is expected that next generation optical access networks will adopt the wavelength division/time division multiplexing (WDM/TDM) technique to increase PONs capacity. Compared with traditional PONs, the architecture of a WDM/TDM-PON requires more transceivers/receivers, hence they are expected to consume more energy. In this paper, we focus on the energy minimization in WDM/TDM-PONs and we propose an energy-efficient Dynamic Bandwidth and Wavelength Allocation mechanism whose objective is to turn off, whenever possible, the unnecessary upstream traffic receivers at the Optical Line Terminal (OLT). We evaluate our mechanism in different scenarios and show that the proper use of upstream channels leads to relevant energy savings. Our proposed energy-saving mechanism is able to save energy at the OLT while maintaining the introduced penalties in terms of packet delay and cycle time within an acceptable range. We might highlight the benefits of our proposal as a mechanism that maximizes the channel utilization. Detailed implementation of the proposed algorithm is presented, and simulation results are reported to quantify energy savings and effects on network performance on different network scenarios.

## KEYWORDS

PON; WDM/TDM-PON; energy-saving; channel utilization

## 1. INTRODUCTION

It is likely that, from a technological outlook, the Future Internet will be an integration of long-haul high-speed optical networks; a number of wireless networks at the edge; and, in between, several access technologies. Among these access technologies, the Passive Optical Networks (xPON) are very suitable to succeed, due to their simplicity, low-cost, and bandwidth [1,2]. PONs can be used to backhaul both fixed and mobile traffic, therefore it is expected that next generation optical access networks will adopt the wavelength division/time division multiplexing (WDM/TDM) technique [3,4] to increase PONs capacity.

One of the main challenges of PONs is the resource (bandwidth and wavelength) management [5,6]. As it has been shown that access networks consume a significant part of the overall energy of the telecom networks [7], the resource management schemes should also include energy minimization strategies.

The main contributors to energy consumption in PONs are the Optical Network Units (ONUs) and the Optical Line Terminal (OLT). In a WDM/TDM-PON we argue here that if we are able to operate with fewer wavelengths, while still serving all bandwidth demands, the energy consumption can be minimized by shutting down idle receivers/ transceivers at the OLT.

Several techniques have been proposed to increase energy efficiency in PONs. Such techniques are regarded as either hardware (or physical), or software (or data-link) approaches [8,9]. Typically, physical approaches introduce new energy-efficient optical device architectures, while data-link proposals require extensions of resource management algorithms to include "energy awareness", but only few of them so far have addressed the case of WDM/TDM-PONs.

In order to strengthen research on energy-saving besides technical studies, standardization works have been carried out by IEEE and ITU-T [10,11] to describe power saving methods. However challenging tasks, such as the parameter values for the timers, are out of standards scope and thus left vendor-dependent.

This paper is an extension of our previous work [12], where we have presented a novel energy-efficient resource management mechanism for WDM/TDM-PON. Additional contributions are mainly oriented to show new results such as the relationship of the channel utilization with energy savings as well as the impact of the cycle time (the cycle is a time period for the transmissions from ONUs, being the cycle length definition an important factor regarding delay issues) on these savings.

The rest of the paper is organized as follows: in Section 2, we provide an overview of existing energy-saving strategies for PONs. Section 3 provides a description of the proposed mechanism for energy savings at the OLT side in WDM/TDM-PON. In Section 4, results are presented to validate our proposal. Finally, Section 5 concludes this paper.

## 2. AN OVERVIEW OF EXISTING ENERGY-SAVING STRATEGIES

The energy-saving strategies in PON are designed primarily for ONUs and consist in exploiting the variations of upstream/downstream traffic load, according to a defined threshold, by enabling the ONUs to switch among different power saving modes (shedding, dozing, or sleeping) [10]. The definition of the threshold is a key issue, as an appropriate values can lead to relevant performance improvements in terms of delay.

With power shedding the ONU turns off non-essential components while a fully operational optical link is maintained. Dozing mode turns off the ONU transmitter function while the receiver function is always on. With sleeping mode both functions, transmitter and receiver are deactivated. Among the power saving modes, enabling sleep mode in ONU is the most studied approach.

The main challenges of power saving modes in PONs are the slow transition from active mode to sleep mode (time needed during the wake-up process), the sleep time definition and the combination with bandwidth and wavelength allocation algorithms. In [13], the authors propose a Sleep and Periodic Wake-up (SPW) regime for ONUs in order to save energy. The OLT activates or deactivates the ONU, depending on the presence or absence of downstream traffic. The OLT determines the presence or absence of downstream traffic by monitoring the average frame interval. Based on the SPW regime, the authors in [14] propose a Sleep-aware Dynamic Bandwidth Allocation (SDBA) algorithm that maximizes the ONU energy-saving while guaranteeing QoS requirements. Unlike other proposals, the SDBA algorithm is experimentally evaluated in an FPGA-based testbed. The SDBA combines cyclic sleep mechanisms with DBA while consider QoS constrains. It implements an online-based scheduling and increases the polling cycle time as much as possible to improve energy-saving. Regarding the sleep time definition, the authors in [15] propose a sleep-time sizing and scheduling framework namely Sort-And-Shift (SAS), the framework comprises an analytical model that provides a closed-form expression of the ONU sleep time based on the maximum delay requirements.

Physical approaches to address energy efficiency in PONs introduce new optical devices architectures. In [16], the authors experimentally demonstrate a sleep mode ONU architecture that enables ONU the transition from sleep mode to active mode in short time using fast clock and data recovery (CDR) circuit. In [17], same authors also addressed the ONU wake-up time through a novel just-in-time sleep control scheme. In [18], the authors use a hybrid ONU equipped with a low-cost, low energy IEEE 802.15.4 module that turns on the ONU when it needs to receive traffic. The PON architecture is formed by one or more coordinator ONUs that periodically receive information from OLT to notify, using a Wireless Sensor Network (WSN), when those ONUs should enter active mode to receive downstream traffic.

The aforementioned studies focus in the ONU energy-saving in TDM-based PON, but in WDM-based PON and WDM/TDM-based PON new strategies, not only in the ONU, but also in OLT, are needed to effectively control the energy consumption. Authors in [19] propose a sleep control function for energy-saving in PON systems where the configuration of OLT consists of multiple PON-interfaces and two layer-2 switches that provide the multiplexing function to combine upstream and downstream signals and the redundant functions to assure redundancy against failures. Based on traffic predictions, the sleep control function allows switching to sleep mode one of the duplicated layer-2 switches in OLT.

A similar proposal can be found in [20], where an OLT comprises multiple OLT line cards, each of which serves a number of ONUs. The authors propose a new OLT hardware configuration by placing an optical switch at the OLT to configure the connections between OLT line cards and ONUs, thus in low load traffic conditions the switches can be configured such that the same OLT line card provides services to multiple ONUs and other OLT line cards can be turning off.

In WDM/TDM based PON, energy-saving has been addressed to effectively control the energy consumption of the OLT according to downstream traffic. The authors in [21] propose an energy-efficient architecture that reserves a single wavelength and a single low speed transceiver to handle all narrow-band downstream traffic to save energy at the OLT.

In [22] the authors propose to insert a remote channel combine/split module (CCS) in the remote active node of an extended-reach WDM/TDM PON system to reduce initial investment and achieve good energy efficiency. The key function of the CCS is to selectively aggregate optical signals from active subscribers such that a minimal number of OLTs and remote amplifiers are operated. The adding of the new CCS module requires some modification in the MAC layer and setting up a communication channel to the remote node. Thus, DBA modules will decide which OLTs remain active, become active, or inactive. To do so a new mechanism to de-register and reregister ONUs from different OLTs is proposed in [23]. In [24] the energy consumed by the OLT is reduced by using fewer wavelengths and setting idle transmitters to sleep. Authors proposed to use thresholds to determine, based on downstream network traffic, when switch on and off one or more wavelength. Unlike the proposal in [24], in our work we base our energy-saving mechanism on the upstream traffic, *i.e.*, we propose a mechanism that uses DBA information to save energy at the OLT receivers.

An overview of studies and works to address the energy-saving issue in the optical access network is presented in [25,26].

In general, while previous proposals have targeted energy savings by using the sleep mode at the OLT transceivers or at the ONUs (based on downstream traffic), in this paper for the first time to the best of our knowledge, we propose a novel energy-aware mechanism to save energy at the OLT in a WDM/TDM-PON based on *upstream network traffic*. Note that by adopting the proposed mechanism, resource management can be performed using any of the existing DBA and DWA algorithms.

# 3. MINIMIZING ENERGY CONSUMPTION AT THE OLT

This section presents our energy-saving mechanism. To implement our mechanism, we exploit the multi-point control protocol (MPCP) in EPON (standard IEEE 802.3ah) that the ONUs employ to report their upstream queue lengths and to transmit their traffic to the OLT. The protocol information exchanged in the control messages between OLT and ONUs is exploited to disclose the use of resources and based on that information to decide how much energy can be saved (*i.e.*, how many OLT receivers can be switch off). Table 1 defines the notation used hereafter.

**Table 1.** Notation.

| | |
|---|---|
| $T_D$ | Time needed by the ONUs to transmit the granted length of data. |
| $T_{cycle}$ | Time needed by the ONUs to transmit the granted length of data together with their associated guard intervals. |
| $T_g$ | Interval of time between any two adjacent timeslots. |
| $N$ | Number of active ONUs. |
| $R$ | Data rate of the upstream link from an ONU to the OLT. |
| $B_i$ | Amount of data requested by $ONU_i$ (bits). |
| $Ball_{i,j}$ | Amount of data granted to $ONU_i$ on wavelength $j$ (bits). |
| $Bmax_j$ | Maximum amount of data among all the ONUs on the same wavelength $j$ (bits). |
| $Bmax_{i,j}$ | Maximum allowed amount of data for $ONU_i$ on wavelength $j$ (bits). |
| $W$ | Number of supported wavelengths. |
| $W_c$ | Number of current utilized wavelengths. |
| $W_a$ | Theoretical minimum number of wavelengths needed. |
| $W_{off}$ | Number of receivers in OLT to be switched to sleep mode. |
| $W_{on}$ | Number of receivers in OLT to be switched to active mode. |
| $T_c$ | Current Time. |
| $U_{high}$ | Observation period for high channel utilization. |
| $U_{low}$ | Observation period for low channel utilization. |

## 3.1. Energy-Aware Wavelength Assignment (EWA)

Let us define as *Energy-aware Wavelength Assignment* (EWA) the energy-saving mechanism performed by the OLT, based on the evaluation of the channel utilization, to decide whether receivers can be put to sleep mode, and consequently perform the bandwidth and wavelength allocation on the remaining wavelengths. So, an OLT implementing EWA performs the functions that are shown in Figure 1.

**Figure 1.** EWA.

Once low channel utilization is detected, a counter is triggered for an observation period $U_{low}$. If along such timespan the level of utilization does not change we infer that one or more receivers at the OLT should be switched to the sleep mode. Similarly, when high utilization is detected, another counter is triggered as well for an observation period $U_{high}$ and if along such timespan the level of utilization does not change, then one or more receivers at the OLT should be switched to the active mode.

In this study we test different values of $U_{low}$, and we also introduce a threshold to define high-load observation period $U_{high}$. In this manner, through channel utilization evaluation, the OLT switches the receivers to sleep/active mode after $U_{low}$ and $U_{high}$ thresholds are reached. Figure 2 illustrates the operation of the energy-saving mechanism.

### 3.1.1. Channel Utilization Evaluation

Our algorithm deduces if has to switch off receivers only if, during a predefined observation period $U_{low}$, the OLT detects low channel utilization. The OLT evaluates the channel utilization by summing the current timeslots requested by each ONU, *i.e.*, by checking if:

$$\sum_{i=1}^{N} B_i/R < (W_c - 1)T_D \tag{1}$$

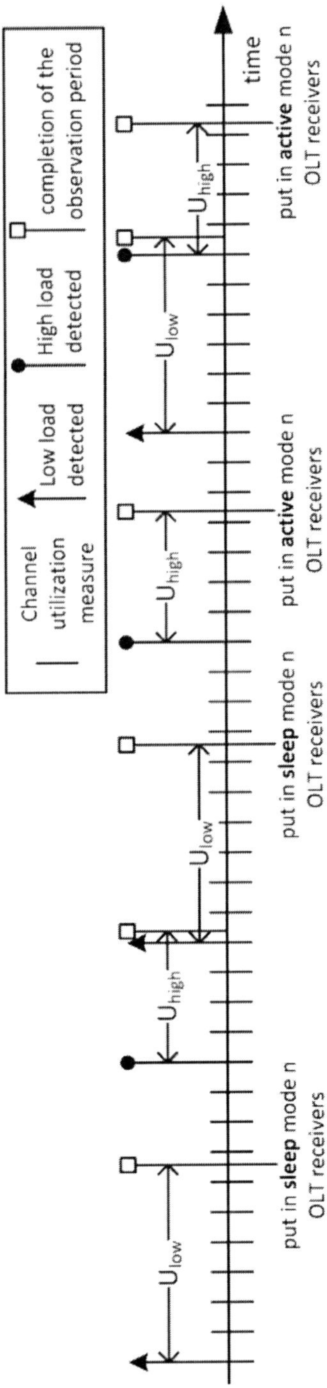


**Figure 2.** Energy saving mechanism.

where TD is the data cycle, *i.e.*, the time needed for the transmission of data in a set of timeslots of different ONUs on the same wavelength *j*, computed as:

$$T_D = Bmax_j / R \quad (2)$$

where Bmaxj is the sum of the maximum transmission length of all ONUs on the same wavelength $\sum_{i=1}^{N} Bmax_{i,j}$. The cycle size with guard intervals is:

$$T_{cycle} = T_D + N(T_g) \quad (3)$$

If $\sum_{i=1}^{N} B_i/R < (W_c - 1)T_D$, then all reported ONU requests might be accommodated in a cycle, and this is regarded as a low channel utilization.

If $\sum_{i=1}^{N} B_i/R > W_c T_D$, then there is not enough room to allocate all the requests and therefore we deduce that the channel utilization is high.

### 3.1.2. Number of Active Wavelengths Required

If the OLT detects low-load channel utilization during Ulow then Woff is calculated as:

$$W_{off} = W_c - W_a \quad (4)$$

where Wa=⌈B/TD⌉ is the theoretical minimum wavelengths needed in the current time Tc and where *B* is the total ONUs request ∑i=1NBi/R.

If, on the contrary, the OLT detects high-load channel utilization during Uhigh then Won is computed as:

$$W_{on} = W_a - W_c \quad (5)$$

### 3.1.3. Wavelength Allocation

As mentioned previously, our resource management scheme is able to work with any DWA. For sake of illustration we refer to two well known algorithms to perform wavelength allocation, called Earliest Finish Time (EFT) and Latest Finish Time (LFT) [27]. We also use EFT and LFT for the computation of the active wavelengths required such that we can then select the OLT receivers to be switched in sleep or active mode.

In EFT the wavelength with the earliest-finish time is allocated to the next transmission, while in LFT the wavelength with the latest-finish time is allocated, as long as this time is not later than a threshold Tlft (Figure 3a). If no channel has a finish time lower than Tlft, then the LFT algorithm works as EFT in the current cycle (Figure 3b).

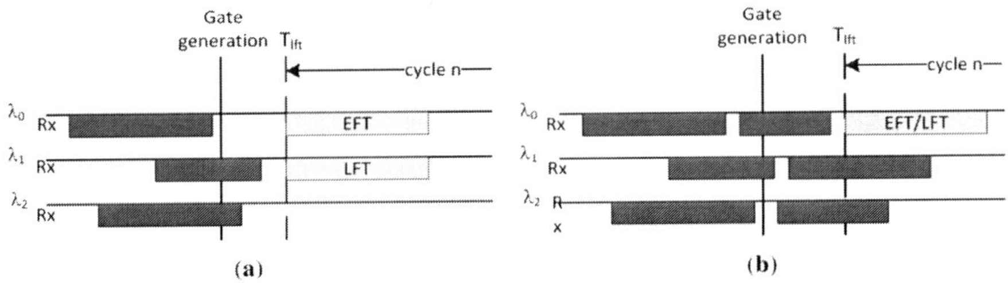


**Figure 3.** Wavelength Allocation. (**a**) EFT operation; (**b**) LFT operation.

### 3.1.4. Bandwidth Allocation

We assume that each ONU is able to transmit only on one wavelength *j* in a cycle, so the OLT updates the value of Bmaxi,j for the next cycle according to the new value of Wc. If all the ONUs have the same traffic load then:

$$Bmax_{i,j} = T_D/(N/W_c) \tag{6}$$

The scheduling in time, *i.e.*, the bandwidth allocation is:

$$Ball_{i,j} = min(B_i, Bmax_{i,j}) \tag{7}$$

where Bi is the ONU queue size in bits.

## 3.2. Summary

Algorithm 1 summarizes the EWA mechanism implemented at the OLT. EWA computes the number of OLT receivers in active or sleep mode.

**Algorithm 1** Energy-aware wavelength assignment (EWA)

1: Channel utilization evaluation

IF $\sum_{i=1}^{N} B_i/R < (W_c - 1)T_D$ THEN low-load

IF $\sum_{i=1}^{N} B_i/R > (W_c - 1)T_D$ THEN high-load

2: Computation of active wavelengths required

IF low-load during $U_{low}$ THEN $W_a = \lceil B/T_D \rceil$; $W_{off} = W_c - W_a$; $W_c = W_a$

Switch OFF $W_{off}$ OLT receivers.

IF high-load during $U_{high}$ THEN $W_a = \lceil B/T_D \rceil$; $W_{on} = W_a - W_c$; $W_c = W_a$

Switch ON $W_{on}$ OLT receivers.

3: Wavelength Allocation (EFT or LFT)

4: Bandwidth Allocation

$Bmax_{i,j} = T_D/(N/W_c)$ according the new value of $W_c$

$Ball_{i,j} = min(B_i, Bmax_{i,j})$

# 4. RESULTS AND DISCUSSION

This section is devoted to evaluate the performance of our proposal in terms of energy-saving and its impact mainly on the packet delay. We carried on simulations using the OPNET Modeler tool. Performance of PONs can be conveniently characterized formulating realistic system models to optimize metrics of interest such as packet delay and energy consumption under different simulation scenarios. We refer to them as balanced/unbalanced input traffic and residential and business user input traffic. Details of traffic models are provided in the following subsections.

We use a single (tunable) transmitter and receiver at the ONU side and multiple receivers/transmitters at the OLT. We assume that each ONU can support all wavelengths (tunable ONU), but it might be only allocated on one wavelength in a cycle. We implemented EWA based on EFT and LFT, for wavelength allocation, and we implemented two different versions of EWA according to the way we switch to sleep/active mode the OLT receivers: *i.e.*, 1-by-1 or n-by-n approach. In the 1-by-1 approach, if OLT decides to save energy, EWA switches to sleep mode only one OLT receiver (even if the number of unused channels is greater than one). In the n-by-n approach EWA can switch to sleep/active mode more than one receiver simultaneously.

Bandwidth allocation is performed according to limited service as described in [28], where $Bmax_i \leq T_D/N$. The energy consumption of each receiver at the OLT was set to 0.5 Watt (W) based on the energy consumption of commercial transceivers/receivers modules [21]. Therefore, the total energy consumption by OLT receivers in our simulation setup is set to 4 W. The input traffic is of the self-similar type and packet length is uniformly distributed based on Ethernet frame size. We assume that the wake-up time for OLT receivers is lower than the Round Trip Time (RTT), which is in the order of 0.2 ms. Note that, in general, the operation of our proposed mechanism is not affected as long as the wake-up time is less than RTT. If the wake-up time is significantly higher then other values such as Ulow and Uhigh should change accordingly. Details of simulations setup are provided in Table 2.

**Table 2.** Simulations Setup.

| | |
|---|---|
| Number of ONUs ($N$) | 64 |
| Number of upstream channels ($W$) | 8 |
| Maximum cycle time ($T_{cycle}$) | 2 ms |
| PON upstream channel data rate ($R$) | 1 Gb/s |
| Guard time ($T_g$) | 2 μs |
| OLT-ONU distance ($d$) | $18 < d < 20$ Km |
| Hurst parameter ($H$) | 0.7 |
| Packet size (uniform distribution) | 64 - 1518 bytes |

Other physical aspects such as transmission power, burst mode reception, optical power losses are already taken care of the physical layer of OLT and ONU by performing the ranging operation as stated by the standard. We assume that physical layer parameters are under acceptable operational levels. However the proposed algorithm does not directly consider physical level parameters. We also assume that the network has been designed properly, and in this paper we focus on the operational aspects, *i.e.*, after the network design has been performed. Since our mechanism is based on Dynamic Bandwidth and Wavelength

Allocation (MAC layer) we only have considered the physical parameters mentioned in Table 2. This is a common approach employed in studies regarding dynamic bandwidth allocation in PONs and it follows the same assumptions as in many other works [5].

The Ulow and Uhigh thresholds values were set in the order of a cycle time (2 ms and 1 ms respectively) because we consider that during a cycle we can have a global knowledge of the channel utilization. If we use more than one cycle is a very long time to make a decision, but if the cycle is smaller, then is a short time to decide. We have tried other values in the range of 1 to 3 ms and the results have not changed significantly.

## 4.1. Balanced Input Traffic

The first simulation evaluates the performance of EFT and LFT algorithms in terms of average number of used wavelengths and average packet delay. We refer to balanced input traffic when the input traffic is equally balanced among ONUs, *i.e.*, the input traffic is the same for all ONUs.

Figure 4a shows the average number of wavelength used in function of the offered network load, which also indicates the number of receivers in OLT that can be put into sleep mode. The energy consumption is directly related to the number of wavelengths used for the upstream transmissions because in our simulative setup each wavelength represents one OLT receiver. We can observe that both algorithms yield similar results, even when not using any mechanism for energy-saving (always on). This confirms that our approach can be used with different DWA approaches and suggests that significant energy reduction can be achieved independently of the DWA approach. Also the fact that LFT and EFT yield very similar results demonstrates that LFT most of times reaches the Tlft and it switches to EFT. Choosing a EWA value of Tlft larger than the one used in our simulations, LFT could allow a better performance of EWA in terms of energy consumption. Nevertheless, with a higher Tlft the average packet delay will be excessively degraded. The Tlft in these simulations was setup as Tlft=RTT+Tcm where RTT is the round trip time and Tcm is the time required for the transmission of control messages.

Figure 4b shows the effect of EWA on the average packet delay. The packet delay is defined as the time elapsed since the packet is generated until it is finally transmitted to the OLT. We noted that by applying an energy efficient mechanism, we increase the delay in the network. However, we consider that these delays remain within reasonable values (below 2 ms, *i.e.*, the duration of a data cycle), acceptable for the majority of current applications.

It is expected that next generation PONs will support various traffic types and services (telephony, TV, mobile backhaul, video streaming) with stringent delay requirements. If these new applications have not their delay requirements satisfied (the 2 ms average delay of our proposal is not appropriate) then EWA should be complemented with strategies based on differentiated wavelength or bandwidth allocation.

The results of channel utilization are shown in Figure 5 (wavelength allocation is performed by EFT following 1-by-1 approach). The channel utilization is the percentage of channel (wavelength) time that is used by ONUs transmissions. If we relate energy savings with the use of the channels then we can highlight the benefits of EWA as a mechanism that maximizes the use of the channels.

Figure 5a shows the average channel utilization for each channel when not using any mechanism for energy-saving ("Always on" with EFT 1-by-1). As expected, the average channel utilization closely reflect the amount of offered load on all the channels. The average utilization for each channel when we use EWA (with EFT 1-by-1) is shown in Figure 5b. In this figure we can clearly see that the different channels have different channel utilization for all network loads.

The effects of introducing a mechanism of energy-saving are also reflected in the cycle time. The cycle time is defined as the time between two successive transmissions of an ONU and can have a fixed or variable length.

To accomplish the delay restrictions that some applications demand, a fixed cycle length should be the most appropriate. However, the coexistence of real-time and non-real-time applications suggests a mechanism with variable cycle length as it applied in our simulations. Note that a large cycle length will result in increased delay for all the packets. The results of the average cycle time that we obtained when we use EWA are shown in Figure 6. Again when EWA is applied, the average cycle time is increased but we consider that they are acceptable values since they are below the defined maximum cycle time corresponding to 2 ms.

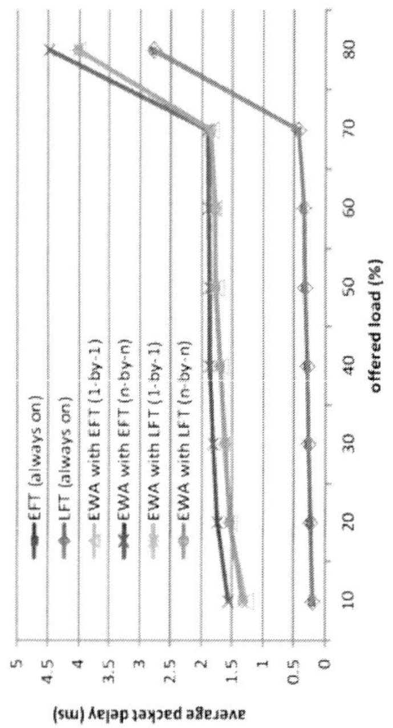


(b)

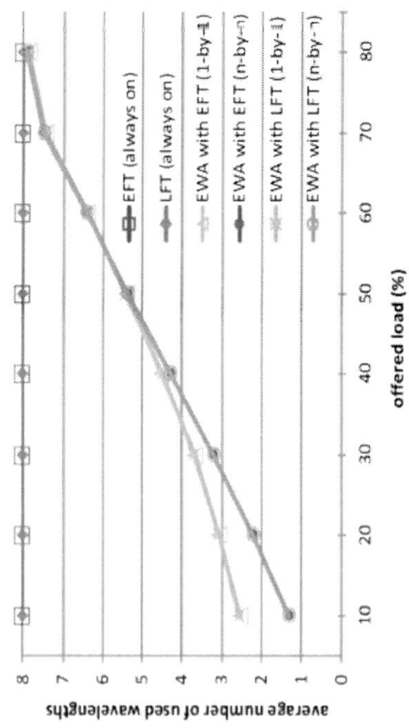


(a)

**Figure 4.** EWA with LFT and EFT wavelength allocation. (**a**) Number of Wavelengths used; (**b**) Average packet delay.

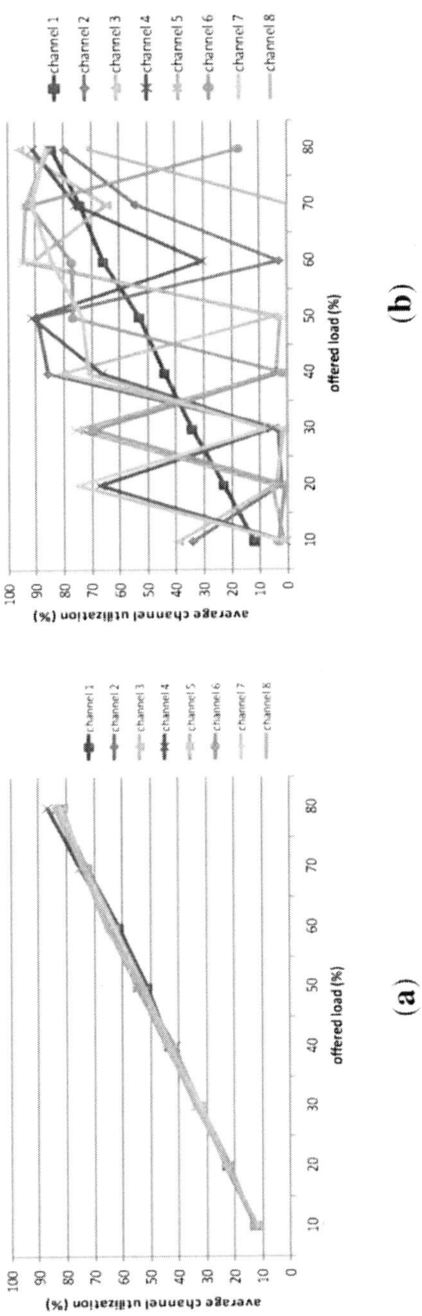


**Figure 5.** Average channel utilization. (**a**) Average Channel Utilization without energy-saving; (**b**) Average Channel Utilization with energy-saving.

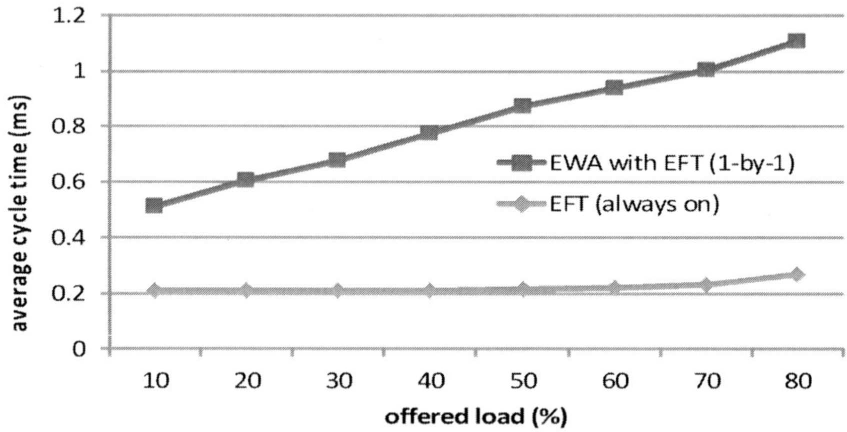


**Figure 6.** Average cycle time.

To conclude we report results directly in terms of energy consumption. Figure 7 shows the average energy consumption for all offered loads. We find that for our proposed energy-saving mechanism is able to decrease energy consumption at the OLT of about 68% low loads, and about 22% for medium and high-loads.

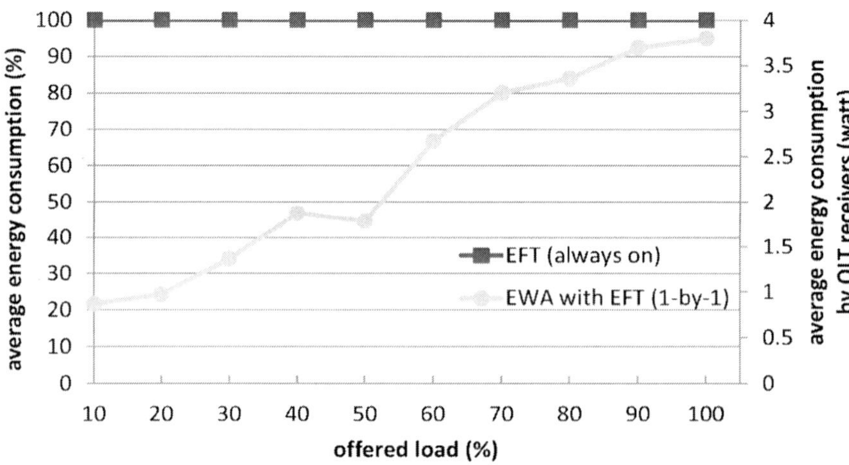


**Figure 7.** Energy Consumption.

## 4.2. Unbalanced Input Traffic

In a similar set of simulations, we evaluated the performance of LFT and EFT, when the upstream network traffic is unbalanced among ONUs, *i.e.*, ONUs are divided in two groups with different input traffic pattern (32 ONUs high-load and 32 ONUs low-load). We consider an ONU as high-load when its load is equal or higher than 0.7, and low-load when its the load is equal or lower than 0.3. Since the results of LFT and EFT are similar also in this case, we will focus hereafter only on EWA with EFT as wavelength allocation method.

First we evaluated the average channel utilization for each channel following 1-by-1 and n-by-n approach (Figure 8a). We found that, the average channel utilization for all channel is about 50% when not using any mechanism for energy-saving (always on). So, we can save energy by maximizing the use of channels, thus energy-saving is about 30%.

Figure 8b shows the average packet delay for high-load and low-load ONUs. In this case we can see relevant variations in the delay between the different approaches, mainly for high-load ONUs. High-load ONUs are negatively affected because the maximum bandwidth guaranteed is the same for all ONUs so several cycles are needed to serve such ONUs. We conclude that EWA affect some applications or services that are delay sensitive in unbalanced traffic scenarios.

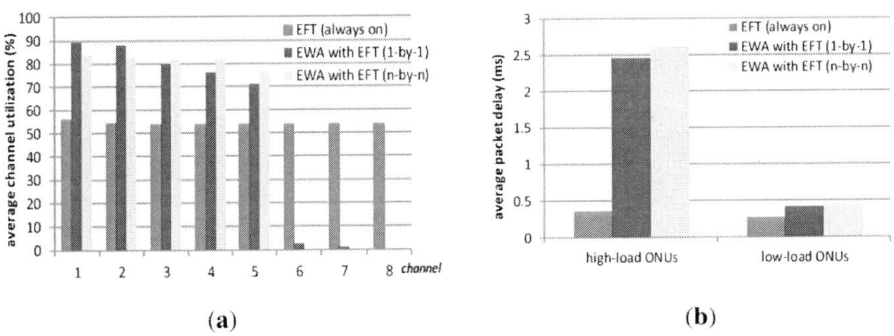


**Figure 8.** Average channel Utilization & Average packet delay. (**a**) Average channel Utilization; (**b**) Average packet delay of ONUs of each group.

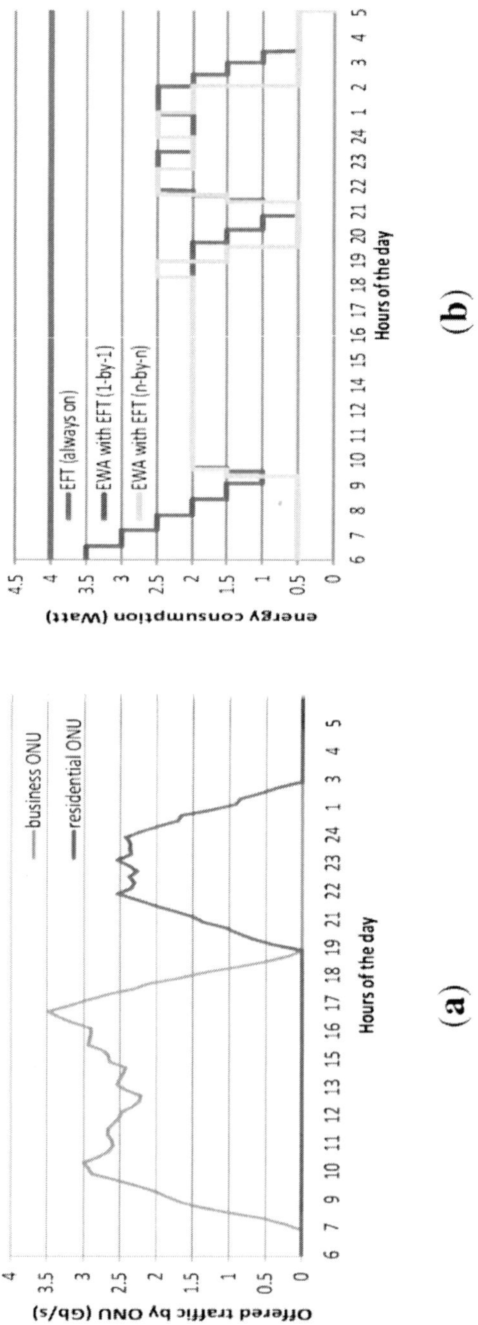


**Figure 9.** Residential and Business usage behaviors. (**a**) Input traffic of each ONU; (**b**) Energy consumption.

## 4.3. Traffic Variation during the Day

We performed a different set of simulations with different configurations settings. In this case we model a traffic with varying intensity and with two classes of users representing residential and business users. Figure 9a represents the upstream traffic behavior along the day of business and residential users.

Figure 9b shows the average energy consumption in watts along the day of the OLT receivers. The scenario based on EFT n-by-n slightly outperforms the 1-by-1 in terms of energy consumption. Differences are seen mainly at the beginning of the simulation period, *i.e.*, at 6 a.m. (in the absence of traffic), because even though the number of receivers that can be shut down is the same, how they are turned off is different. Following the 1-by-1 approach at the beginning of the simulation, energy savings are achieved slowly because EWA starts considering all the available wavelengths switched on.

Figure 10a shows the average channel utilization for each channel. In this case, by using EWA we maximize the channel utilization, and we only need half of the available channels, and thus we are able to save energy. Figure 10b shows the average packet delay of business and residential traffic. The EFT n-by-n works slightly better than EFT 1-by-1 mechanism.

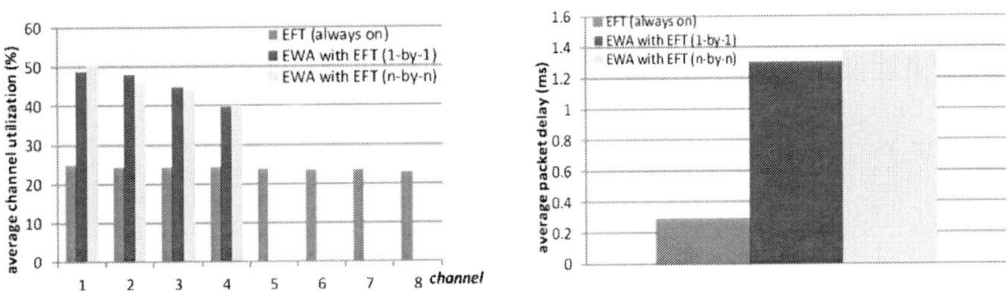


(a) (b)

**Figure 10.** Average channel utilization & Average packet delay. (**a**) Average channel utilization; (**b**) Average packet delay.

# 5. CONCLUSIONS

In this paper we have presented a novel energy-aware wavelength assignment mechanism to save energy at the OLT in a WDM/TDM-PON based on upstream network traffic. Results have shown that energy consumption in OLT receivers can be reduced up to 30%. We have shown new results, compared with our previous work, where the average utilization of the available wavelengths reflects the amount of offered load on all the channels and the average cycle time is below the defined maximum cycle time.

Simulations results show that our proposal is able to save energy at the OLT, based on a periodic evaluation of channel utilization. The new results for the channel utilization show that significant energy savings can be achieved at the OLT by the proper use of upstream channels. We find that, for our simulation setup, if we consider a balanced scenario where the input traffic for all ONUs is the same, our proposed energy-saving mechanism for low-loads is able to decrease energy consumption of about 68% while for medium and high-loads about 22%.

In our study we have assumed some specific values of the duration of OLT receivers in sleep or active modes as well as the length of the observation periods Ulow and Uhigh. Such values are crucial parameters that impact directly the delay performance. Regarding the wake-up time for elements in OLT, we found that our proposal does not introduce any penalty whenever the wake-up time in OLT receivers is less than the RTT.

As future works, we notice that the identification of effective values under different scenarios as well as a generic and suitable definition of Tlft for the LFT algorithm, might provide higher energy-saving and more efficient resource management.

# ACKNOWLEDGMENTS

A partial version of this work was previously presented in NOC 2013 conference [12]. The research leading to these results has received funding from the European Community's Seventh Framework Programme FP7/2013-2015 under grant agreement no.317762 COMBO project and from the Comissionat per a Universitats i Recerca del Departament d'Innovaciò, Universitats i Empresa de la Generalitat de Catalunya, and the European Social Fund.

## AUTHOR CONTRIBUTIONS

All authors had equivalent contribution in the proposal, analysis, and revision of the work. The first author provided additional contribution in writing most of the manuscript.

## CONFLICTS OF INTEREST

The authors declare no conflicts of interest.

## REFERENCES

1. Ghazisaidi, N.; Maier, M.; Assi, C.M. Fiber-wireless (FiWi) access networks: A survey. *IEEE Commun. Mag.* **2009**, *47*, 160–167.

2. Sarkar, S.; Dixit, S.; Mukherjee, B. Hybrid Wireless-Optical Broadband-Access Network (WOBAN): A Review of Relevant Challenges. *J. Lightwave Technol.* **2007**, *25*, 3329–3340.

3. Kani, J.-I. Enabling Technologies for Future Scalable and Flexible WDM-PON and WDM/TDM-PON Systems. *IEEE J. Sel. Top. Quantum Electron.* **2010**, *16*, 1290–1297.

4. Yuanqiu, L.; Xiaoping, Z.; Effenberger, F.; Xuejin, Y.; Guikai, P.; Yinbo, Q.; Yiran, M. Time- and Wavelength-Division Multiplexed Passive Optical Network (TWDM-PON) for Next-Generation PON Stage 2 (NG-PON2). *J. Lightwave Technol.* **2013**, *31*, 587–593.

5. Zheng, J.; Mouftah, H.T. A survey of dynamic bandwidth allocation algorithms for Ethernet Passive Optical Networks. *Opt. Switch. Netw.* **2009**, *6*, 151–162.

6. McGarry, M.P.; Reisslein, M.; Colbourn, C.J.; Maier, M.; Aurzada, F.; Scheutzow, M. Just-in-Time Scheduling for Multichannel EPONs. *IEEE J. Lightwave Technol.* **2008**, *26*, 1204–1216.

7. Lange, C.; Gladisch, A. On the energy consumption of FTTH access networks. In Proceedings of the Optical Fiber Communication Conference, San Diego, CA, USA, 22–26 March 2009.

8. Dhaini, A.R.; Pin-Han, H.; Gangxiang, S. Toward green next-generation passive optical networks. *IEEE Commun. Mag.***2011**, *49*, 94–101.

9. Valcarenghi, L.; Van, D.P.; Castoldi, P. How to save energy in Passive Optical Networks. In Proceedings of the 13th International Conference on Transparent Optical Networks (ICTON), Stockholm, Sweden, 26–30 June 2011; pp. 1–5.

10. ITU-T, G.Sup45: GPON power conservation: 2009. Avaiable online: https://www.itu.int/rec/T-REC-G.Sup45-200905-I(accessed on 11 August 2014).

11. *IEEE Standard for Service Interoperability in Ethernet Passive Optical Networks (SIEPON)*; IEEE Std: 1904.1-2013.

12. Garfias, P.; de Andrade, M.; Tornatore, M.; Buttaboni, A.; Sallent, S.; Gutierrez, L. A novel traffic-aware mechanism for energy-saving at the OLT in WDM/TDM-PON. In Proceedings of the Conference on Network and Optical Communications (NOC), Graz, Austria, 10–12 July 2013; pp. 225–232.

13. Kubo, R., Kani, J.; Fujimoto, Y.; Yoshimoto, N.; Kumozaki, K. Proposal and Performance Analysis of a Power-saving Mechanism for 10-Gigabit-class Passive Optical Network Systems. In Proceedings of the Conference on Network and Optical Communications (NOC), Valladolid, Spain; 2009; pp. 87–94.

14. Van, D.P.; Valcarenghi, L.; Chincoli, M.; Castoldi, P. Experimental evaluation of a sleep-aware dynamic bandwidth allocation in a multi-ONU 10G-EPON testbed. *Opt. Switch. Netw.* **2014**.

15. Dhaini, A.R.; Ho, P.-H.; Shen, G.; Shihada, B. Energy Efficiency in TDMA-Based Next-Generation Passive Optical Access Networks. *IEEE/ACM Trans. Netw.* **2014**, *22*, 850–863.

16. Wong, S.; Valcarenghi, L.; Yen, S.; Campelo, D.R.; Yamashita, S.; Kazovsky, L. Sleep Mode for Energy Saving PONs: Advantages and Drawbacks. In Proceedings of the IEEE GLOBECOM Workshops, Honolulu, HI, USA, 30 November–4 December 2009; pp. 1–6.

17. Wong, S.; Yen, S.; Afshar, P.; Yamashita, S.; Kazovsky, L. Demonstration of Energy Conserving TDM-PON with Sleep Mode ONU Using Fast Clock Recovery Circuit. In Proceedings of the Optical Fiber Communication Conference, San Diego, CA, USA, 21–25 March 2010.

18. Newaz, S.H.S.; Cuevas, A.; Lee, G.M.; Crespi, N.; Choi, J.K. Improving Energy Saving in Time-Division Multiplexing Passive Optical Networks. *IEEE Internet Comput.* **2013**, *17*, 23–31.

19. Shimazu, S.; Kani, J.; Yoshimoto, N.; Hadama, H. Novel Sleep Control for EPON Optical Line Terminal Employing Layer-2 Switch Functions. In Proceedings of the IEEE Global Telecommunications Conference (GLOBECOM), Miami, FL, USA, 6–10 December 2010; pp. 1–5.

20. Zhang, J.; Wang, T.; Ansari, N. Designing energy-efficient optical line terminal for TDM passive optical networks. In Proceedings of the 34th IEEE Sarnoff Symposium, Princeton, NJ, USA, 3–4 May 2011; pp. 1–5.

21. Tadokoro, M.; Kubo, R.; Nishihara, S.; Yamada, T.; Nakamura, H. Adaptive bandwidth aggregation mechanisms using a shared wavelength for energy-efficient WDM/TDM-PON. In Proceedings of the 10th International Conference on Optical Internet (COIN), Yokohama, Kanagawa, 29–31 May 2012.

22. Feng, H.; Chae, C.; Tran, A.V. Cost-effective and power-efficient extended-reach WDM/TDM PON systems. In Proceedings of the 2011 and the National Fiber Optic Engineers Conference on Optical Fiber Communication Conference (OFC), Los Angeles, CA, USA, 6–10 March 2011.

23. Feng, H.; Chae, C.; Nirmalathas, A. Seamless handover scheme in energy-efficient long reach WDM/TDM PON systems. In Proceedings of the 17th Opto-Electronics and Communications Conference (OECC), Busan, Korea, 2–6 July 2012.

24. Shi, L.; Lee, S.; Mukherjee, B. Energy-Efficient Long-Reach Passive Optical Network: A Dynamic Wavelength Allocation Scheme. In *Integrated Photonics Research, Silicon and Nanophotonics and Photonics in Switching*; paper PTuB1; 2010.

25. Kani, J. Power Saving Techniques and Mechanisms for Optical Access Networks Systems. *J. Lightwave Technol.* **2013**,*31*, 563–570.

26. Valcarenghi, L.; Yoshida, Y.; Maruta, A.; Castoldi, P.; Kitayama, K. Energy saving in TWDM(A) PONs: Challenges and opportunities. In Proceedings of the 15th International Conference on Transparent Optical Networks (ICTON), Cartagena, DC, USA, 23–27 June 2013.

27. Kanonakis, K.; Tomkos, I. Improving the efficiency of online upstream scheduling and wavelength assignment in hybrid WDM/TDMA EPON networks. *IEEE J. Sel. Areas Commun.* **2010**, *28*, 838–848.

28. Kramer, G.; Mukherjee, B.; Pesavento, G. IPACT a dynamic protocol for an Ethernet PON (EPON). *IEEE Commun. Mag.* 2002, *40*, 74–80.

# INDEX